A Dialogue of the Senses

A Dialogue of the Senses

Designing Interaction with the Body in Mind

Sile O'Modhrain

The MIT Press
Cambridge, Massachusetts
London, England

The MIT Press
Massachusetts Institute of Technology
77 Massachusetts Avenue
Cambridge, MA 02139
mitpress.mit.edu

The MIT Press would like to thank the anonymous peer reviewers who provided comments on drafts of this book. The generous work of academic experts is essential for establishing the authority and quality of our publications. We acknowledge with gratitude the contributions of these otherwise uncredited readers.

This book was set in Stone Serif and Stone Sans by Westchester Publishing Services. Printed and bound in the United States of America.

Library of Congress Cataloging-in-Publication Data

Names: O'Modhrain, Sile author
Title: A dialogue of the senses : designing interaction with the body in mind / Sile O'Modhrain.
Description: Cambridge, Massachusetts : The MIT Press, [2026] | Includes bibliographical references and index.
Identifiers: LCCN 2025028944 (print) | LCCN 2025028945 (ebook) | ISBN 9780262052399 paperback | ISBN 9780262052405 pdf | ISBN 9780262052412 epub
Subjects: LCSH: Human-computer interaction | User-centered system design | Perception | Reaction time | Virtual reality | Augmented reality
Classification: LCC QA76.9.H85 O46 2026a (print) | LCC QA76.9.H85 (ebook)
LC record available at https://lccn.loc.gov/2025028944
LC ebook record available at https://lccn.loc.gov/2025028945

EU Authorised Representative: Easy Access System Europe, Mustamäe tee 50, 10621 Tallinn, Estonia | Email: gpsr.requests@easproject.com

Contents

Acknowledgments

This project has been a long time in the making. Its seeds were planted during my time as a postdoc at MIT, where I taught a class on interaction design in the Media Arts and Sciences program at the MIT Media Lab, and later at the University of Michigan. I am very grateful to the students in these courses for helping me shape and refine the book's content. Tod Machover and Carol Strohecker then persuaded me to submit the project to Doug Surrey and Noah Springer at the MIT Press, who have patiently overseen its development.

Much of the thinking reflected in these pages grew out of my participation in the EU-funded Network of Excellence "Enactive Interfaces," and I owe a debt of gratitude to its members for setting me on the intellectual path that has shaped my research career. I am also indebted to my former PhD students Matthew Rodger, Nicholas Ward, and Nicholas Gillian, whose work reflects aspects of the embodied approach to interaction design that I was developing while at Queen's University Belfast. My colleagues Cathy Craig, Fred Cummins, Marek McGann, Val Morash, and Terry Winograd patiently answered endless questions relating to embodiment, phenomenology, and the finer points of the enactive approach to cognition. Beyond this group, I am also grateful for the support I have received from the haptics, NIME, TEI, assets, and CHI communities, whose members have provided timely and critical feedback on earlier publications. In particular, my colleagues Brent Gillespie, Mark Cutkosky, Roberta Klatzky, Susan Lederman, Lynette Jones, and Hong Tan have been the pillars of support on which my academic journey has been founded. Were it not for the grounding in psychoacoustics I received from John Chowning and Max Matthews, and for the thesis advice I received from Chris Chafe (at Stanford's Center for Computer Research in Music and

Acoustics), I would not have found this path at all. I am also grateful to my colleagues at the University of Michigan, in both the School of Music, Theatre, and Dance and the School of Information, who have provided the practical and scholarly environment that enabled this work to progress. I especially want to thank Michael Gurevich and John Granzow, who stepped up to take over the position of department chair at various points to give me space to write; Mark Newman and Beth Yakel for their continual encouragement; and Marie McCarthy for making time in her busy schedule to go for walks and to cook dinner on numerous Sunday afternoons.

If the book is readable at all, it is because of input from those who have taken time to read and comment on the manuscript. Chief among these are David Pratt and Elena Abbot, who helped me develop the structure of the text and carefully edited its various versions. Original illustrations were drawn by Hans Anderson, who patiently listened to my random musings about how to represent many of the more abstract concepts he has captured so eloquently in his drawings. I am also grateful to Lauro Ojeda for helping me capture and analyze footsteps, and to Emelia Paine in her role as chief wrangler of copyright permissions, alt texts, and references. Financial support for their work was provided by the Carnegie Fund and by the University of Michigan. I also wish to acknowledge the support of the Smith-Kettlewell Eye Research Institute, in particular John Brabyn and James Coughlan, who hosted me for sabbatical in fall 2017, enabling me to start work on this book.

To Matthew Rodger and Lynette Jones I owe a huge debt that I can never repay. They generously gave of their time and patience to read many iterations of the book's content; without their input, I could not have brought the project to completion.

There are so many friends and relatives who have stuck with me along the way, far more than I am able to mention here. Lynda Marlow has been my ever-present cowriting buddy, turning up online for three hours every day to keep me accountable, while Kim Warren has continued to put my numerous writing panic moments into perspective. Thank you, Mikel, for turning up on so many mornings to go running by the river, and thanks to Scott for Wordle, Byrdle, and organizing so many memorable gatherings with friends, on my balcony, to keep me sane. Frankie James has been a constant source of support, distracting me to go shopping and to visit the fish in Monterey Bay Aquarium when I became too stressed. Anil, Zeynep, Ella, Joe, Renee, and Laura, you have made Ann Arbor such a wonderful place to live.

Alex, Brent, Hrishi, Thomas, and the rest of the NewHaptics team, you have brought our shared dream of a multiline braille display to life. Annie, Richard, Drew, Liz, John, Niamh, Fiona B., Fiona P., Piers, and Jan, I couldn't have made it through COVID and this book project without our various weekly chats and your constant encouragement. Loretta and Darragh, thank you for all the times you turned up with food and wine, regardless of continent or time zone.

But my greatest support network has been my family. They are the ones who have had to witness all the ups and downs of the project and who never gave up on me. Eamonn, Cormac, Lughaidh, and Maeve, you have been my strongest allies and closest friends. Robert, Susan and Ali, you have quietly supported the people whom I most care about, my siblings, my nieces and nephews, and now their children too.

This book is dedicated to the memory of my parents, Lughaidh and Mairead O'Modhrain. Their love, commitment, and tenacity are the reason I received the education and training that made it possible to embark on my journey as a researcher and, ultimately, made it possible for me to write this book.

1 The Senses in Dialogue

1.1 Introduction

"As the firefront approached, the sky blackened. Spotfires broke out as blazing embers fell from the heavens. The heat intensified and a blistering wind, whipped up by the change of air pressure caused by the fire sucking in oxygen, sounded like a freight train."[1] This firsthand account of the 2009 Australian bushfires is a vivid reminder that perception of events is acquired not through one sensory modality but rather from the synthesized experience of simultaneous perceptions associated with an event. In the case of a brush fire, these are smoke (olfactory), heat (tactile), crackling flames (auditory), and the motion and color of flames (visual). As the psychologist J. J. Gibson points out, a fire "provides information for the ears, the nose, the skin, and the eyes. The crackling sound, the smoky odor, the projected heat, and the projected dance of the colored flames all specify the same event, and each alone specifies the event. One can hear it, smell it, feel it, and see it, or get any combination of these detections, and thereby perceive a fire."[2]

The body is exquisitely equipped to function in a complex, ever-changing environment that presents a continuous series of puzzles to solve for the sake of survival. Our highly evolved sensory systems, coupled with our moving, acting, thinking bodies, allow us to continually make judgments about how to obtain needed information. That information helps us pursue and achieve the individual goals that, together, form the sequences of events that make up our lives. When doing anything—from waking, rising, and dressing in the morning to negotiating traffic or public transport to accomplishing highly skilled mental and physical activities to earn a living—people routinely accomplish tasks that even the most sophisticated computational devices cannot. Or, to frame this observation slightly differently, people can switch between complex tasks and adapt to new situations in ways that are challenging for computers. Experiences are rarely evaluated as isolated sensory events. Rather, they are integrated with prior experiences accumulated over a vast range of situations, enabling assessment of the present in the context of the past. Having lost balance once while hefting a large but lightweight box, one might approach future large boxes with the awareness that they too may be light.

This book is about the relationship among perception, action, and design. It is intended as a resource for designers of interactive digitally augmented systems and artifacts that test the ability to learn through experience, just as in everyday life. Such learning is more than remembering a sequence of tasks. It goes down to the muscular level, so that seeing, hearing, or feeling the body's responses to an action will guide performance of a subsequent action. Anyone who has ever played an augmented reality (AR) or virtual reality (VR) game that tracks player movements and gives visual, auditory, and/or haptic feedback will recall the process of figuring out what the system expects. In a virtual golf game, for example, a player might have to deduce what the system considers a "good" swing. This is subtly different from the memory of a good swing with a real golf club, but the body nonetheless quickly learns how to do what the game expects. This book investigates our human ability to learn how to perform actions in a computationally enhanced environment.

This process of learning has two aspects. First, in the case of golf, what perceptual cues must a partly or wholly simulated game provide for the player to grasp what counts as a good swing? Second, what information must the game extract from user movements and the broader environment to determine what is being attempted? In the early days of computing, the user had to speak the binary language of the machine. As computers advanced, interfaces began to allow users to phrase commands in familiar language and to use typing, handwriting, and, more recently, gestures to communicate intentions to their devices. Computer scientist Alan Kay points out that computers have effectively taken a developmental path opposite that of a growing child. Children learn by exploring the world with their mouths and hands, then learn to associate objects with images, as in picture books, and finally negotiate the symbolic world of writing and reading. By contrast, computers "evolved" from machines that operate through highly abstract languages into machines that communicate through visual imagery such as icons, and they eventually became sensitive to gesture and touch.[3] The actions, commands, and iconography that form the basis of our interactions with computational devices such as phones, TVs, cars, and immersive games are carefully designed and evaluated to ensure a good fit between what we need to accomplish and the constraints of a given task environment. The in-vehicle infotainment system in your car, for example, is designed to support quick access to maps, phone calls, and music with minimal distraction for the driver. Meanwhile, the interface on your handheld phone provides access to similar functionality but draws on a wider variety of gestures (e.g., typing instead of voice commands) and is designed to focus your attention on its small screen.

The subfield of design that encompasses the development of the interface between people and computational devices is called either "interaction design" or "user experience design," and its practitioners are referred to as either "interaction designers" or "user experience (UX) designers." These practitioners work at the nexus of hardware design, software design, user needs assessment, product evaluation, and even marketing to deliver a product that allows a user to accomplish a task while satisfying budgetary constraints and delivering a branded user experience. Consequently, UX/interaction designers have come to play a crucial but often invisible role in the development of all of the interactive products we use today. As the range of ways systems can track human movement has increased and the variety of ways computers are embedded in our environment has expanded, interaction designers are now required to become conversant with the world of interaction beyond the office desk.

This book focuses on how people move and act in the world. Notably, it does not address socially situated interaction and design, a topic deserving a book of its own. Rather, its goal is to introduce interaction designers to the fields of embodied cognition and embodied interaction design and to research on perception and action in real and virtual environments. The premise of the book is that developing digitally augmented artifacts and environments to support physically engaged exploration of augmented and virtual worlds requires turning the interaction design process inside out. It argues that the designer of virtual worlds should start with the seeing, hearing, feeling, moving body that will accomplish a task rather than with the performance capabilities of the machine. Successful design of interactive systems should create environments sufficiently complete that they allow users to acquire knowledge as seamlessly as in a physical environment. The hope is that the framework introduced in this book will lead to better interactive experiences by showing how users extract information from real environments and how they acquire a physical understanding of the virtual world.

This book thus departs from previous texts on interaction design in that it is a theoretical and more grounded approach to rebuilding from first principles that need to be considered when we look at designing for digital spaces. It is not a typical design book telling practitioners what to do; it is a rich exploration of the theories, principles, and research findings that can help them think more expansively and more critically about what they do. Although it is thorough, it is not exhaustive. It follows the pathways that I have found most illuminating and useful in my own design practice. The book owes much to the work of influential practitioners such as Donald Norman and Paul Dourish—with whom many readers may be familiar—but I have chosen to emphasize the fields from which their work also emerged. In the twenty-five years since Paul Dourish published his groundbreaking book *Where the Action Is*, the fields of embodied cognition and embodied interaction design have advanced considerably: There are now conferences and publications specifically focusing on the interaction between these fields. This book is not intended to summarize or replace these early and significant sources. Instead, my hope is that I can provide a way forward for those who want to deepen their knowledge of this increasingly important body of work.

The book is divided into seven chapters. This chapter establishes the purpose and outlook of the book, inviting students and professional interaction designers to consider their work with a fresh perspective. It asks you to join me on a journey that includes theoretical and practical exploration of human perception, movement, and interaction to enable us to think, imagine, and experiment more creatively in our work and develop more innovative, immersive, and effective artifacts. Chapter 2 sets out the theoretical frameworks crucial to developing purposeful and effective artifacts. Expansive but not exhaustive in its survey, this chapter draws on work in the fields of embodied cognition and the ecological approach to perception, situating the field of human–computer interaction (HCI) in its behavioral and phenomenological contexts. It offers an intentional, but not necessarily typical, exploration of the field. Students of interaction design will develop their own frameworks of theoretical perspectives, whereas seasoned practitioners will be asked to reengage with material from a refreshed critical perspective.

Chapter 3 sets forth the case for treating the moving, acting body as the source of lived experience, proposing that interaction design begins with understanding how bodies move. Our perception of the world arises from our actions upon it, and subsequent actions are, in turn, guided by the response of our surroundings to those actions. Moving the head to shift one's gaze or to listen more closely to a conversation is as much a part of acquiring information about the world as discovering patterns present in light falling on the retina or sound reaching the ears. This is even more evident with touch, because one must literally "reach out" to bring one's hand into contact with the object of interest. Chapter 3 thus lays out approaches to capturing and analyzing movements, and it introduces three practices that place the body at the center of the process of interaction design.

The next three chapters, chapters 4–6, reposition our perception of structural aspects of the environment. They build heavily on the theoretical groundwork set forth in chapters 2 and 3 and gradually become more technical, so students and practitioners alike can focus on the sections most suitable for where they are in their learning and practice experiences. Chapters 4 and 5 are devoted to the interrelated concepts of time and space, respectively. There are many kinds of time. Clock time runs at a constant speed, presented as the hours and minutes that divide the days or the microseconds that pace the streaming of media. Then there is time as it is experienced. Most activities have their own rhythms and ways of dividing time. Artists, for instance, must be attuned to audience perceptions of time. The timing of a joke, the suspense created by a playwright or filmmaker laying out a narrative, or the more abstract manipulation of time in music—these all depend on the principle of creating an expectation and then choosing when (and how) to fulfill or subvert that expectation and lead to a resolution. Chapter 4 proposes a third type of time, action time, in which purposeful actions generate the sequencing, adaptation, and duration of lived experience. Action time is an arc along which we plan actions in the context of knowledge of what went before, awareness of the present, and anticipation of what is to come. It is increasingly important to include the backdrop of time in designing interactive artifacts and environments.

Space, meanwhile, is not an abstract geometric concept for the moving, sensing being. It is often measured in units whose origin is derived from bodily dimensions such as feet and hands.[4] In moving about in the world, what matters are not abstract distances or dimensions but rather awareness of our bodily capabilities, the surfaces and objects in our surroundings, and the actions to perform. It is important to know, for example, whether something is within reach, whether we can pass through a gap, or if we will collide with a stationary or moving object. In chapter 5, I propose that in simulating virtual spaces, the relevant information to render is not the geometry of a virtual environment but rather a global energy array populated with visual, auditory, and tactile information that allows users to discover patterns of light, sound, or mechanical energy, patterns that can guide their actions in partly or wholly simulated worlds.

Chapter 6 discusses the design of digitally augmented interactive objects in partly and wholly rendered forms. Rendering objects that retain certain salient properties, no matter how they are seen, heard, or felt, is one of the greatest challenges for designers of interactive systems. Although holograms provide the illusion of three-dimensional objects, and robotic devices can create illusions of touch, these illusions fall apart when the user attempts to treat virtual objects as material ones. For example, one cannot use the whole hand to pick up a virtual object. There is a better outcome from "augmenting" a real object by embedding it in a virtual world. Now the user can manipulate something with real mass and materiality, with a fixed form. However, incorporating the manipulation of objects into an interaction with a virtual or augmented environment is a trade-off between the tangibility of malleable objects and the malleability that allows objects to be felt and wielded. Chapter 6 discusses current techniques for rendering objects virtually, as well as corresponding approaches to interacting with objects in real and augmented environments. The chapter also introduces the design of hybrid or blended worlds from the perspectives of users who must integrate digitally augmented artifacts into lived experience.

The final chapter makes a case for designing with bodies in mind—that is, with awareness of how bodies of all kinds and capabilities will interact with the artifacts and systems we build. I argue that if we are conscious of building for "bodies" (not one singular "body"), our designs are more likely to reflect the needs of the broadest range of embodied user experiences from the ground up. In this way, I suggest, we can avoid falling into the one-size-fits-men chasm that marginalizes women, children, people with disabilities, and people who, in so many ways, do not have bodies that fit the dimensions of the devices through which we interact with the online world—phones, head-mounted displays, fitness trackers—or that are even reflected in the voice- and motion-tracking algorithms on which these devices depend. Taking to heart the embodied approach to interaction design at the center of this book, I argue, will require designers to respect the needs and lived experiences of all potential users in every aspect of their work. This process begins with selecting team members who can represent multiple design perspectives and pervades every stage of the design cycle, from the choice of design methodologies for initial prototyping, through hardware and software development, to product evaluation. In short, no member of a design team is off the hook.

1.2 Questioning Our Assumptions

At the outset of this journey, I am going to ask you to turn onto a road you may never have traveled before. I provide a map with landmarks, but the individual waypoints may be unfamiliar. In short, you may have to let go of ideas you assumed were unquestionable in terms of your relationship with the environment you expertly negotiate every day.

The first assumption to question is that perception is absolute—that what we perceive is an entire and accurate representation of our physical environment. Because our interactions with the world are predictable in most instances—and because the objects we encounter very seldom dematerialize or shape-shift without explanation—we seldom doubt what we perceive. Further, we never notice that we, in fact, have access to a limited subset of the physical properties of our surroundings. Our ability to see relies on our visual system being sensitive to electromagnetic radiation with wavelengths between 400 and 700 nm. But many birds and insects see colors above the wavelengths we can perceive, in the ultraviolet range (figure 1.1).

Other animals, such as some snakes, can perceive energy in the infrared spectrum by means of a pair of pit-like receptors on their heads, which allows them to pinpoint moving prey accurately in space, in much the same way as we would use our two ears or two eyes. Our hearing too is sensitive to a narrow band of frequencies of sound waves, which we usually sense as changes in air pressure. We typically hear vibrations between 20 Hz and 20 kHz. Elephants can hear much lower vibrations, so-called infrasound, whereas bats and most other small animals can hear sounds above 20 kHz in a range we call ultrasound. There are sources of energy to which we are not sensitive at all, such as fluctuations in electric fields that many fish can sense, or in magnetic fields that birds and insects use for navigation. Of course, we can now

Figure 1.1
The range of color in a crocus flower as seen by a bee, which can see the ultraviolet spectrum. Top: A flower as seen by a human eye. Bottom: The same flower as seen by a bee. Note the greater contrast, in grayscale, between parts of the flower on the bottom image, which represents the more extensive bandwidth of color visible to the bee. *Source:* Adapted by Hans Anderson, reprinted with the permission of the Science Photo Library, spring crocus (*Crocus vernus*) and spring crocus in ultraviolet light, photo by Bjorn Rorslett.

build artificial sensory receptors, such as night-vision goggles to see in the infrared range, or compasses that use the earth's magnetic field.

Beyond the fact that there are signals in the physical environment that our bodies don't detect, there are also ways our perception of the world is less reliable than we might think. For instance, there is the experience of sitting on a stopped train when another train very slowly passes by, and being unsure whether your train or the other one is moving. There is a visual perception of movement but not enough speed for the body to detect motion and thus for the brain to know which train has moved. This disconnect between two senses results in a momentarily ambiguous situation.[5]

In a well-known example, Simons and Chabris demonstrate that even obvious properties of our environment can change before our eyes without our noticing.[6] Their study involved asking subjects to watch a video of two basketball teams passing a ball, counting the number of passes made by one of the teams. Partway through the video, the experimenters added a moving image of a gorilla. Most people in the experiment missed the gorilla because their attention was focused on counting ball passes. This phenomenon, called "attentional selectivity," arises because attention has been selectively directed toward a task that requires so much concentration it causes other aspects of the visual scene to go unnoticed.

This is why performing any task that redirects our visual attention while driving is so dangerous. We may believe we are aware of everything happening around the vehicle but can easily miss hazards, even if, as Simons and Chabris demonstrate, they are large and obvious.

The second assumption to be questioned is that perception is passive, that sensations just *arrive* at our sensing organs. When we think about perceiving something, it is typically from the perspective of an observer; we see the plane in the sky, feel the heat of the sun, hear the approaching car. And it can be true that we sense external events outside our sphere of control. Much of the time, however, our sensory receptors are stimulated by our own actions. Notice, for example, what it feels like to move your foot in your shoe. This becomes even more obvious in a task such as pouring water into a glass. We continuously watch and listen to the water and likely feel the changing weight of the pitcher. We are in a dialogue with the pitcher, the glass, and the environment (including gravity), during which action and perception are calibrated together. The philosopher Marek McGann elegantly encapsulates this dialogue as follows:

> Cognition is a bit like a handshake. You can't make a handshake by yourself, and you can't really do it by accident. Handshakes involve reaching, and they involve a rather subtle form of negotiation: how firmly do you grip? How many pumps? How long do you hold on before you let go? Once you know how to shake hands though, you can do lots of interesting things, such as decide whether the person you are shaking hands with is confident or timorous, friendly or suspicious, interested or disdainful. So it is between you and the world in the case of cognition.[7]

It is therefore better, McGann suggests, to understand interaction with the environment not as an abstract process occurring in an isolated mind but rather as a skilled activity involving both the individual and the environment. The fundamental shift this requires in our interactions with our environments is realizing that perceptual cues meaningful to an individual gathering information from the world are not static but instead continuously changing as the goals of the perceiving organism change. In McGann's terms, the cues that guide actions such as pouring water or shaking hands can exist only in the process of interacting with the glass and pitcher or with the person who has offered their hand.

When it comes to your relationship with the environment, the third assumption to be questioned is that sensations are signals we receive via individual and separate sense organs. As the example of the bushfire at the beginning of this chapter illustrates, we do not need to see fire to know it is coming. The roar of flames, the smell of smoke, and the heat are sufficient. Any one of these cues would probably alert us to danger, but their combined presence leaves no doubt. Even without danger, the examples of moving trains and pouring water demonstrate that we rarely rely on information from a single sense modality. As you read this book, you are relying on different sources of information to perceive the presence of the book, to maintain a stable body posture, and to visually track the lines on the page while reading.

Letting go of old assumptions about the senses might be disconcerting, but it is also liberating. It allows us to consider:

- What perception might be if it is not merely a veridical representation of our physical environment
- What perception might be as an active engagement with the environment
- What perception might be if it concerns more than the processing of signals from sense organs

1.3 Theory, Experience, and Experiment Design

Let's take an important step back and review our subjective understandings of perception. In doing so, let us review the developments in theoretical understanding of experience, and consider studies designed to validate these theories. The idea that the senses are in dialogue with the moving, acting body aligns with thinking that has gained traction in recent years in the fields of perception, cognition, and interaction design. On the one hand, the ability to render stimuli in virtual environments has allowed researchers to design new kinds of stimuli for studies that systematically examine the role of movement in guiding action, a central tenant of the embodied approach to perception. For example, by tracking people walking toward closing gaps in virtual environments that may or may not follow the lawful behavior of real-world objects, such as closing elevator doors, researchers have observed consistent behaviors in participants as they try to determine whether they can pass through the gaps.[8] On the other hand, interaction designers are increasingly turning toward the field of embodied cognition to help them understand how people will interact with objects when moving about in augmented and virtual environments. Measures of "presence," someone's sense of being in a virtual environment, are heavily dependent on how participants feel that the environment responds to their actions. Developing and using measures of presence is therefore predicated on understanding the link between perception and action for people moving about in the real world.[9]

Interaction designers such as Kristina Höök go further by proposing that the body must be central to the design process before any hardware, software, or virtual world is even developed. Höök argues that one can arrive at a successful interaction design only by finding its movement vocabulary through a practice that involves exploring the shapes of movements, much as a writer finds the right way to express their thoughts by exploring the shapes of sentences.[10]

It is unfortunate that, at the same time as theories of embodiment from cognition and the multimodal sensing and display capabilities of interactive devices and environments are converging, the field of interaction design itself is undergoing a seismic shift, with an increasing move toward automating the process of interaction design. At the very point where we need to center the process of design on the capabilities of the human body, the human designer is being removed from the design process altogether. As noted in the ninth edition of the "State of UX Design" report from the UX collective, published in 2024, such automation can go only so far: "We'll always have to find the right use cases for new technologies, and we'll continue to see design expanding beyond flat screens with the popularization of augmented reality and multimodal AI Agents."[11] As we move away from interaction design centered on flat screens toward designing interactions for devices that track our movements and respond to our interactions in multimodal, often nonvisual ways—designs that can account for the fact that we are moving, thinking human bodies living in a multidimensional world—it helps to be a moving, thinking, embodied human designer. This is the argument at the very center of this book.

Like all fields that pursue a scientific method, fields that address human perception are engaged in negotiating tension between theoretical frameworks to guide research on a topic and designing empirical studies to generate data that test those theories. Asking whether experimental findings generate theories or whether theoretical frameworks guide scholars toward key findings is a chicken-or-egg exercise. Clearly, the ability to abstract a problem and formulate it as a framework for structuring a series of research studies is a useful way to constrain a body of work and avoid overlooking hypotheses to be tested. The social psychologist Susan Fiske points out, "Good theories posit causal relations, attempt coherence, form a good narrative, aim for parsimony, are testable, prove fertile, and solve problems."[12] This is not to say that two apparently contradictory theories cannot coexist. Many of the theoretical frameworks discussed in this book will appear to overlap and even to present contrary perspectives. This is tenable because the scientific endeavor is founded on the premise that a theory is not set in stone but rather a structure for thinking about a problem until another theory supplants it. Overlapping and contrary perspectives are useful because they signal places where open questions remain and where further assumptions need to be tested. Readers should also keep in mind that laboratory experiments are designed to test a specific hypothesis by ensuring that all confounding variables are controlled. This means that a laboratory experiment is always an abstraction, meaning that it is rarely representative of a situation that we might encounter in our everyday lives.

Any theory of perception, however fully evaluated, must also take into account the individual's subjective experience. For the designer of a perceptual study, the problem is that we are so attuned to our habitual environments that the processes by which we discovered the cues for negotiating everyday life remain in the background. Occasionally, we may be forced to foreground or discover these cues due to an encounter with a novel situation such as learning a new skill—how to ride a bicycle, how to play a musical instrument, and so on. These instances force us to rediscover the explicit process of connecting the actions performed with the sensory feedback received from them, reminiscent of early childhood, when we deliberately worked to figure out the connections between action and sensation. Most people will not remember what it felt like to learn to walk but can likely recall the sense of vulnerability of learning how to drive, swim, or ski, when the relationship between actions and associated sensory feedback was not automatic. There was no recognition of what aspects of the sensory milieu were salient, nor any separation of relevant cues from all the available sensory stimuli. At other times, these cues are uncovered by tears in the fabric of our perceptual assumptions, when there is a breakdown experienced as ambiguity, such as the one in the moving train example above.

The challenge faced by the designer of a perceptual study, or by the designer of an interactive system aimed at drawing on the bodily knowledge of unknown end users, is that bodily knowledge is, by definition, knowledge acquired by a *particular* body. A person brings to the interaction with the environment not only the unique anatomy and physiology of their body but also their individual experience and skills, their emotional state, their current level of fitness, their attention—all of the physiological, cognitive, and behavioral resources at their disposal. These resources are brought to bear on a particular task goal within a particular context. From the point of view of the individual perceiver, their experience of an interaction is thus subjective, immediate, and context dependent.

It is difficult for any study of perception, performance, or behavior to account for these complex and interdependent aspects of a user's experience. Many studies have sought to reduce complexity by narrowing and simplifying the tasks used in evaluations. Typically, an evaluation is constrained in scope, time, and space to rule out as many potential "confounds" as possible. But those same confounds may provide the very sensory and motor cues that shape the action being evaluated. Think again of the train example. The ambiguous visual cue in this context is linear vection, which relates to our ability to judge our own motion with respect to the position and motion of objects around us. When a large portion of the visual field appears to move while a small portion is stationary, a viewer will perceive that they, and not their surroundings, have moved. After all, this is what would normally be true.

Early studies that explored this ambiguity used the device of a "moving room." The observer was stationary with respect to the ground, and the walls of the room moved past them. Like the movement of a neighboring train, this resulted in changes only to the visual scene.[13] Constraining this study to a manipulation of the visual component of vection was useful in allowing researchers to isolate the effect of "visual flow" on people's perception of the direction of motion. In most real-world situations, however, the ambiguity that arises in the case of the train would be resolved by the presence of other cues indicating self-motion, including perhaps the movement of air on the skin, acoustic cues, and cues from the vestibular system. Thus, by simplifying experimental design, researchers can foreground the contribution of one sensory cue, but often at the expense of understanding the effects of cues from other sensory channels. When those constraints are removed, if the observers themselves are actively engaged in moving, other cues about self-motion that reveal aspects of interaction with the environment can be uncovered.

In a recent study, Zadra and Proffitt demonstrated that people's perception of effort during walking is calibrated to optic flow.[14] Participants walking on treadmills who were shown videos with higher rates of optic flow relative to their walking speed perceived that they were exerting greater effort. Zadra and Proffitt propose that people's perception of walkable distances may be measured not in stride length or walking speed, as was previously assumed, but rather in bioenergetic units, such as the magnitude of effort required to produce a given magnitude in optic flow. Although this study does not account for other sensory cues relating to self-motion, it demonstrates that active engagement by the observer reveals relationships between actions and responses from the environment that are ecologically relevant for a living, moving being.

In addition to constraints meant to reduce experimental confounds, what we learn from studies is shaped by what those studies set out to discover. We have learned answers to the questions the studies posed, but gaps remain because other questions simply were not asked. There can be a gap in the technological resources needed to ask these questions, to create needed stimuli, or to display stimuli appropriately. Research on spatial hearing, for example, advanced significantly once it became possible to generate, store, and reproduce multichannel sound. And research on cross-modal perception of space is now being propelled by the availability of head-mounted displays that offer coordinated visual and auditory scenes with appropriate depth and directional cues. These disparities in methods for designing and presenting stimuli for different sense modalities have resulted in disparities in our knowledge of different sensory systems. We are only now beginning to understand the chemical senses (olfaction and taste), whereas our knowledge of visual perception is vast, building on almost 150 years of continuous work by thousands of researchers. It was not until 2004 that a Nobel Prize was awarded to researchers studying olfaction,[15] whereas work on vision has been recognized since the Prize's inception.[16]

In short, what we know about perception is influenced by the questions researchers posed, by the technology at their disposal for providing appropriate stimuli, and by the theories that framed the work. The theoretical frameworks that inform this book approach perception from the perspective of a moving, thinking person with objectives to accomplish, some of which involve interacting with computational devices.

1.4 The Origin Story

The genesis of this book was a course I taught twenty years ago at the MIT Media Laboratory, introducing graduate students to basic facts about how human bodies interact with their environments and giving students a view into the literature on sensation, perception, and action. Thus, in addition to being a resource for designers, this book is intended as a starting point for readers interested in the interaction between our bodies and our real or simulated environments and who are unfamiliar with current theories and methods in the field. Like the course from which it stems, the book blends theory and practice to explore how the field can move forward. Participants in the original class included computer scientists, poets, painters, filmmakers, musicians, clothing designers, and interaction designers. All of them enriched their disciplinary perspectives through the design projects they undertook.

There are robust and important fields of scholarship addressing critical themes in interaction design such as disability, neurodiversity, and the politics and identity implications of embodiment. Although I highly encourage students and practitioners to engage with these evolving conversations, they are not a focus of this book. Here, I am concerned with introducing interaction practitioners to theories of embodied cognition and to ways in which these theories can be, or have been, incorporated into the design of interactive systems. It is a sad reflection on these fields that the bodies that have been studied are primarily those of able-bodied students at universities, and that studies have been designed to rule out any factors that might introduce "confounds," which are the factors that might actually reflect the range of abilities of the people who live in the world. As a person with a disability myself, I can only wish for the day when we will be included in such research as a matter of course and not relegated to siloed studies that focus on our disabilities. True, we may move differently, but in observing such difference, surely there is something valuable to be learned, some data to be gathered that can expand our knowledge and change how we design.

My hope is that this book, by re-centering the role of the moving, acting body in the design process, can encourage designers to build for all bodies. As the tools for tracking and analyzing movement expand to include artificial intelligence, there is no need to design for the movements of a "default" user. Systems can learn how users move and then calibrate their software to respond, just like voice-recognition systems do. My hope is that the material presented here will be a starting point to enable design practitioners to fundamentally rethink how interactive systems are designed, and I hope this book will reach interested readers, from an array of communities, who will be challenged to create new experiences, technologies, and artworks that reflect something meaningful about our interactions with the world we live in.

2 The Body in the World

2.1 Introduction

We are moving, thinking human bodies living in a multidimensional world. It should therefore come as no surprise that successful interaction design relies on being able to capture the movements of those who will use our systems and to respond to those movements with richly multimodal feedback. The argument at the center of this book is that the process of designing interactive systems and artifacts must be grounded in the capabilities of our bodies and in a deep understanding of how the world of our experience is brought forth. This is because the artifacts and systems we design are encountered by our users in their daily lives, and the experiences our systems bring forth are embedded within the context of everything else they are doing. Think of answering your phone while you are in the midst of doing another task, such as cooking or walking your dog. The experience of the world may drop into the background as you focus on your phone conversation, but the experience of interacting with the phone does not encompass everything you are doing or feeling.

In chapter 6, I will introduce design methodologies that seek to blend the experiences of interacting with physical artifacts and digital environments by centering the role of the lived experience of beings with bodies in blended worlds. But here, my goal is to present current research on how we perceive the movements of ourselves and others, and how such perception guides our actions. I then introduce interaction design methodologies grounded in movement practices, such as dance, to argue that designing for the moving bodies of our users requires us to become conversant with how our own bodies move.

In most traditional accounts of perception, the processes of sensing, interpreting, and responding are described as distinct steps to be carried out in sequence, almost like lines of code in a computer program. In this model, we sense light, sound, or touch (sensation), then we process sensory stimuli in order to perceive some property or object in the environment (cognition), and then we respond (action).[1] Philosopher Susan Hurley refers to this approach to perception and cognition as the classical sandwich, and it stands in sharp contrast to McGann's notion of cognition as a handshake. The sandwich model suggests that perception and action are separate, that they are peripheral to cognition, which acts as an interface between them, and that they are also distinct from the "higher processes" of cognition deemed necessary for perception to be possible. In this formulation, "The mind is a kind of sandwich, and cognition is the filling."[2]

This modular and somewhat static model of cognition has been questioned from several quarters. Significant among these are the ecological approach to perception and action based on the work of psychologist James Gibson in the mid-twentieth century, as well as a growing body of work grouped under the term "embodied cognition" that has emerged over the past fifty years. Below, I offer a discussion of these approaches, which are all predicated on the assertion that the ability to act in and understand the environment is dependent on the ongoing interaction between the moving body and that environment. The relationship between the structure of the human body and that of the environment determines what actions are possible and what responses are detectable. It is through the dynamic interaction between the body and the environment that we gain experience and become skilled inhabiters of our world.

The work presented in the following review deviates from cognitive theories suggesting that the representation of the world is a mental construct independent of sensorimotor information. Rather, the body helps us make sense of the world. This theoretical foundation underpins the material presented in the later chapters of this book. For readers familiar with the theoretical foundations of interaction design, this chapter will serve as a useful reminder. For those unfamiliar, this chapter offers critical understanding of how philosophers, psychologists, and other researchers have sought to understand how human agents (and other agents) perceive and interact with their environments. The chapter ends by offering an assessment of how interaction designers can move forward effectively in light of what these theoretical frameworks have to offer us.

2.2 The Ecological Approach to Perception and Action

The ecological approach to perception and action is a key theoretical starting point because of its emphasis on how an individual's actions are guided by and dependent on the structure of both the individual and their environment. After all, the designer of any artifact, from a chair to a computer mouse, is ultimately determining the user's experience of an interaction in an embedded relationship with the broader context of the world.

The ecological approach to perception arose primarily from the work of Gibson in the mid-twentieth century. Gibson claimed that an animal's ability to act in the environment is a function of the relationship between the structure of its body and the structure of the environment. An animal can act because it can perceive how the environment supports, or *affords*, certain opportunities for action. The feet of lizards afford opportunities for climbing because they possess adhesive properties sufficient to support their weight so that they can cling to vertical surfaces such as walls and rock faces. More radically, Gibson claimed that animals are guided not by energy in the form of light, sound, or mechanical forces but rather by observable higher-order properties in these energy sources such as proportions and ratios—properties that are "invariant" even as the animal's movements modify their relationship to their surroundings. In Gibson's world, there might exist many possibilities for action, only some of which are relevant for a given animal performing a given task. The individual may choose from the available affordances by selecting those most aligned with their current task goal.

Gibson arrived at this conclusion while working as a scientist with the US Army's Aviation Psychology Program during World War II. He observed that complex coordination tasks such as takeoffs and landings, or pursuit and evasion in flight, not only could be mastered by aircraft pilots but were routinely accomplished by insects and birds.[3] Whatever enabled organisms to accomplish skillful control of flight, Gibson surmised, must not depend on a high degree of cognitive processing of sensory stimuli. As a boy, Gibson had spent summers traveling through the Midwest on trains with his father, a railroad engineer. The young Gibson noticed that if he stood at the front of the train, the world flowing past him appeared to expand, whereas if he was looking out the back of the train, the flow converged. Pilots, Gibson later deduced, must use similar perceptions to control their planes in flight. He concluded that animals and people are aware of the world, not the sensations the world gives rise to.[4]

In his book *The Senses Considered as Perceptual Systems*, Gibson proposes that the senses make available the conscious qualities of experience, whereas perception is the means by which we pick up information from the environment. He points out that stimuli from individual sense organs are variable and susceptible to external conditions. But observable higher-order properties such as stimulus energy ratios and proportions are unvarying, and these invariants carry information for the perceiving animal. They remain constant with respect to movements of the observer and changes in stimulus intensity. Therefore, although energy levels at the receptor surfaces (retina, cochlea, skin) may vary, the invariant properties of stimulus energy correspond to properties of the environment that are permanent, such as the invariant patterns in light that encode information about an approaching object. "The active observer," Gibson argues, "gets invariant perceptions despite varying sensations."[5]

Gibson thus claims that perception depends upon what he terms "perceptual systems," in which sensory receptors can provide meaningful information only when they are coupled to motor capabilities that provide control over what is sensed. He identifies five such perceptual systems:

1. The basic orienting system, which provides information about the forces applied to the body, acceleration of the body, or parts of the body, and orientation to gravity.
2. The auditory system, which picks up information concerning the location and nature of vibrating sources.
3. The haptic system, which provides information about the physical properties of objects and surfaces through the stretching and compressing of receptors in skin, muscles, and joints.
4. The combined chemical senses of taste and smell, which, through chemical interactions with receptors, provide information about the chemical composition of odorants and tastants.
5. The visual system, which is sensitive to ambient light and, as such, provides access to information present in the structure of optic flow.[6]

What is important here is that Gibson distinguishes between sensations arising from perceptual systems and modes of perception involving the pickup of invariant patterns in the structure of sensory stimuli. It is by perceiving the invariant structure within, for example, optic flow, Gibson argues, that people and animals can distinguish their own motion from that of their surroundings. In either case, the image on the retina changes. Gibson proposes that the structure of such change would depend on the relationship between the observer and the moving image. As we noted in the train example in chapter 1, overall or global change in optic flow arises from the movement of the individual with respect to the environment, whereas local change (against a stationary background) arises from the movement of some object in the environment with respect to the observer. What is relevant for the observer is not the changing pattern of light but rather the lawful way in which that pattern changes. For example, the global optic flow associated with forward motion through the environment is characterized by regularity in how the image on the retina changes. Technically, the structure of the image is described in terms of a velocity vector field radiating from a focus of expansion—the center of the image expands outward as the viewer approaches the object (figure 2.1).

If the expansion is smooth, the assumption is that the observer is approaching a flat surface. Faster expansion inside a contour indicates the presence of an obstacle, whereas faster expansion outside a contour indicates an approaching opening such as a door.[7] Under normal circumstances, these lawful changes in the visual field provide cues that are invariant for people moving toward objects in their environment.

It is important to note here that such perceptual invariants are present in stimuli other than light. For example, the lawful patterns illustrated here for optic flow can also be observed in the ambient acoustic array. Higher-order variables related to the position, velocity, and time until arrival of a moving sound source, for instance, are specified by lower-order invariants such as interaural time delay, changes in average sound intensity, and, for faster-moving sources, pitch changes due to Doppler shift.[8] Rosenblum et al. demonstrated that both interaural time differences (ITDs; the difference in the times of arrival for a sound at each ear) and monaural intensity change (change in the level of a sound approaching one or another ear) are salient cues that, when correlated, define the point of closest travel for a source passing in front of a listener.[9] When the source is directly in front of the listener,

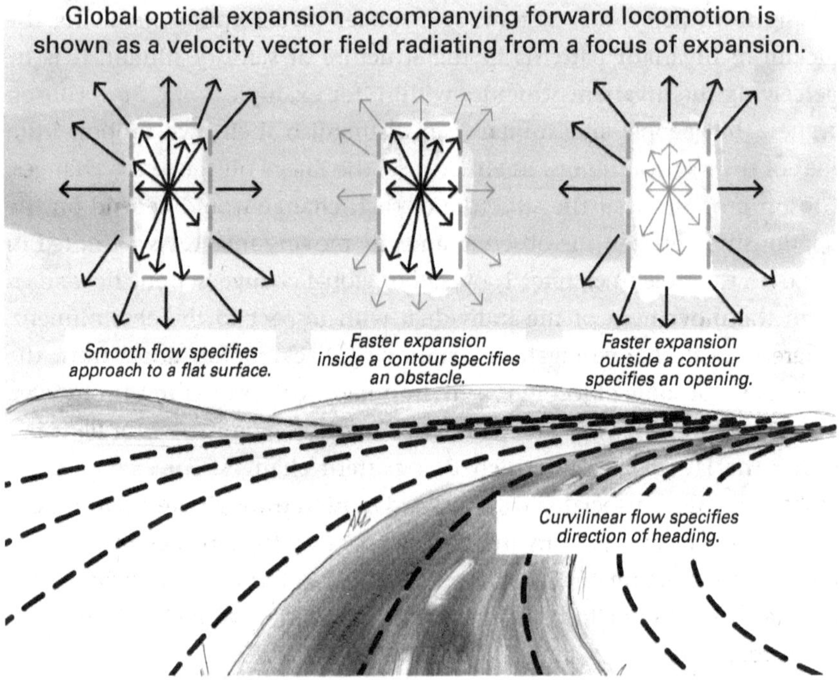

Figure 2.1

Visual flow accompanying forward motion. Global optical expansion accompanying forward locomotion is shown as a velocity vector field radiating from a focus of expansion. *Source:* Original drawing by Hans Anderson.

the interaural time delay is zero and monaural sound intensity is at a maximum. It is this lawful relationship that is invariant for an observer tracking the movement of an approaching sound source.

The key to Gibson's formulation of this ecological approach is that such lawfulness depends on the relationship between an organism and its environment—that is, on the ecology that encompasses them both in an animal–environment system. Gibson proposes that animals are not born with a priori knowledge of such relationships; they discover them through interacting with the environment. His "theory of affordances," meanwhile, asserts that an animal's environment presents meaningful possibilities for action, perceivable as relational properties to be discovered, regardless of an organism's intent to use them. In time, prior experience is brought to bear, so that animals do not exist in a perpetual exploratory mode; that is to say, they do not have to rediscover the lawful relations in the stimuli they encounter that reveal, for example, the presence of obstacles.

Later proponents of the ecological approach have refined some of Gibson's assertions. Turvey and others, for example, have challenged Gibson's description of affordances of an environment as simply being "what it offers the animal."[10] They claim that Gibson failed to connect his dots—that he did not adequately account for how invariant structures in light, sound, or touch could indicate to the animal what opportunities for action the environment might afford. Gibson himself realized it would be necessary to account for the process by which the physical world of surfaces surrounding an animal and the meaningful world of relevant properties of those surfaces could be related. He asked, "If there is information in light for the perception of surfaces, is there information for the perception of what they afford?"[11]

Turvey suggested that a better approach would be to define what he termed the "dispositional properties" of an individual object that, when encountered by an animal with a complementary dispositional property, would constitute an affordance.[12] In our example of the lizard, the dispositional property of the wall is that it is a vertical surface, whereas the dispositional property of the lizard is that it has feet that can adhere to a vertical surface. For the lizard, the wall "affords" climbing. Others have suggested that affordances are relational properties that can arise in an encounter between an animal and its environment.[13] In the case of the lizard climbing the wall, the "climbability" affordance only comes into being as an outcome of the relationship between the lizard's feet and the wall when the lizard's goal is to climb. The subtle distinction here is that, framed as relational properties, affordances can be both generally available to a species and specifically available to an individual member of that species in a given context.

Baggs and Chemero suggest a further distinction between the environment inhabited by a whole species (its habitat) and the environment that surrounds an individual member of a species (its *Umwelt*).[14] Affordances in this case serve a different purpose. With respect to habitat, affordances can be thought of as dispositional properties that exert evolutionary pressure on a species, such as the structure of feet that allows lizards to climb vertical surfaces. With respect to the *Umwelt*—the world surrounding an individual—affordances are relational properties relevant for an individual in their environment. The stickiness of a particular lizard's feet is relevant when that particular lizard encounters a wall. These affordances are not static but rather change as the animal changes: when it grows, becomes more fit, or acquires new skills.

A final distinction between habitat and *Umwelt* is the role action plays in guiding perception. In the case of a species and its habitat, affordances are defined in terms of the generalized opportunities for action among members of the species that may or may not be taken. In the *Umwelt* of an individual, actions are performed and reveal lawful relationships within the global energy array that can guide that individual's future actions. The global energy array, as defined by proponents of the ecological approach to perception, is composed of all the sources of energy available to an individual animal—electromagnetic energy usually perceived as light, mechanical energy such as force, vibration picked up by receptors in skin and joints, and so on. For a lizard, for example, the change in the direction of force applied to its body by gravity as it climbs is a signal that can guide its actions, in that it will change how it moves its legs and feet to adhere to the surface.

In the ecological approach to perception and action, the opportunities for action afforded the individual are functions of their ability to discover what they can do, learn from past experience, and adapt to changes that may disrupt their progress toward accomplishing a goal. Applied to the design context, all of these qualities are present in the interaction between a user and any technological artifact. The designer of the artifact is ultimately responsible for determining the user's experience of an interaction, embedded in the broader context of the world. As designers, we build components of the *Umwelt* that our users inhabit, and we provide at least part of the structure that will guide a user's actions toward accomplishing goals. The digital map that guides a driver and the electronic stud finder that guides a carpenter are examples of new ways that digital artifacts can assist their users in discovering possibilities for action in their environment. Such technologies introduce new sources of information that a user can track as they move—the lines converging on the map or the sounds that home in on the position of a stud in a wall. Throughout this book, I will introduce design methodologies informed by concepts that inherit from the ecological approach to perception and action, many of which find their home within the field of embodied interaction design.

2.3 Embodied Cognition

Building on Gibson's work, proponents of embodied cognition have developed a richer understanding of the interdependence of body, mind, and environment in meaning-making. While Gibson was discovering how the moving, acting body could pick up and act upon perceptually salient information from the array of ambient energy around it, scholars from other disciplines were advancing the broader notion that, beyond perception, understanding one's environment must depend on having a body to actively engage with that environment. Cognition is said to be embodied when the understanding an organism can acquire of its environment is dependent on the physical features of the body it brings into that environment, and when a range of sensory features of the body directly affect cognitive processing.[15] This perspective contrasts with the Cartesian view of cognition, which proposes that perception is purely an intellectual pursuit independent of the senses.[16]

The embodied cognition framework emphasizes the interdependence of body, mind, and environment in precisely the relationship that McGann described when he characterized an interaction with the world as a handshake. In this framework, both the body and the environment are necessary for meaning to emerge. Embodied cognition as a research area is grounded in perspectives and methods from scholars in disciplines as diverse as phenomenology, philosophy, psychology, robotics, and HCI. Early attempts at unifying language and methodological approaches from these fields resulted in contention over exactly what embodied cognition means, but as the research has matured, some central concepts have emerged. Shapiro, in particular, has provided a useful framing of three emergent themes.[17] He acknowledges doing this not to arrive at an exhaustive account of the diverse perspectives on these concepts but rather because any account of embodied cognition that omitted these topics would be incomplete.

The first of Shapiro's themes, conceptualization, captures the idea that an organism's ability to conceptualize its environment depends on the kind of body that it has. Altering the structure of that body would therefore alter its understanding of how the world functions. Evidence for this can be found in a landmark study by Adolph et al., who compared the behavior of fourteen-month-old toddlers and eight-and-a-half-month-old infants faced with slopes of differing steepness.[18] When encouraged to descend steeper slopes, most of the toddlers switched from walking to sliding, whereas crawling infants plunged down, even though they might fall headfirst. The authors concluded that the transition from crawling to walking resulted in a need for toddlers to rediscover the affordances of slopes and thus to reassess which slopes were navigable. Toddlers don't acquire new bodies when they learn to walk, but they do acquire new locomotor skills that reorient their relationships to their environments guided by the capabilities of their postures.

In a still more compelling series of studies, Smith and Thelen and colleagues showed that newborn stepping, an involuntary behavior that infants perform when they are held upright and their feet are touching a surface, disappears at around the age of two months; the behavior is a precursor to walking behavior that is first disrupted and then reestablished as the baby gains the strength and muscle mass to support its rapidly growing body.[19] In support of this hypothesis, the authors noted that the loss of this newborn stepping behavior was most rapid in babies who gained weight more quickly. The researchers demonstrated this behavior modification by varying critical environmental conditions for the infants. The authors determined that at around the age of three months, infants pass through the first of two transitions: the disassembly of coordinated stepping behavior in the face of rapid changes in their bodies. The second transition, at around one year, marks the reestablishment of stepping behavior once the infant's body is strong enough to support its weight. The authors found that by changing the environmental constraints for these infants—by supporting their bodies in water, effectively removing some body weight—they could prompt the second transition about six months before infants would normally support their own weight in upright walking. In short, Smith and Thelen's work provides supporting evidence for the first of Shapiro's three themes, conceptualization, by demonstrating that, in these cases, infants were using their coordinated interactions with their environments to reconceptualize their altered bodies.

Shapiro's second theme, replacement, concerns the positioning of the field of embodied cognition with respect to established approaches. Formulating the relationship between an organism and its environment as an interactive and interdependent exchange of energy replaces accounts of cognition that depend on representational states, abstract the process of cognition from the mechanisms of sensory processing and motor control, or rely on algorithmic processes to account for the interaction between the body and the mind.

Hurley and McGann are among the many proponents of the embodied perspective that argue for such replacement.[20] Moreover, the work of Smith and Thelen, described above, gives concrete evidence that an embodied approach to cognition can empirically account for behavior in real-world contexts through a systematic methodology for observing cognition in the wild.

The third theme that Shapiro draws attention to is constitution. He refers here to the claims for embodied cognition that the body is a constituent in cognition, not just a causal influence. To illustrate this distinction, he points out that oxygen is a constituent of water (with two molecules of hydrogen) but a causal component of an explosion after a fuse has been ignited.[21] Thus, for bodily knowledge to be acquired, the body must be a constituent in the process of acquiring knowledge. In learning to ride a bicycle, for example, the body constitutes a part of the rider–bicycle system in a more engaged way than just sitting on the saddle, pushing the pedals, and waiting to see or feel what happens. The experienced rider anticipates future states of the system, such as tilting their body to prepare to turn a corner. The constitution hypothesis suggests that the body as a whole is a constituent component of cognition and that, without a body, nothing can be known.

Although Shapiro's themes are helpful in highlighting the central tenets of the embodied cognition approach, they are insufficient on their own. It is also imperative to understand the schools of thought that led to its emergence as a framework. The following pages address approaches that are relevant for interaction design, beginning with a brief side step to talk about phenomenology, as many approaches to embodied cognition build on this work.

2.3.1 Phenomenology and Embodied Cognition

Phenomenology, founded on the work begun by philosopher Edmund Husserl in the 1890s, is the philosophical study of what experience is like. Husserl's phenomenology rests on the central idea of time consciousness, our awareness of the continual stream of time in which we encounter objects, perform intentional acts, and reflect on our consciousness. His radical claim was that in order to understand experience, we must step outside of conscious awareness of the world and focus awareness on the experiences to which the world gives rise. This reductionist approach was considered too radical by later phenomenologists, who believed that Husserl was returning to the kind of Kantian transcendental idealism that they were trying to escape.[22]

Among those who sought to move beyond Husserl's approach was Martin Heidegger, who was Husserl's research assistant and ultimately his successor as a professor of philosophy at the Albert Ludwig University, Freiburg. Heidegger's seminal work, *Being and Time*,[23] has had an enduring influence on European philosophy and on fields as diverse as architecture, literary theory, theology, and, of course, cognitive science.[24] In *Being and Time*, Heidegger departs from Husserl's thinking in important ways. Husserl's insistence that objects are perceived on the basis of a priori categories of sense impressions becomes, for Heidegger, a search for more fundamental categories of experience that determine what it is *to be*. Husserl's notion of self, the transcendent ego, is rendered by Heidegger as Dasein, a social being with prior knowledge of how to exist in a given social context (such as how to be a student in a classroom or a sibling at home). Finally, Husserl's idea of intentionality, the mode of interacting with objects in the world, is supplanted by Heidegger's "being-in-the-world" (*In-der-welt-sein*), an attitude of openness to the world of potential opportunities.[25]

Being-in-the-world is of interest here because it has come to the attention of researchers in the field of HCI, most notably through the work of Winograd and Flores and of Paul Dourish.[26] Dourish draws attention to Heidegger's proposal that the things around us are, for the most part, objects that we use to accomplish tasks—tools or things we co-opt as tools. Even our office becomes a workshop, tooled to allow us to accomplish the work that must happen there. It has a desk, a computer, and perhaps some pens. A different workshop, such as a kitchen, will have different tools, appropriate for performing different work. The point is that in each context, the tools that are there are given meaning by what they are to be used for: "their 'being'—that is, what they are—is understood in terms of what they are for."[27]

Consequently, Heidegger argues, if tools are to be useful, we must approach them as engaged users. The alternative mode of encountering such equipment, Heidegger argues, is to *gawk* in an attitude of detached observation. If you think about something you use every day, such as your phone, you can imagine the difference between your awareness of the phone while you are practically engaged with it and your awareness of it when you are gawking at it, thinking of it as a desirable product to be admired. In engaging with the phone, Heidegger would suggest, we are using it for a purpose; when we are gawking at the phone, we see it as an object rather than experiencing it as a tool. In each case, what we experience is different. As equipment, the phone is experienced as a communication device that functions well or not so well; as an object, the phone is experienced as a thing with a shape, a color, a design—qualities that you might not be consciously aware of when you are using it as a phone. For Heidegger, these two attitudes represent different ways of being: *zuhanden* and *vorhanden*. When you are making a call, engaged with the phone as a piece of equipment, the phone is for you *zuhanden*, or ready to hand. When the phone is an object of study, it is *vorhanden*, or present at hand.

Heidegger argues for a further distinction between what a thing is and what it is available or usable for. Most philosophical accounts of objects describe them in terms of what they are—their basic properties, such as shape and material composition—and these objects derive their purpose from this basic description. Heidegger claims that this is the wrong way around—that in purposefully engaging with the things that surround us, we encounter equipment in Dasein, as "skillful, familiar, disposed, purposive, caring" beings deeply situated in our world.[28]

Skill, for Heidegger, is the ability to know how to purposefully engage with equipment. If nothing goes wrong, the tool remains transparent, and we can focus on the task—the tool is *zuhanden*. If, however, something goes wrong, and the equipment fails to function correctly, it is no longer available, no longer ready for use, and it reverts to being an object of our attention that is *vorhanden*. It can be difficult to independently observe this process of breakdown happening because it may occur suddenly, experienced only by the individual using the tool. This is the aspect of Heidegger's phenomenology that most interested Winograd and Flores and later Dourish, each of whom pointed out that users of technology are often frustrated because some aspect of an interaction forces them into this distanced attitude because something has not functioned as they expected.[29]

What is striking about Heidegger's account of purposeful, skilled engagement with the world is that he rarely mentions perception of that world or the role of the body in coming to grips with its equipment. The phenomenologist who most directly addresses this oversight, who takes on the challenge of extending Heidegger's concept of being-in-the-world to include the capabilities of the body, is French philosopher Maurice Merleau-Ponty. He proposes that the attitude or intentionality implicit in being-in-the-world is primarily a motor intentionality, bound up in the skillful actions through which we encounter the world. For Merleau-Ponty, action and perception are not separate. Perception is a bodily process only possible through action. For many proponents of embodied cognition, Merleau-Ponty's *Phenomenology of Perception* represents a point of departure and is considered a pioneering work in this field.[30]

Merleau-Ponty proposes that phenomenology is both a study of essences, such as perception and consciousness, and a method for understanding experience, particularly as it relates to our interactions with the encountered world. Central to this proposal is a distinction first discussed by Husserl between the body as an object (denoted by the German word *Körper*) and the body as we experience it (denoted as *Leib*). For Merleau-Ponty, *Leib*, or the lived body, is the body of experience: the body that can see, hear, touch, move, and so on. He writes, "I can only understand the function of the living body by accomplishing it and to the extent that I am a body that rises up towards the world."[31] In other words, it is the perception of one's own lived body that makes perception of everything else possible, and it is impossible to develop a theory of perception that does not account for this first-person perspective. The world in which the lived body exists is not simply an arrangement of objects in space but rather a space of possibilities.[32] Through the habits and skills of the lived body, we are open to these possibilities.

Merleau-Ponty refers to our readiness for, or openness to, deploying our skills toward the world as the "body schema." The body schema makes available the space for actions, a space that situates experience; this is a lived space rather than a distanced one. The way we reach for and grasp a cup, for instance, gives form to that cup as an object we can reach and grasp.[33] Foreshadowing the work of Gibson, who was certainly familiar with his writing, Merleau-Ponty also emphasizes that the concept of the body schema demonstrates that the context for movement and sensation is governed not by basic sensations or sensible qualities but rather by the way movement structures the environment. Our ability to sense, see, hear, or touch is inseparable from our ability to act; we learn to sense as a part of acquiring the habits and skills that we need for coping. As our bodies grow, these skills and habits are reorganized in coordination with our changing relationship to the environment—think again of the stepping baby. It is this malleability, the body's ability to be open to any of many possibilities for coping in a given context, that provides evidence for experience that is rooted in bodily skill. Therefore, as we climb stairs that we have never encountered before, we bring to this action prior bodily knowledge of stair climbing that is adaptive and can be brought to bear on skillfully coping with this particular set of stairs.

Although there are many more ways *Phenomenology of Perception* is considered a fundamental precursor to embodied cognition, I will focus on just one other point: Merleau-Ponty's discussion of how we understand the relationship between bodily space and our representation of objective space. Because our bodies are always with us, he claims, we do not need to represent them in objective space, and it makes no sense to think of them with respect to a set of spatial coordinates. Nor is objective space simply the space outside our bodies. As Kaufer and Chemero state: "When we act skillfully, the objects we use skillfully become extensions of our own bodies, or the body and its tools become part of the same system."[34] In familiar situations, our tools become extensions of our bodies, incorporated into our bodily space of perception and action. To illustrate this, Merleau-Ponty uses the case of the blind man walking with a cane. The tip of the cane is the means by which the world is touched, and its extent is quite literally "incorporated." The cane is, for that task, part of the body schema.

In situations of skillful use of familiar tools, there arise familiar experiences or signs to guide further actions. Note here that Merleau-Ponty goes beyond both Heidegger's notion of coming to grips with equipment, which still requires the equipment to be separate and available to be gripped, and Gibson's notion of affordances, which are opportunities for action that are present, regardless of whether there is an intent to act. For Merleau-Ponty, accomplishing the goal of a task is a process of being meaningfully directed toward the object of that goal, and this, in turn, is a result of the bodily skills and habits we can bring to bear on that task. So, the hammer we use to drive a nail into the wall is not simply an object that affords hammering, nor is it just the hammering equipment that is familiar and transparent in use. It is a task-dependent extension of our body schema that provides a means of coupling the action of hammering the nail to the perception of the progress of that action. Like the blind man's cane, it extends the body not only by extending the individual's reach but also by providing an appropriate structural coupling between the individual's body and the environment, a coupling that can perform tasks that would be inappropriate or impossible for the body to accomplish on its own.[35]

It is no surprise that many ideas central to Merleau-Ponty's thinking have reemerged in the work of proponents of embodied cognition. His emphasis on perception as being possible only through action, his concept of the body schema as a readiness or openness to deploying skills adaptively in the world, and his proposal that the tools we use can be incorporated as we skillfully cope with the world are all aspects of his thinking that are reflected in most writings in this area. Moreover, these concepts increasingly find their way into practical approaches for the design of the digital tools and devices we bring into our environments today. As technologies become more immersive, the ability to design systems that enable users to accomplish increasingly complex interactions within these environments also increases. It is therefore a challenge to craft the kinds of environments where users can experience the adaptive, skillful coping they draw upon in the lived world.[36]

2.3.2 Phenomenology Through the Modeling Looking Glass

So far, our discussions of Gibson's ecological approach, embodied cognition, and phenomenology have been abstract. This fact is not lost on either the proponents or the opponents of the embodied cognition program. Opponents say that embodied cognition has little concrete evidence to support its claims, in contrast to more traditional, representation-oriented approaches characterized by emphasis on empirical studies of behavior. Proponents of embodied cognition, meanwhile, have sought concrete, formal methods to characterize or model the unfolding interactions between organisms and their environments as observed in real-world, ecologically grounded situations. They have sought to observe and model phenomenology at work by applying techniques from fields of study as diverse as dynamical systems theory (DST), biology, robotics, and human motor control. All of their approaches start from the premise that an organism and its environment can be formally modeled as a coupled dynamical system, and that there is a way to describe how the relationship evolves over time. But it is important to distinguish between a model a scientist might use to describe the behavior of a system in the abstract and an internal mental model that traditional cognitive science proposes must exist in the individual mind to control behavior in the world. In the discussion that follows, we are concerned with the former, with modeling as a methodology for describing observed or predicted behavior that is not generated by an internal controller but rather arises from the interaction between an individual and the environment. This, after all, is where our work as designers is situated. If we can develop an understanding of how people discover possibilities for action in their environments when it includes the artifacts we design, then the likelihood of supporting them in accomplishing the tasks that depend on these artifacts will be greatly increased.

Notably, the following approaches to embodied cognition also solve a problem arising from Gibson's theory of affordances. In the environment Gibson describes, infinite potential affordances or opportunities for action are available to individuals as they go about their activities, but only some are relevant to the task at hand. How, then, does the individual determine which among all possible affordances are relevant? Cutting pointed out that although Gibson emphasized an animal's agency in navigating and accomplishing tasks in its environment, he failed to provide a parallel theory to account for the organism's motivation to behave in a particular way[37]—or, in Cutting's words, "My behavior is virtually unconstrained by its affordances"[38] (see also section 5.4).

In their 2012 paper, Withagen and colleagues provide a framework to address Gibson's oversight.[39] They suggest that although an environment may offer an agent many opportunities for action, the agent does not perceive all opportunities in the same way. They cite a study from the field of industrial design on the movement of automatic doors.[40] They suggest that when doors opened slowly or paused, people felt the door movements were inviting them in, but when doors opened and then quickly closed, people perceived the doors as reluctant to let them in. Withagen et al. suggest that manipulating the way the doors moved changed the likelihood that people would or would not enter. This, they suggest, demonstrates that the environment is not a neutral collection of action possibilities. Rather, the environment can "invite" an individual to take certain actions or can discourage them. It is up to the individual to respond or not.

Understanding the behavior of an individual in an environment, therefore, is also a matter of understanding how the agent chooses to respond to the environment's invitations to action. This, they suggest, is how the agent manages to attend to the affordances in the environment that are relevant to an immediate task goal. If my goal does not involve entering a store, I will pass by its automatic doors, no matter how inviting they may be. As we shall see, this notion of attraction toward an action possibility will prove to be a key feature that appears in many models of the coupled agent–environment system reviewed below.

2.3.2.1 Radical embodied cognitive science Kugler et al. propose another way to solve the problem of selecting among affordances without the need for the brain to act as a central planner.[41] They suggest that the actions of an individual, coupled with their environment, are self-organizing. That is, the actions of the coupled organism–environment system exhibit patterns of behavior that emerge from the way the parts of the system interact rather than being governed by some central plan. Self-organizing systems are typically modeled using methods from DST. A system, in this case a cognitive agent coupled with its environment, is described in terms of quantitative variables that are interdependent and concurrent, that change continually, and that can be framed in terms of the laws of dynamics expressed as nonlinear differential equations. The goal is to model perception, action, and cognition by finding parameters and variables describing the patterns observable in a given task.

In their 1980 paper, Kugler et al. suggest that when two limbs are engaged in a coordinated task such as swinging, they can be considered to be oscillating in a coordinated or coupled manner.[42] This behavior will persist only so long as there is a source of energy to prevent them from slowing and stopping. Kelso demonstrated this idea using a finger-wagging task, where people were asked to wag their two index fingers from side to side in time with a metronome. He noticed that, given no further guidance, people would wag their fingers either in the same direction (in phase) or in opposite directions (180 degrees out of phase). As the speed of the metronome increased, the out-of-phase option seemed to fall apart, and people converged on the in-phase behavior. This pattern was later captured in a dynamical model, the Haken–Kelso–Bunz model. This model and its subsequent iterations capture three aspects of such movement patterns: the initial tendency of coordinated movements to be attracted toward either in-phase or antiphase coupling, the transition of antiphase to in-phase with increasing frequency of cycles, and the asymmetry, or hysteresis, observable in this transition. Hysteresis arises because once antiphase coordination has transitioned to in-phase behavior due to increased frequency, there is not a transition back to antiphase behavior if the frequency of movement is slowed down again. In describing the motion of wagging fingers in terms of the relative phase of coupled oscillators, Kelso and his colleagues elegantly provided a means of modeling all sorts of coordinated behaviors involving the symmetric movement of limbs, from waving arms to swinging feet.[43]

Models from DST are useful for embodied cognition for several reasons. First, they provide a means to describe observed behavior using quantitative variables and parameters and mathematical models that can account for a given behavior. Second, such models can serve not only when we observe current behavior but also when we seek to predict the conditions under which such behavior might be sustained or might break down, as described in Smith and Thelen on newborn stepping.[44] Finally, because DST can account for variables that might belong to the agent, the system, or a combination of both, DST provides a formal method for explaining the ongoing dynamical behavior of an agent coupled to its environment. As Kaufer and Chemero point out, DST satisfies Gibson's two main concerns: that perception should be describable in terms of how it functions to guide action, and that neither perception of the world nor action in the world should be derived from an internal representation of the world.[45]

Applying models from DST to observed and predicted patterns of behavior in self-organizing body–environment systems has engendered a thread of work in embodied cognition that Chemero calls "radical embodied cognitive science."[46] By combining theories from ecological psychology with methods derived from DST, it is possible to explain the distinction between affordances of an environment and invitations to action within that environment in a way that ecological approaches cannot. These kinds of attractors toward particular action opportunities, as in the example of automatic doors, can be modeled similarly to wagging fingers because the tendency for fingers to become coordinated is itself an attractor, an invitation to act in a particular way. Thus, Chemero argues, the combination of these methodologies can address the problem of selecting among infinite affordances in a way that the ecological approach cannot. Finally, whereas DST seeks to describe behavior in terms of parameters, variables, and models, radical embodied cognitive science rejects any suggestion that the individual is engaged in any sort of computational processing when selecting among available affordances to guide actions toward a goal.

The ability to observe and model behavior is a fundamental skill for any interaction designer. Sometimes, such observations reveal patterns of behavior at a macro level, such as people's attitudes toward automatic doors. At other times, being able to observe and model movement patterns in data picked up from sensors that track a user's actions at a much more fine-grain level, such as finger tapping, can help guide the design of algorithms to more accurately predict a user's intent. These predictive algorithms often function to increase the usability and accuracy of interactive systems that rely on tracking a user's movements. I will return to this topic in chapter 3.

2.3.2.2 The enactive approach to cognition Whereas the radical approach to embodied cognition seeks to extend the ecological approach by employing methods from DST to model the behavior of individuals and their environments, the enactive approach to cognition finds its inspiration in the behavior of the underlying biological systems themselves.[47] The term "enaction" and the first substantive explication of the enactive approach to cognition appeared in *The Embodied Mind* by Varela et al.[48] The enactive approach goes beyond radical embodiment not only by proposing that individuals select from available affordances that invite attention as guidance for action but also by asserting that cognition is itself the action by which living beings *bring forth* a meaningful world. Without action, there is no meaning, no cognition; without cognition, there is no action.

Like proponents of radical embodiment, the enactivists seek a way to model the behavior of living systems that accounts for their coupled interactions with the environment. They claim that a living organism has to exhibit two properties: self-creation and self-maintenance. This led to the theory of autopoiesis as a means to mathematically model life and living systems. Autopoiesis is defined as "the property of a living system (such as a bacterial cell or a multicellular organism) that allows it to maintain and renew itself by regulating its composition and conserving its boundaries."[49] Therefore, the autopoietic system must be operationally closed. Kaufer and Chemero provide the example of a system involving two chemical reactions. The first is a catalyst for the second, which, in turn, is a catalyst for the first. The system is said to be operationally closed because it can keep catalyzing itself indefinitely, assuming no loss of energy.[50] Further, the autopoietic system must be structurally coupled to its environment. An important extension to this original formulation of autopoiesis was to state that such a system should be not only autonomous but also adaptive. Given some perturbation to the system that pushes it toward failure, it should be able to adapt to its changed conditions and return to a stable state.[51] This, Di Paolo argues, is what distinguishes a living system from one that is merely autonomous.

The enactive approach has given rise to subdomains of scholarship, one of which focuses on further developing the theory of autopoiesis.[52] Autopoiesis, proponents suggest, is central to the enactive approach because it is how enactivists can address the problem posed by Merleau-Ponty: How can we account for the relationship between the objective body that we see in the mirror and the lived body of experience? The answer, enactivists claim, is autopoiesis. The objective body is, for all intents and purposes, operationally closed in that it can sustain its own activities and regenerate itself independently from its environment. Operational closure also implies a separation between the organism and the environment. As Kaufer and Chemero explain, this implies the emergence of a self, which in turn implies a corresponding world of nonself. The fundamental point about the enactive approach is that the world of the nonself emerges only through the actions of the autopoietic system.

In a paper discussing the self-organizing origins of agency, Kelso quotes Piaget's description of a twelve-week-old baby, Laurent, who, having been surprised that the rattle in his hand made noise when he moved, then purposefully moved to reproduce the sound.[53] Kelso argues that, by repeating the action, the baby was figuring out how the changing auditory and visual stimuli were related to the felt dynamics of the movement of his arm. The rattle temporarily became an extension of the baby's body, coupled to his arm through his grip on its handle. By shaking it, the baby made a sound that was under his control and followed the dynamics of his movements. The sound is distinct from other sounds around him and provides auditory proof that he is a self-motivating being, distinct from the world of the nonself in which he is situated. In this way, the self and the world are said to co-emerge—a process enactivists refer to as "sense-making."

So, sense-making is the process by which an organism maintains a meaningful relationship to its environment. Meaning does not exist independently of action; it coevolves in accordance with the activities of the organism. Or, as Di Paolo puts it, in the enactive approach, "cognition is sense-making in interaction."[54] The properties of the environment that are salient for an organism also have valence. In the process of sense-making, the organism will be drawn toward conditions that support success and will be repelled by conditions likely to lead to harm or failure. These basic action tendencies of approach and avoidance are the simplest forms of emotional response and the basis for any intentional behavior. If I am repelled by something in my surroundings, I will avoid it.[55]

The key assertion of the enactive approach is that the world of experience is the world *brought forth* by the actions of the biological organism. If I put on skis and stand on a snowy mountain, the world that is brought forth for me is the world that is skiable, a world of slippery slopes that has meaning only when my feet are equipped for sliding and I can experience what it is to ski. My ski-clad feet now form the structural coupling to the skiable world of the mountain.[56] Note how this assertion differs from Gibson's notion of affordances—affordances are meaningful possibilities for action available to an organism to be discovered, but the skiability of the mountain has meaning only once a skier puts on skis and brings forth the skiable world, which will include the snow, the wind conditions, and so on.

A final point to make about the enactive approach concerns what enactivists term "incorporation," which relates to Merleau-Ponty's notion of body schema and refers to the ability of a system to co-opt resources beyond its body so that they constitute part of the coupling between the autonomous biological entity and its environment. For a skier, the skis are a case in point. The boards on their feet are no longer objects attached to their body but rather provide a transparent coupling to the skiable world. For Merleau-Ponty's blind man, too, his lifelong experience exploring the world elevates his cane from a simple tool co-opted in the moment to the means through which he feels the path he is traveling.

In summary, the enactive approach provides a way to solve Merleau-Ponty's question of how the objective body and the lived body can coexist. For the enactivists, they coexist because a biological body, the autopoietic entity, brings forth its world of experience through its actions, its structural coupling to the environment as a lived body. The world can, in fact, only be experienced or brought forth through action, which implies that only a living being is capable of having experience. In this way, the enactive approach claims to provide an account of the lived body that can accommodate both its objective biological structure and its first-person phenomenal experience. Although there are many more ways the enactive approach addresses the connection between biological life and lived experience, the fundamental idea is that the world of experience for a living being is brought forth by its actions. Experience arises from the sense the organism can make of its environment by means of its actions upon that environment.

2.3.2.3 The sensorimotor approach As the enactive approach began to grow away from its original biologically inspired roots, some tears began to appear in the fabric of the original formulation. Some proponents continued to pursue biologically inspired models based on autopoiesis, whereas others, such as O'Regan, Noë, and Hurley, moved away from biodynamics and notions of the coproduction of organism and environment. Instead, they began to focus on the phenomenal and goal-directed nature of perceptual experience.[57] They developed a theory of action-dependent perception, which they term the "sensorimotor approach." They claim we are sensitive to correlations, or regularities, in the relationship between our bodily movements and the sensations they arouse.[58] They refer to these regularities as "sensorimotor contingencies" (SMCs). For example, when one runs a hand from left to right across a textured surface, the texture slides beneath the hand moving from right to left. Similarly when one turns the head to the left, the image of one's surroundings will move across the retinae in the opposite direction. Perception, in this approach, is defined as the ability to recognize these sensorimotor relationships as regular or, in the language of SMCs, to master the sensorimotor contingencies for a given situation. As McGann points out, "The regularities that constitute these SMCs are not simply part of the world, nor part of our sense organs, but arise because both the world and our sense organs are relatively stable, and therefore the interactions between the world and our sense organs will have relatively stable characteristics."[59]

The sensorimotor approach stands apart from the biologically inspired enactive approach in another important way. Its proponents claim that, in any experience, there must be awareness of the SMCs that give rise to that experience. Because different senses rely on different motor behaviors, researchers argue that the sensorimotor approach provides evidence for senses being different from each other. As evidence, they cite experiments in sensory substitution, where information picked up using the actions associated with one sense modality is presented to the body through a second, substitute sense. They use the example of Bach-y-Rita et al. who mounted cameras either in front of or on the bodies of blind participants.[60] As the camera was moved, or as participants actively moved the camera by moving their heads, the images it picked up were "translated" into tactile stimuli presented to another body site, the participant's back or stomach, a technique they called "tactile visual sensory substitution" (TVSS). They claimed that, after training, participants could "see" what was in front of them and began to use patterns of head movement usually associated with acquiring information through vision to generate sensory stimulation (now presented through touch).

Auvray et al. have reviewed studies on TVSS more rigorously.[61] They observe that participants are more likely to report that the sensations experienced arose from the exploration of an external object when they could precisely control the relationship between their movements and the stimuli from the tactile device directly in contact with their bodies. From their survey, Auvray et al. concluded that the key to TVSS was that participants could master a new set of SMCs associated with this new sense of touching objects they could not reach, by learning to observe the systematic covariation between their own movements controlling the camera and the changes in tactile stimulation on the skin. For any form of sensory substitution to be successful, they propose, the sensations arising from SMCs must be attributed to the presence of a distal (in this case untouchable) object. In turn, if distal attribution is to occur, three conditions must be met:

1. The observer's movements must be coupled to the sensory stimuli they are receiving.
2. There must be a distal object present to be perceived.
3. There must be an external space that is separate and distinct from both the object and the observer—what Auvray et al. term the "perceptual space."

Perceptual space is that in which both viewer and object are situated—the space that defines the relationship between the two. As the viewer moves to observe the object from a different angle, the patterns of light falling on the lens of the camera in the TVSS device will change. But certain structures within those patterns will be invariant. Auvray et al. argue that these invariant structures in light falling on the camera's lens and then the user's back are as meaningful for the user of the TVSS system as they would be for someone perceiving the same patterns falling on their retinae, even if these patterns are sensed by the skin. This raises the question of whether the experience of the observer is visual or spatial. If the former, one could say that sensory substitution has occurred; if the latter, one must say that distal attribution (the attribution of sensations to properties of the distant object) has occurred.

Thus, the sensorimotor approach to perception inherits from the enactive approach the central concept of action-dependent perception: the proposal that perceptual experience is shaped by the actions we perform. But it departs from the enactive proposal that our actions bring forth the world of experience and meaning. Rather, it focuses on how we come to master the relationship between actions and sensations, so that these relationships can meaningfully guide us in our interactions with the world. Enactivists, following the thread of autopoiesis, think about structural couplings such as that of the skis bringing forth a skiable world. The sensorimotor approach suggests that, in skiing, we must learn to master new SMCs, a new relationship between our ski-clad feet and the snowy mountain. Chapter 3 discusses the implications of these subtle differences for the design of interactive systems, as the various aspects of embodied cognition become most relevant when we think about changing the structure of our bodies by either physical augmentation or representing ourselves in different forms in virtual environments.

2.3.3 Embodied Cognition in Action and Word

The greatest challenge in the embodied approach to cognition is finding ways to observe its processes in action. The very fact that it is predicated on embodied, first-person experiences of people performing daily tasks in their everyday environments means that we have to develop methods of observation that are suitable for understanding cognition in the wild. It would be easy to design highly controlled experimental situations informed by principles of the embodied approach, but we would be falling right back into the trap that supporters of embodied cognition railed against—namely, classical studies of cognition using artificially simplified and highly constrained experimental tasks, what we might term "cognition in the lab." The best evidence in support of embodied cognition is therefore found when we observe people in real situations performing everyday activities. And there are now many examples of such studies. Here, I will focus on two bodies of work—skill acquisition in infants and embodied metaphors in language—both of which are useful for interaction designers to familiarize themselves with.

2.3.3.1 Skill acquisition in young infants Smith and Thelen's work on the development of movement patterns in infants represents some of the most compelling evidence in support of embodied cognition. Their observation of infants interacting with objects illustrates not only how infants learn to coordinate and control their body posture and limbs in reaching and grasping[62] but also how thought leads to action. In a recasting of the classic A-not-B perseverative reaching task, Smith and Thelen challenged Piaget's interpretation of object permanency.[63] Piaget formulated object permanency as the ability to grasp the idea that an object that disappears from view is not gone forever. Piaget suggested that an infant's ability to understand this concept is a distinct developmental milestone.[64] In the task he devised to test permanency, infants watch as an experimenter hides an object under one of two boxes. After a short delay, the infant is allowed to search for the toy (by either lifting a lid from one of the boxes or removing a cloth cover). Initially, the experimenter puts the toy in box A and repeats this action a number of times. Then, they put the toy in box B. At this point, eight- to ten-month-old infants make what Piaget termed the "A-not-B" error, continuing to reach for box A, even though they saw the toy being placed in box B. Piaget explained this error in terms of the infant's stage of maturity, claiming that infants within this age range did not yet have a concept of objects as permanent entities.

However, Smith et al. propose an alternative explanation.[65] They point out that the developmental time frame is critical. Infants younger than about seven months do not manually search for objects, and infants older than twelve months easily perform the task without error. They also note that, during this developmental time frame, infants also learn to reach, grasp objects, and locomote independently. And there is evidence that infants who can crawl or walk are more successful in performing the A-not-B task.[66] Smith and Thelen propose that the persistent or "perseverative" reaching toward box A is an outcome of a body-centered, visually guided process that directs the hand toward a location in visual space. They further propose that, rather than being a top-down failure of a cognitive process related to object permanency, this behavior is an outcome of the body-centered encoding of reachable space, the strong coupling between looking and reaching, and recall of prior goal-directed reaching tasks. Their argument against Piaget's interpretation is further strengthened by the fact that poor performance in the A-not-B task can be reversed by even slight changes to either the task dynamics (e.g., adding a wrist weight to one arm) or some aspect of the visual presentation of the boxes (e.g., changing the color of one of their lids). When some property of the task is subtly altered, the infant's memory of repeatedly reaching toward box A is sufficiently disrupted to reset their motor memory of the task, and subsequent reaches become part of a new task—reaching toward a box of a different color, reaching with an arm that is heavier, and so on.

The most critical component of the A-not-B task, Smith and Thelen argue, is where the infant's gaze is directed because this defines the goal of the reaching task, which, in turn, defines the trajectory of the reaching movement. But the goal and the movement toward it, in turn, influence where the infant will look. The current action is therefore a product of the locus of the infant's gaze, the trajectory of the current movement, and the past history of both. Smith and Thelen therefore propose that the A-not-B error emerges as a consequence of complex interactions among the components of this bottom-up, self-organizing system and the infant's lack of skill in managing its internal dynamics. So, what changes for infants when they figure out where to search for the hidden object? Smith and Thelen propose that, during the developmental time frame within which the A-not-B error can be observed, infants gradually gain sufficient control of their reaching behavior to plan more easily—and therefore more predictably succeed in accomplishing reaching and grasping tasks. This, they suggest, is when the infant has reached the "moment of knowing," when thinking about grasping an object and acting upon that thought become one and the same. And, they argue, this moment of knowing is dependent on both the current task-specific actions and their immediate predecessors.

Their conclusion with regard to Piaget's object permanency hypothesis is radical: "There is no such thing as an 'object concept,' at least not in the sense of a causal mechanism that generates a thought or a behavior. It is merely convenient to say that adults 'have' an object concept because there is great regularity and stability across individual moments of knowing, across individual acts of reaching, looking, thinking, and talking about objects."[67] The fact that infants ultimately master the A-not-B task at around twelve months provides evidence, they claim, for the developing link between thought and action, which becomes more robust as the infant matures. Ultimately, this capacity to stabilize reaching behavior reflects the process by which adults acquire new bodily skills through new task-specific moments of knowing.

In addition to their work on the A-not-B task, Smith and Thelen also studied how infants developed successful strategies for reaching and grasping objects.[68] In a longitudinal study involving weekly observations of four infants over a period of eight months (from three weeks to nine months old), experimenters tracked the way reaching movements evolved. Each of the infants came to the task with different pre-reaching skills and ultimately exhibited different movement strategies before developing a stable movement pattern that would result in repeated success. The infants also tested different movement strategies along the way. Smith and Thelen concluded that each infant had to solve the problem in a unique way related to their body mass, their level of energy, and the strategies they had developed for attempting to reach prior to the start of the study. Smith and Thelen point out that the different paths the infants take to solving the reaching problem make sense when it is viewed as a bottom-up challenge for a self-organizing system in the process of moving from one stable state to another. For the infants, this meant losing the stability of a set of learned but unsuccessful pre-reaching strategies before acquiring new strategies that could guarantee success. As we have already noted, with respect to newborn stepping, the disassembly and reassembly of bodily skills is an adaptive strategy to enable the developing baby to cope with changing constraints in their body–environment systems—changes that arise because their bodies are rapidly changing too.

Proponents of embodied cognition consider this work to provide convincing evidence that action can shape processes of understanding (the A-not-B task); success in skill acquisition depends on embodied hypothesis testing—that is, on bodily exploration of action possibilities within a task domain (the reaching task); and, in learning new skills, we engage in a long-term process of assembling and disassembling task-relevant behaviors (the newborn stepping studies). As we shall see in section 2.4, HCI scholars such as Kirsh have recently begun to draw the attention of designers to the fact that, in order to acquire new skills, interaction with the environment, whether computationally augmented or not, depends upon this ability to construct ad hoc concepts to cope adaptively with changing dynamic constraints.[69]

2.3.3.2 Embodied metaphor in language More evidence in support of embodied cognition in the wild can be found in the role that embodied experience plays in shaping metaphor. There are many ways our bodily actions and experiences structure language. And language, in turn, can shape what we do. For example, in telling a friend about a book we have just finished, we might say it's a light read. Although we are communicating the idea that the book will not demand too much emotional energy, we are invoking the metaphor of weight: Heavy things require more energy to lift and carry. For Lakoff and Johnson, however, metaphor is much more fundamental to connecting our experiences to our thoughts and our thoughts to our actions.[70] All, they claim, are linked by language because we conceptualize, plan, and reflect upon our actions through the inner language of our thoughts. Their approach, which has since been termed the "conceptual metaphor theory," posits that there is no separation between our ability to conceptualize abstract events, such as planning a party, and how we plan actions situated in our current space and time based on percepts immediately available to our senses. They argue that the conceptual systems fundamental to defining everyday experiences of reality are mediated by language and are largely dependent on metaphor. In other words, language can reveal how thought becomes action and how both are connected through metaphorical concepts. Metaphors in language are therefore intrinsically linked to embodied knowledge and experience.

Conceptual metaphors also transcend physical boundaries. Think of how images of icons, such as the trash can icon on a computer desktop, communicate their underlying functions. And such metaphors can even transcend their original conceptual framing. Most people now use laptops that have solid-state drives, yet they speak of opening their "disk drive" to find their "files." Such conceptual metaphors have become a mainstay of design within the field of HCI.[71] As the devices we interact with begin to engage more and more of our bodily skills, the conceptual metaphors that underpin the design of these interactions also change.

Harrison et al., in highlighting the evolution of interaction design for HCI, point out that there have been three distinct paradigms to date, each characterized by its own conceptual framework.[72] The first paradigm, human factors, derives its metaphorical framing from the engineering approach to interaction design, where the interaction is conceived as the coupling between the "user" and the "machine," function dominates form, and safety and efficiency are prioritized over usability and transparency of process. There is little attention to the meaning of an interaction unless the user cannot understand what to do and a problem arises that might impede the successful completion of an action. The second paradigm focuses on information processing and builds on a conceptual metaphor in which the computer and the mind of its user are now coupled information processors. Meaning within this paradigm is characterized as the two-way flow of information between the user and the computer.

The third paradigm, which Harrison et al. call the "phenomenologically situated paradigm," is firmly rooted in embodied understanding and situated action. The central tenet of this paradigm is that meaning is constructed on the fly by people in interaction with their surroundings, in the context of current goals and past experiences. The paradigm is dependent on embodied knowledge and conceptual metaphor and has its roots in the same phenomenological origins as many of the approaches to embodied cognition previously discussed. For this reason, the third paradigm defined by Harrison et al. is an important antecedent to the approach to interaction design that is central to this book; we shall thus return to it in some detail in section 2.4. The important point here is that conceptual metaphors provide a useful window into how people frame their current experience and how they can carry concepts from one context (throwing paper into a trash can) into a second situation (deleting virtual documents from their computer).

Before leaving the discussion of the relationship between language and bodily movement, it is important to acknowledge the work of the choreographer and evolutionary biologist Maxine Sheets-Johnstone. In her book *The Primacy of Movement*, published in 1998, she argues that our primary means of understanding the world is by moving in it.[73] She suggests that when we are babies, the world only becomes accessible to us as something to be perceived either because we move or because there is movement in the world around us. Language, she argues, arises from movement because babies first learn to perceive action either by observing others or by actively engaging with the world themselves. She argues that discovering the world through movement is not pre-linguistic, but that language is post kinetic—in other words, without movement there is no language.[74] Further, she points out that if learning and understanding could happen only through language, many bodily skills such as walking, swimming, dancing, and playing musical instruments would be impossible to acquire. After all, babies walk before they can talk, and no one would suggest that they learn to walk only because their parents laboriously describe the bodily movements necessary to achieve upright locomotion. She writes, "Indeed, corporeal concepts are the essential without which language cannot arise and could not ever have arisen. There would be nothing to hold language together, no body of thought which would anchor it. Language is not experiences; it is the means by which we describe experience—or try to describe experience, for the gap between experiential and the language is not easily bridged."[75]

In this view, metaphor is an outgrowth of corporeal experience, for which we come to find expression through language. Think of the book that is a light read or the crossword that is difficult—both describe the amount of effort required to accomplish the task. However, although Sheets-Johnstone draws heavily upon the work of Heidegger and Merleau-Ponty in proposing that movement is the means by which we can come to be present in the world, she rejects the whole notion of embodiment, pointing out that talking about "embodied interaction design" implies that there are designs, devices, and even users that are disembodied. This, she suggests, can never be the case.

In rejecting embodiment as a conceptual framing for the interaction between people and their environments, Sheets-Johnstone also points out that the role of movement and kinesthetic perception has been overlooked in the design of interactive systems, even those that claim to draw on embodied knowledge. This valid critique will be addressed in the next chapter, which will draw more heavily on disciplines such as movement science and musical performance to illustrate how kinesthetic proprioception plays a central role in bridging the gap between physical stimuli in the environment and bodily knowledge that can guide action. Without understanding the mechanisms of perception and action, it is difficult to design for the bodies that will use these mechanisms to interact with the devices and systems we build.

2.4 Embodied Cognition and Interaction Design

And so we arrive at the question at the core of this chapter and, indeed, of this whole book: What can we, as designers of interactive systems, take away from a deeper understanding of the interactions between our bodies and their environments? In this section, I argue that because the systems and artifacts we design are interactive—that is to say, predicated on the presence of a user who will perform actions for which they expect perceivable responses—we are designers not just of artifacts but also of the phenomenal experiences they give rise to.

Is it too bold to claim that, as designers, we can shape phenomenal experience? As described above for the case of TVSS, it is not possible to design a system that creates sensory substitution. The best approximation is to design a system that creates the conditions under which sensory substitution might arise. The phenomenal experience of the user must be one in which the invariant properties relevant for one sense modality (in the case of TVSS, the structure of light) are perceivable through a different sense modality (in this case, touch). An action-dependent transformation of the structured invariants of light into structured invariants in mechanical skin stimulation can decouple the phenomenal experience of structured light from the visual modality altogether. In other words, what the creators of the TVSS system designed is a phenomenal experience for the tactile sense that is driven by invariant structures in light, which are normally apprehended visually. The learning curve for the participant is in understanding how their actions relate to these redirected invariants. They must learn the SMCs associated with this new sense modality. If the TVSS is successful, these invariants can come to guide action in the same way as they would when picked up from the structure of light arriving at the retina. In short, what is substituted here is not the sensory experience of the visual world but rather the mechanism by which the phenomenal experience of the structure in light is made available for guiding action.

The example of sensory substitution highlights the degree to which a designer must commit to designing the *conditions* under which a phenomenal experience might arise. Although sensory substitution is not part of mainstream HCI, it raises many of the same questions designers of virtual and augmented environments already face. The designer's aim is to create, through computational mediation, conditions under which a participant will set aside the sensory stimuli normally associated with the physical surroundings in favor of those that are generated by software and displayed using specialized hardware such as head-mounted displays, gloves that provide haptic feedback, and so on. This phenomenal experience is often referred to as "presence."[76] As with sensory substitution, technology itself cannot generate presence. The experience can only be designed to promote the suspension of disbelief required for a sense of presence in the virtual environment to occur. As Thompson et al. point out, "Presence is a psychological phenomenon that occurs in the human mind and not in the specific technology. The technology is merely a means used to arrive at this state of mind."[77]

Although designers of virtual environments are committed to creating an entire world of cues to promote phenomenal substitution, proponents of embodied interaction design (such as Dourish) and the third paradigm in HCI (such as Harrison) would argue that no interaction designer is off the hook. It is impossible to ignore the role that phenomenally situated action and the acquisition of embodied knowledge play in making sense of any interaction that may be designed, even if it is with a phone. Therefore, I argue that interaction designers are not designers of computational artifacts. We need to consider ourselves designers of conditions *under which* phenomenal experiences can arise, but now within real environments that happen to contain digital artifacts.

As Dourish points out, such a phenomenologically informed approach to design does not require a change in what we build.[78] It suggests that we need to fundamentally change how we think about what we build. Stewart et al. make the same point by emphasizing that computational devices are not privileged but rather should be treated the same as other devices. Certainly, they are devices of a special sort—and the worlds that are brought forth when a human being uses computational devices are of a special sort—but the interaction that occurs is between the human being and this world, mediated by the machine; it is not an interaction between the human being and the machine.[79]

What means do designers have at their disposal to make such a shift in thinking? Kirsh suggests that, as designers of tools, we are also designers of what he calls "enactive landscapes"—structural couplings an actor brings forth or cocreates with an environment when engaged in goal-directed action.[80] Enactive landscapes determine for the user of a tool the possibilities for action available to them when they, the tool and the task are brought together in the task environment. For example, a musical instrument provides the physical landscape for a musician's actions. Kirsh writes, "The more skillful the musician, the larger the enactive landscape they inhabit, because skills combine with instruments to constitute a broader world of possibilities."[81] Instrumental performance, in Kirsh's terms, depends on an enactive landscape meant to bring forth music, the making of which is a goal-directed task that reveals the activity-dependent nature of perception. Kirsh points out that designers must therefore be continually aware that the tools we design and the skill level for mastering them will combine to reshape the enactive landscape for each user in each context of use.

In addition to highlighting how new tools bring forth their own enactive landscapes, Kirsh points out that many aspects of our environment are defined by rules and cultural constraints. He gives the example of someone preparing a meal in a kitchen. He notes that the same kitchen will become a different enactive landscape for each cook because of their skill level, their particular interests, the requirements of the recipe, the tools at hand, and so on. As they work through the tasks associated with preparing a dish, each chef will become aware of different details of the surroundings, depending on the tools they use. If they are holding a knife, they focus on things to be cut. If they put down the knife and pick up a spatula to stir ingredients, they see only places where they can use the spatula.[82] Their enactive landscape is adaptive, dynamically reshaped as their goals and subgoals change and as the tool that forms the structural coupling to the environment (knife or spatula) also changes. If we are multitasking, a number of enactive landscapes may be superimposed with differing task constraints. Kirsh argues that skill in designing the environment where the user's actions are situated, whether a kitchen or an operating room, "resides in blending the many enactive landscapes of its probable inhabitants to minimize error, maximize effectiveness, reduce fatigue and delight aesthetic sensibilities."[83] Understanding how tools bring forth enactive landscapes for their users is, Kirsh suggests, the first stage in such a design process.

The second step for the designer is understanding how new tools shape new concepts. When we acquire new skills, our actions cocreate an enactive landscape that is initially specific and highly situated. For the musician learning a new piece, for example, that landscape is constrained by the nature of the structural coupling between the body and the instrument and by very specific musical goals (to play this note at this time). Kirsh suggests that each such act represents an ad hoc concept, a way in which actions are arranged around subgoals but are still adaptive to changing environmental constraints. In manipulating a laparoscope, a surgeon might never perform the same maneuver the same way twice but is always aware of which aspects of the movement are negotiable in a given context and which are not. The general concept of performing the maneuver is therefore an ad hoc concept that is modified on the fly, in service of a particular goal. This, Kirsh argues, is evidence that design is never complete and that universal design can never be truly universal. Users are always learning and adapting, and their enactive landscapes are shaped by their idiosyncratic behaviors and are continually evolving in accordance with their individual task goals.

By suggesting that enactive landscapes arise in response to goal-directed actions and that skills arise from ad hoc concepts, Kirsh provides interaction designers a means to address Merleau-Ponty's assertion that it is only through the habits and skills acquired as beings adaptively coping in our world that, when we are faced with new tools, we are open to the emergent possibilities for action.

2.5 The Third Paradigm in HCI

Framing approaches to HCI design and evaluation in terms of three distinct paradigms was first proposed by Harrison et al., who connected the distinct intellectual waves of influence shaping the field to the design approaches that mirrored each wave.[84] As noted in the previous discussion of conceptual frameworks within HCI, the first wave of influence (and its associated paradigm) grew out of early work on man–machine systems or human factors engineering. It focused on the need to optimize the fit between the requirements of a control system and the capabilities of its operator. At that time, computational systems were hard to use because it was assumed that users would learn to interact with the machine, which would be designed to function as efficiently and safely as possible.

The second intellectual wave to influence HCI was centered in cognitive science and shifted from the idea of an individual operating a machine with a single purpose toward interactive computers that serve more diverse needs among specific populations, such as office workers and graphic designers. This second wave adopted information processing and other cognitive science approaches that modeled how information *flows* between user and computer.[85] It used theories of action that modeled the steps a user would take to accomplish a goal.[86] The second wave thus provided clear structures and guidelines for HCI design and evaluation, which gave rise to the field of usability engineering. This field focused on testing interaction designs to see whether they met the requirements of stakeholders, including both the purchaser of a system (the client) and the end user.[87] Along with increased emphasis on structured approaches to design and evaluation, the second wave also saw greater value placed on the participation of users in the design process.[88]

Third-wave HCI represents another shift, this time deriving its intellectual framing from what Harrison et al. call the "phenomenological matrix." This work emphasizes the phenomenologically situated nature of user experience and focuses on the user—not just in the process of designing systems to accomplish a task but also as an active participant in constructing meaning within the context of that task. Third-wave approaches acknowledge that the design of technologies and how they are evaluated must reflect the diversity of users, their goals, and, above all, the social and cultural contexts in which computing is embedded. Notably, all three interaction design paradigms overlap. In safety-critical applications, for example, the practices associated with human factors engineering still dominate design approaches. Concepts central to the third wave, such as sense-making and embodied cognition, have existed alongside earlier approaches for some time, particularly in the work of Dourish, Gaver, Hornecker and Buur, Fallman, and others.[89]

Rather than representing a distinct methodological approach, the third wave is characterized by a convergence in thinking around common themes, including situatedness, embodiment, and meaning construction, and a clearer articulation of design values as a part of the design process.[90] These themes are discussed below. First, however, it is useful to review some of the pioneering work that set the stage for what is generally referred to as the "third-paradigm" approach to HCI.

Third-paradigm and related approaches to embodied and phenomenally situated interaction design have emerged almost in parallel with embodied approaches to cognition. This is not surprising, as early computers mirrored in their architecture the sandwich approach to cognition that Hurley and others discredited. Just as human cognition was described in terms of sequential steps of sensing, cognitive processing, and action generation, so computer architecture was described in terms of devices for accepting input, processing data, and generating output. As computing has become embedded in the devices around us and as the means by which we interact with them have come to include a greater range of actions such as whole-body motion, gesturing, and speech, our understanding of what computers are has also changed. We no longer assume that our interactions with these devices require us to type or use a mouse. We no longer expect their output necessarily to appear on a screen. Furthermore, we assume that devices are networked, and we are often unaware of where exactly data is processed or stored. Computation has become embedded in the ecosystem for much of the industrialized world. In parallel, approaches to interaction design have moved away from computational models based on rules and scripts for HCI toward models that seek to predict user intent for a given task.

Winograd and Flores were among the first to anticipate this shift, suggesting that the most successful interactive computer systems would not depend on a priori models of a given task domain (e.g., a precise model of a road network for autonomous vehicles).[91] They suggested that robust systems would instead be those concerned with the fundamental structure of that domain (e.g., the constraints associated with driving in a dynamic environment). Such systems could allow for modification and evolution to generate new structural coupling that adapts to changes in this environment. In this, they predicted the adaptive algorithms that are the basis of many of today's most advanced computational systems, from self-driving cars to smart devices—systems that refine their performance in response to both user behavior and changing environmental conditions.

In emphasizing the situated nature of interaction with computers, Winograd and Flores also anticipated the emergence of the third paradigm in HCI. They drew attention to breakdowns—those moments when a failure of the system to respond as expected causes the "equipment," or the interface, to break into the user's foreground attention and disrupt their task flow. If there is a moment of breakdown, then there are also preceding stretches of time when the interface is transparent to the user, when the conditions for the experience of having a tool ready to hand do exist. Winograd and Flores used the concept of breakdown to demonstrate that a computer, like Heidegger's hammer, could disappear from foreground awareness, that it was indeed possible to create conditions under which the phenomenal experience of directly manipulating the symbols on the computer screen could arise, conditions that become apparent only when breakdown occurs. Although Winograd and Flores were the first to bring the work of phenomenologists such as Husserl, Heidegger, and Merleau-Ponty to the attention of HCI scholars, others have since applied these principles to design interactive systems across many application domains, some of which will be discussed.

2.5.1 Situatedness

The emphasis of the third paradigm on a phenomenally situated design perspective owes much to the work of Lucy Suchman. In her 1987 book *Plans and Situated Actions*, she says that if we are to talk about interaction between people and machines, it is important to understand conversational interactions between people. This insight arose from viewing video recordings of people attempting to use a new photocopier developed at Xerox PARC. As she watched people approach the machine and try to figure out how it worked, she noted where they kept getting stuck, and she wanted to give them spoken directions.[92] The problem, she surmised, was that the designers had failed to consider the many contexts in which the machines would be used. By applying techniques from ethnomethodology, Suchman developed an analytical framework that revealed the resources available to both the user and the machine in a manner that exposed the successes and failures of the interaction.[93] She summarized her framework in a pair of tables, one of which captures the capabilities and resources of the user and the other those of the machine, as illustrated in table 2.1.[94]

The two columns on the left, when annotated, represent the resources (or actions) of the human, but only the second column is also available to the machine. The columns on the right denote the resources available to

Table 2.1
The user versus the machine. Summary of resources available to the user and the machine during a canonical human–machine interaction. Source: Adapted from Suchman, 1987.

The Machine		The User	
Design rationale	Effects available to the user	Actions available to the machine	Actions not available to the machine

the machine, which Suchman terms "effects" and "design rationale." (Robertson and Loke, discussed later, suggest that the term "effect" refers to the machine's sensing and display capabilities, whereas "design rationale" is better thought of as the internal resources available to the machine, such as programmed routines and algorithms.)[95]

Suchman's central claim was that, just as the user is situated in a world brought forth through their actions, the machine is also situated. The world brought forth for the machine is determined by what it can sense and how it can act. The photocopier had limited capabilities to sense. It could identify when trays were being moved in and out, when doors were opened, and when buttons were pressed. It also had a limited repertoire of actions. It could initiate a copying protocol determined by the algorithm and defined by a particular sequence of button presses, or it could inform the user of a malfunction. In acknowledging that both users and machines act on the basis of the resources available to them in a situation, Suchman's approach shifts focus from the design of the interface between a user and a machine situated in an environment to consider the broader context in which both user and machine are situated or embedded. Suchman reframed the interaction between the user and the computer as a conversation in which both should be considered as active partners. Moreover, she claimed that both should be equal in their assessment of the situation if the interaction is to succeed.[96]

Among the researchers who adopted Suchman's analytic framework were Robertson and Loke.[97] They incorporated it into a comprehensive methodology for designing interactive systems based on sensing bodily movement. In analyzing interactions with the Sony EyeToy, a digital camera for the PlayStation 2, the researchers noted that Suchman's framework could contribute to the design of interactive systems in three ways:

1. The framework can reveal asymmetries between the resources available to the user and those available to the machine for the perception of action. The EyeToy, for example, can detect motion within a narrow spatial area and a short time frame, constraints that do not apply to the human player.
2. In the case of the EyeToy, feedback from the system enables players to have a shared understanding of success within the game.
3. The framework can enable designers to describe the movements of users relative to the context of an interaction. For Robertson and Loke, this means being able to situate the actions of EyeToy users within the spatial, temporal, and gestural constraints created by the game designers.[98]

In subsequent work, Robertson and Loke used Suchman's framework to articulate requirements for Bystander, an immersive multiuser interactive environment intended for museums.[99] Their breakthrough realization was that in designing the situation of the interactive system, they were also designing the resources available to the system within a given situation and the degree to which the system could assess the situations of its users. Their goal was to provide points of interaction where the situations of the machine and the user could meet. In short, they wrote, "We need to understand what it is we are actually designing when we design the potentials for interaction between people and technologies."[100]

In the years since its publication, Suchman's framework for interaction analysis has been applied to the design of technologies in fields such as HCI, computer-supported collaborative work (CSCW), and participatory design. The emphasis placed in these fields on the situated nature of interaction has, in turn, played an important role in the articulation of the third paradigm of HCI. As awareness of the emerging field of embodied cognition began to impact HCI, Suchman's assertion—that both users and machines bring forth a world of possibilities and that interaction between them is predicated on a shared understanding of the resources they have at hand—came to shape not only CSCW but also the emerging fields of embodied interaction and tangible computing.

2.5.2 Meaning Construction

A further application of the third paradigm focuses on what Harrison et al. call "meaning-making." They suggest that the first paradigm took a pragmatic approach to meaning-making, assuming that a user's ability to make sense of an interaction was a given unless a problem arose. The second paradigm framed the users' construction of meaning in terms of their ability to construct a mental model of interaction that could account for the flow of information to and from the machine. If they performed an action such as clicking on an icon, they could assume that any change in the system state was accounted for by a metaphor that connected the symbol associated with the icon to a given system response, such as removing a file from a hard drive by placing it over the trash can icon. The third paradigm offers a different perspective, holding that "meaning is constructed on the fly, often collaboratively, by people in specific contexts and situations."[101]

Like Suchman, Harrison et al. argue that, in the third paradigm, interaction is a central component of meaning construction that is deeply intertwined with the cultural perspectives, past experiences, and current resources available to a user trying to make sense of the interaction. As such, "The process by which meaning is constructed in this paradigm is likely to lie beyond the reach of formalization."[102] This was Suchman's first realization on viewing the video recordings of photocopier users. Her solution was to develop a framework that made no attempt to define the correct way to interact with the machine. Instead, she recognized that accounting for the resources available to both the user and the machine and their shared context of interaction would provide the necessary flexibility for ad hoc strategies for meaning construction. Putting this approach to the test, Gaver et al. proposed that we consider such ambiguity as a design resource, suggesting that the designer can shape how ambiguity arises in the relationship between a user and a digital artifact.[103] As Fallman, in discussing Merleau-Ponty's perspective on ambiguity, says: "The world will always be ambiguous and inexhaustible, as there is always something one cannot see from where one stands."[104]

Gaver et al. propose that designing for ambiguity has many advantages.[105] It allows a designer to engage users without constraining their responses. It allows a designer to express their own views while remaining accessible to people with differing sociocultural backgrounds. And it allows a designer to creatively explore the limitations of a given technology. Think of how ambiguity plays a crucial role in facilitating interaction with voice-activated home assistants. If it were not for the system's ability to cope with ambiguous voice patterns, cultural expectations, and environmental conditions, these systems would be useless. Ambiguous interactions promote greater levels of engagement from users because this very ambiguity introduces a level of unpredictability that requires continuous attention.

Recall from the discussion of sense-making in the context of embodied cognition that meaning does not exist independently of action; it coevolves with situated activities. With respect to the third paradigm, Harrison et al. note that if meaning-making relies on the immediate, local context of an interaction, then the knowledge the user brings to the interaction must be situated knowledge informed by physical, social, and cultural experience. Consequently, designers can no longer rely on one set of metrics to account for all users and situations. They must find an approach that accounts for the situated activities of users while avoiding favor toward any of the varied and perhaps conflicting perspectives.[106]

In addressing this concern, Hummels and van Dijk suggest that this is more challenging than designing with the information-processing perspective of the second paradigm, where the tasks and actions associated with the human and the machine can be broken into a sequence of component steps and processes and presented as design guidelines.[107] The work of Hummels and van Dijk has its roots in tangible, embedded, and embodied interaction. Their approach, embodied sense-making, draws upon embodied cognition and the concept of sensorimotor couplings to support social coordination between people engaged in what they call "face-to-face embodied sense-making technology"—people engaged in a shared activity grounded by a technological artifact such as an interactive table. Their goal is to assist designers of these interactions by providing them with design principles for moving between multiple perspectives as they address the challenge of designing interactions that draw on the embodied and situated knowledge of their users. Citing Merleau-Ponty, they emphasize that such knowledge exists neither in the world nor in our heads but instead arises from meanings derived from interacting with the world. Their principles are therefore intended to provide a practical way for designers to use Merleau-Ponty's phenomenology to inform embodied interaction design. The principles include supporting a first-person perspective, coordinating social spaces for shared action, and eliciting user engagement grounded in the technological artifact before them.

Central to this approach is the concept of "embodied couplings": ways to describe interactions between people and artifacts in a shared environment that account for physical, social, sensory, and action dimensions and respond to the changing requirements of a situation. Embodied sense-making, in this view, is an ongoing, embodied, and situated interaction in a shared action space that is grounded by a technological artifact. Rather than rely on a protocol of communication between individuals or between an individual and a machine, embodied sense-making relies on the social coordination Suchman highlighted. It is how people in social situations coordinate their actions with the actions of those around them.[108] If one of the participants in this action space is a technological artifact, it too must be situated. It must be able to engage in coordination by eliciting action and contributing appropriate responses. It must embody the behaviors that will support all participants in making sense of their interactions within a shared environment.

2.5.3 Embodiment

In third-paradigm HCI, Harrison et al. assert, the embodied approach to interaction design is nonnegotiable.[109] Thus, if we commit to Suchman's proposal that both users and machines are situated, then interaction with computing artifacts should be framed as interaction with environments in which both the bodies of users and the artifacts they act upon are embedded. This is not to say that concepts related to embodied action were absent from the human factors approach of the first paradigm or the information-processing approach of the second. Rather, as Harrison et al. state: "Embodiment in the third paradigm is based on a different stance: drawing on phenomenology, it takes as central that the way in which we come to understand the world, ourselves, and interaction derives crucially from our location in a physical and social world as embodied actors."[110]

Indeed, the fact that our actions in the world are shaped by multiple forces, including physical engagement with objects, social interaction, cultural experience, and task context, has led to some confusion as to what embodiment means for HCI. In the work of designers such as Hornecker and Hummels, the term refers to the embodied knowledge that a user brings to bear on interaction with objects, technically enhanced or otherwise. This work emphasizes the role the body plays in meaning construction and how interfaces act as structural couplings to the environment, parts of which may be digitally rendered. A second meaning of embodiment refers to how computational processes are embodied in artifacts present in the environment. A smart speaker, for example, embodies the functionality of the digital assistant to which it provides access. This interpretation is more prevalent in the work of Ishii and others on tangible user interfaces (TUIs). Still another interpretation relates to the embodied actions of conversational agents in social interaction, whether as physical gestures or as inflections of speech.[111]

In *Where the Action Is*, Dourish reinforces the emphasis on phenomenology for HCI. He points out that the multiple facets of embodied interaction with the world must allow for multiple perspectives. He claims that interactions with physical objects should not be isolated from social interactions or from gestures. Indeed, in a 2013 review of the book's impact, Dourish acknowledged that his emphasis on TUIs as exemplars of interactions that privilege bodily action with computational artifacts gave rise to the notion that embodied interaction is tangible computing. He states:

> My goal in articulating how tangible computing could be understood within the frame of the ongoing phenomenologically inspired program of research was precisely to draw connections between tangible computing and other forms of interactive systems and to suggest that they were amenable to common forms of analysis. . . . If there was one thing that I would like to make clearer in *Where the Action Is*, it is the relevance of an embodied account of interaction to traditional user interface design and analysis.[112]

Dourish has suggested that even the most basic forms of HCI draw upon the kinds of situated and embodied knowledge that phenomenologists such as Heidegger would recognize. Keyboards and mice are, like the hammer, tools that become ready to hand for the skilled user of a word processor or a digital drawing application.

The problem, as Klemmer et al. point out, is that, for a long time, the keyboard and mouse were the only means of interacting with computers, even as the tasks to be performed became increasingly diverse.[113] The unintended consequence, they point out, is the increasing homogeneity of physical work. In their 2006 review, they sought out approaches to interaction design that emphasized the role of bodily skill and situated action, and they identified five themes that could guide the design of interactive systems toward a more appropriate convergence of bodily capabilities and computational resources.

The five themes are thinking through doing, performance, visibility, risk, and thick practice. Thinking through doing captures much of Suchman's framing of the way thought and action coproduce understanding and reasoning because body, mind, and objects of interaction are colocated or situated in the lived environment. In the language of the embodied approach to cognition, we can say that by manipulating objects, we bring forth possibilities for action and, in so doing, shape our understanding of what we can accomplish. If those objects are augmented with computational capabilities, then our actions upon them can bring forth opportunities for understanding that could not be present in the non-digital world. For example, Klemmer et al. cite Underkoffler and Ishii's "Illuminating Light," a TUI that allows architects to place a scale model of a building on a table and simulate seasonal and diurnal movement of the sun to see shadows cast by planned buildings.[114] Rotating the model can reveal how small adjustments could prevent glare from windows disturbing drivers on nearby roads.

Klemmer et al. define their second theme, performance, as the ability of an interaction to support the acquisition of bodily skill. They note the overemphasis in the field of tangible interaction on what they call "walk up and use" scenarios. In focusing on the objects of interaction and not on interactions with those objects, such approaches overlook the power of TUIs to support the tacit knowledge on which skill acquisition depends.[115] By contrast, they cite the work of Djajadiningrat and colleagues, who assert that assigning different actions to functions within an interactive system promotes the bodily or kinesthetic memory necessary for learning a new skill.[116] In Kirsh's terms, by associating specific actions with particular functions, a designer leaves open the possibility of the construction of task-specific couplings to the environment and of the construction of task-specific enactive landscapes such as the kitchen for the chef.[117]

The themes of visibility and risk address how bodily actions shape social interaction in collaborative augmented environments. Visibility is the degree to which the performance of an activity is visible to collaborators or onlookers who need to maintain awareness of task progress. The example cited is air traffic control, where the progress of flights through an airspace is noted on paper strips that are moved around on a board to reflect the relationship between the planes as they fly through the air space. This shared representation ensures that the activities of all controllers who interact with the strips are visible at all times, so that everyone maintains an awareness of the current status of the airspace.[118] In addition to awareness, Klemmer et al. claim that interactions that are visible promote peripheral or bystander participation in problem solving, situated learning through observation of others, and coordination of action through manipulation of artifacts common to a given work practice.

The suggestion that embodied interaction design should acknowledge risk grows out of the assertion that all physical activity is inherently characterized by risk.[119] Unlike actions in virtual environments, actions in real environments cannot be undone. Choosing an action comes with the possibility of making the wrong choice and being harmed. Similarly, some social blunders in face-to-face interaction cannot be undone. Understanding the degree to which a given context for computer-mediated interaction supports physical or social risk is central to promoting embodied social interaction in shared and collaborative environments.

The final theme, thick practice, suggests that while it is impossible to simulate the real world in all its dimensions, being able to draw on embodied knowledge and embodied social and physical skills can diminish this limitation by facilitating interdependence of physical action, embodied reasoning, and abstract thought, which are the foundations of our everyday interaction with the physical world. As Svanæs states, "All human interaction with digital technology is embodied, in the sense that the technology is physically omnipresent in our everyday lives."[120] In short, embodiment, as a central tenet of third-paradigm HCI, acknowledges that actions have consequences beyond the response of a digital device, encompassing the physically, socially, and culturally situated actions of the individual, mediated by the computational tools they use. The next section, "Reconsidering Design Values," shows how this has resulted in a reappraisal of what it means for a design to be considered "good," since there is little benefit in a design approach that accounts for multiple perspectives if there is no parallel effort to evaluate designs in a way that takes these multiple perspectives into account.

2.5.4 Reconsidering Design Values

Finally, in their description of third-paradigm HCI, Harrison et al. acknowledge that a design must be sensitive to the values and experiences of its intended user community. As noted, third-wave approaches to HCI are defined by a shift in focus, from designing systems that allow people to accomplish specific tasks in a particular context to designing artifacts that enhance everyday experiences of a lived environment. This shift requires acknowledging that people and their environments vary, and the way artifacts will work for individuals in diverse contexts is shaped by social and cultural conditions. As a consequence, there is a growing understanding that "good design," as a practice, must be deconstructed to account for the perspectives of both designers and users of the technologies many now depend on. Fallman states, "In the transition from usability metrics to user experience, what appears lacking is a more explicit characterization of what it is HCI now strives for as a discipline—i.e., what constitutes a 'good' user experience?"[121] The assertion here is that whatever the new good is, it must go beyond the concept of task-specific evaluations of system usability to encompass culturally and socially situated evaluations of user experience: "If interactive artifacts are knowingly designed to provide users with specific user experiences, then guiding visions must be developed, which provide the means for revealing, analyzing, and discussing the obvious implications (human, social, cultural, ethical, moral, ecological, and political) of these experiences and how they foster particular relationships between users, designers, artifacts, and contexts."[122]

Although the emphasis in this book is on the action-dependent nature of perception and the role of the body in creating meaning through a mediated sensorimotor coupling with the environment, it is worth noting that this perspective is only one feature of a user's experience. It is a vital part of enabling the construction of meaning, but it does not include the social, emotional, or cultural shaping of that experience. Evaluating how good a design is at promoting sensorimotor coupling to the environment that allows for new phenomenal experiences is one step in a process that has to account for socially and culturally meaningful situated experiences. Although the goal is for technological artifacts to become, in Heidegger's terms, transparent in use, the fact is that technology can never be neutral. Fallman, citing the work of Ihde, reminds us that any form of technological mediation alters how we experience the world. Using Ihde's example of the telescope, which allows us to see the stars and planets, Fallman notes that its invention also reshaped our understanding of our position in the universe.[123]

Fallman also points out that technologies themselves are multi-stable, that the relationship between us, our technologies, and the world where both are situated is not fixed. Citing the example of someone using a digital camera to take a picture of a dog, Fallman points out how the photographer alternates between the unmediated view of the dog in the world and the mediated view of the dog and the world through the viewfinder. In the first case, the camera is an object of no relevance to the dog or the viewer; in the second, it mediates and constrains the photographer's view of the dog.[124]

Determining what "good" means for these overlaid experiences is complex because the layers are interdependent. As a possible starting point, designers can draw on Suchman's notion that technologies are situated to direct their attention away from evaluating technological artifacts themselves and toward understanding the experiences they mediate. Approaches such as participatory design and the socio-technical design methodologies that emerged from Scandinavia in the 1970s have provided qualitative tools that can also uncover the dimensions of people's interaction with new technologies in the workplace. Fallman argues that only by stepping back from individual design perspectives can we develop new methods for determining what "good" means for the design of experiences that depend on technological mediation. He proposes that philosophers of technology can help by engaging designers in continued critical reflection on the values imbued in our designs and by guiding the search to understand the experiences of a diverse population that depends upon the systems and artifacts that we build.

In sum, I have suggested that it is impossible for designers to ignore the role that embodied knowledge plays in allowing people to interact meaningfully with and through technological artifacts because our experience of the world is grounded in actions upon the world and in what we perceive as the world's responses. The more phenomenal experience is mediated by tools, the more our understanding of the environment is shaped by the world that is brought forth by those tools.[125] This situated knowledge is not in the mind of the user, nor in the technological artifact; it arises from meaning created by a user's interaction with the world in which both are situated. The interaction is between the human being and the world, not the human being and the tool.[126]

Tools form structural couplings to task environments. When we design new tools, we design the means by which users form new structural couplings with task environments and the means by which they bring forth new worlds of possibilities. Think again of the skier and the snowy mountain. Kirsh terms these task-dependent structural couplings "enactive landscapes," suggesting that they are adaptive to real-world constraints and coevolve as skill is acquired and as the goals of a task change. For this reason, he claims, design must be an open-ended process, a continual dialogue between the user, the technological artifact, and the environment in which both are embedded. Culkin states in a review of McLuhan, "We shape our tools and thereafter, our tools shape us."[127] The third paradigm in HCI affirms that the move toward embodied approaches to cognition, emphasizing the coupling between perception and action in situated interactions with the world, is now shaping the computational tools we design. The emphasis placed on the ability of an organism to seek meaning actively through structural coupling to the environment is mirrored in embodied approaches to interaction design that can provide the conditions for phenomenal experiences arising from adaptive coping with an environment. Because users are always learning and adapting to changing environmental conditions, Kirsh argues that the design process is never complete, and there can be no truly universal solutions.

Third-wave HCI and the associated third-paradigm approach to design acknowledge the complexity of designing for experiences that depend on the situated and embodied understanding users bring to their interactions with technology. Proponents of third-wave HCI acknowledge that actions have consequences that go beyond the responses of digital devices to encompass the physically, socially, and culturally situated actions of the individual, mediated by the tools they use. Third-wave HCI thus demands a redefinition of what the field considers to be a good user experience,[128] as it embraces a commitment to diverse experiences and perspectives.

Notably, not all proponents of the embodied approach advocate rejecting formal usability evaluation methods. Van Dijk considers embodied, situated approaches to cognition as theories of behavior that contribute in important ways to understanding the actions of users as they establish structural couplings with environments with technological artifacts. As we shall see when we return to a discussion of van Dijk's work in chapter 6, this sets the stage for an embodied, embedded usability practice that enables designers to disentangle how local constraints and affordances, past experience, and ad hoc plan formation each shape the behavioral couplings between users and technical artifacts.[129] The challenge is to find a way to understand the complexity in these structural couplings without falling back on disembodied task representation. However the field of HCI may define the new good, it is clear that new methodologies must encompass multiple perspectives and account for physically, culturally, and socially situated evaluations of user experience.

2.6 Conclusion

We will return throughout the book to the approaches previously reviewed as we consider how designers can support these premises, knowing that digital artifacts are just one part of the world users inhabit. Digital artifacts are by no means privileged in this regard. Like all technical artifacts, from the wheel to the steam engine, they are designed by humans to help accomplish goals beyond the powers of the body. So, before embarking on the remainder of this journey, let us ask three salient questions that arise from the corresponding assumptions about perception introduced in chapter 1:

1. What is perception if it is not absolute?

   It should be clear by now that the properties of our surroundings available through sense organs are a subset of all the available properties in those surroundings, and that these properties manifest themselves to us through our actions. If this is the case, then the illusion that perception is absolute might arise from the fact that, in a stable environment, the same actions bring forth the same world most of the time. But there is no reason why this should always be the case. Nor is our perception of the world complete compared with the scope of energy sources available in the physical world at any given place and time.

   In his final book, Gibson suggests that the world of the perceiving being is the environment that is relevant to what it needs to accomplish.[130] It is not the totality of the physical world, although they overlap. Thus, perception of infrared energy in the form of ambient heat is meaningful to a snake that uses this to locate prey, whereas perception of Earth's magnetic field is useful to migrating birds. Humans lack direct access to either of these cues because they aren't informative to human capabilities.

2. What is perception if it is not passive?

   Here, I address the assumption that we do not wait for sensory receptors to be stimulated then "process" this stimulation to figure out salient elements of incoming stimuli. From the preceding review, it should be apparent that little meaningful interaction with the surroundings occurs without active participation in acquiring relevant information. Although proponents of the ecological approach and embodied cognition may disagree on the extent to which our actions shape our understanding of the surroundings, they all agree that cognition is dynamic, goal directed, and context dependent.

   Designers of interactive systems must view themselves not just as designers of artifacts but also as designers of experiences for people who interact with those environments and devices. Stewart et al. remind us that digital devices are not privileged.[131] A user brings to interaction with these devices the same body, experiences, and ability to acquire skills as for other objects they interact with. Just as bodily skills are intrinsic to sense-making in interaction with physical artifacts, there is an active sense-making process in engagement with digital artifacts. We actively perceive the visual, auditory, or haptic feedback they deliver as responses to our actions. These devices extend our capabilities to engage with an environment beyond our reach. They provide a coupling between our responses to the physical world and tasks that must be accomplished digitally. Perception of the environment, therefore, depends on action in the environment. It cannot be passive.

3. What is perception if it is not purely signal processing?

   Proponents of the ecological approach to perception, following Gibson, have proposed that, in moving about in their environments, animals attend to invariant structures in stimuli, such as ratios and proportions, because energy arriving at individual sense organs is too variable and too susceptible to external conditions to be a reliable source of information to guide their actions. An animal's ability to respond to invariant structures in the global energy array arises from their being embedded in a body–environment ecosystem, where their movements impart structure to the energy array that is relevant for their particular sensory apparatus. Perception is concerned with the pickup of invariant patterns of information from an array of energy sources rather than with processing signals arriving at different sense organs.

   Proponents of embodied cognition have often rejected the ecological approach to perception, saying it fails to account for perception being structured not just by the environmental invariants a species is sensitive to but also by individual goals and skills. McGann points out that if we follow approaches such as enaction to their logical conclusions, then perceptual modalities are structured not just by invariant properties of the energy array but also by the stability that arises from the actions of the skilled individual seeking information to guide them toward a goal.[132] In the next chapter, we will review how invariant properties within the global stimulus array that might otherwise be completely inaccessible to a non-skilled individual are revealed when perception and action are coupled in service of a skilled activity such as intercepting a thrown ball. McGann points out that it is this context-dependent, goal-directed sensitivity to properties of the energy array in support of meaning-making that distinguishes embodied approaches, in particular enaction, as more than just signal processing.

At the outset of this book, I asked you, the reader, to join me on a journey through unfamiliar territory and to question basic assumptions about perception of the surrounding world. The most unintuitive step is to give up the idea that perception is veridical and that there is a direct correspondence between the physical world and the ability to perceive it. The proposal elaborated in this chapter is that experience of the world is a phenomenal one dynamically shaped by our bodies, our goals, and the habits and skills brought to interactions with the world. The environment we live and work in now contains devices that connect us to an invisible world of bits. Although these devices bring forth a world of new possibilities, they are nevertheless just new tools—new structural couplings to a world in which we have things to do and places to go. Like all tools, they rely on our ability to harness their capabilities in service of our goals. I have drawn upon theories of ecological perception and embodied cognition to propose that the bodily knowledge we acquire in learning to walk, ski, or play musical instruments is no different from the bodily knowledge we draw on when we become skilled users of technical devices that couple us to the world of bits, such as cameras, game consoles, and laparoscopic tools.

The theoretical frameworks in this chapter are rooted in a view of cognition that places the body at the center of the interaction between an individual and the environment. The premise is that cognition depends upon our ongoing and coordinated interaction with that environment, a skilled activity through which we continually gain an understanding of how that environment will support us in attaining our goals. In this view, the body helps us make sense of the world. This is different from cognitive theories that suggest the representation of the world is a mental construct and that actions are planned with respect to goals independent of the sensorimotor information that supports the execution of those actions.

As section 2.4 shows, the relevance of embodied cognition to HCI is already established as an intellectual foundation in third-wave HCI. This "third-paradigm approach," as proposed by Harrison et al.,[133] helps close the gap between third-wave attempts to account for a broader range of user experiences and designer needs for new methodologies to build and evaluate systems for a heterogeneous population of users in varied contexts. The threads that tie these theoretical approaches to emerging third-paradigm practices in HCI are bound by three central claims:

1. Our interaction with the world is embodied and shaped by the physical dispositions of our bodies and the capabilities we acquire as skilled inhabitants of that world. These bodily dispositions are not static, nor are the environments in which we move.
2. Interaction is situated. Our actions take place in a context shaped by the surroundings. The contribution of third-wave HCI is to acknowledge the wide range of influences that can shape the context in which users experience technology. These influences include social circumstances, abilities, and cultural norms that define how technology will fit into people's lives.
3. Interaction brings forth meaning. Proponents of third-wave HCI assert that technically mediated interaction with the environment draws on people's ability to discover regularities between what they do and how a system responds and make sense of their interactions with the world of bits brought forth by a technical artifact.

The remaining chapters offer a practical guide for designers of digital artifacts and augmented or virtual environments that draws on the embodied knowledge people bring to bear in negotiating the everyday physical world. The next chapter addresses how bodies move and how, in turn, the information revealed by movement guides action. Then we will look at the different purposes of movement and how these can impact data present in signals captured by systems that track movement. The remaining chapters address challenges that the capture and analysis of movement pose for designers of interactive systems with respect to time, space, and the design of interactive objects.

3 Movement

3.1 Introduction

We move even before we are born. The philosopher Maxine Sheets-Johnstone goes so far as to say that movement is our native language.[1] It is how we come to know our bodies and, through our bodies, how we come to know the world.

As designers, we populate the world with interactive artifacts, and the people who use these artifacts discover how they work by moving, using their bodily resources to discover what our devices can do. In this chapter, I propose that bodily movement is a material resource central to the process of designing an interactive system. This might seem obvious, because the term "interaction" assumes that someone is moving to perform some action to which the system will respond. But people move for all sorts of reasons. The first task of the designer of an interactive system is therefore to understand *why* someone is moving. An interaction designer's second task is to determine *how* someone is moving. People move all the time—to adjust their posture, to multitask, even unconsciously as they breathe. Only some of the movements they make are intentionally directed toward their devices, even when they are fully immersed in virtual worlds. Part of a system designer's work is therefore to observe how users move as they interact with their devices so that they can determine relevant movement patterns to serve as templates for systems that will ultimately track and analyze sensor data.

There is also the question of *where* people are moving. In everyday life, people's actions are shaped by their environments, the tools they are using, and the goals of the tasks they are performing. As the devices we design are integrated into people's everyday lives and activities, they have come to shape how people act. But devices too are shaped by the requirements of their users and the worlds in which they are encountered. In chapter 2, I laid out the theoretical frameworks informing our understanding of perception, focusing on how researchers have approached the role our bodily skills and capabilities play in "bringing forth" the world of our lived experience. This chapter, which concerns how movement is tracked, captured, analyzed, and integrated into the design of interactive systems, represents the other side of the coin—a practical foundation that sets the stage for discussing how, through the movement of our bodies, we can bring forth parts of our lived environment that are augmented with new digitally enabled behaviors. In the case of virtual environments, we must, through our motion and actions, bring forth entirely new worlds. To do this, it is necessary to understand how our bodies move and for what purposes. One goal of this chapter, then, is to synthesize current work that can inform and guide designers in determining how to provide adequate and appropriate perceptual cues for users of their systems. In chapter 6, I will return to many of the points that I will focus on in this chapter to discuss the practical implications of designing interactive systems and artifacts from a practice that is grounded in observing, capturing, and analyzing the movements of those who will ultimately use the things we build. Here, I am first concerned with how movement and the perception of movement guide our actions in our physical environments. The purpose of this discussion is to motivate an introduction to motion-capture systems and analysis methods that designers can use to produce better artifacts that align with their design goals.

Historically, the role of movement in interaction design has been considered either from the perspective of the designer, who observes people to understand their actions as part of the design process, or from the perspective of the machine, by those choosing the sensors and the movement analysis algorithms when the system design is implemented. But as Loke and Robertson point out, we often forget to consider the felt perspective of the user, the everyday experience they have in coming to grips with the tools and devices we have built for them.[2] In the final part of this chapter, I thus present three approaches to interaction design that aim to redress this imbalance by centering the felt perspective of those who will ultimately engage, through their bodies, with the interactive artifacts and environments we design.

3.2 How Bodies Move

All movement begins with a need or desire to move—whether it is to find food, avoid danger, or call a friend on the phone. Whatever the goal, movement is accomplished through our bodily engagement with the environment and guided by information picked up from the sensory milieu of our world. Sheets-Johnstone suggests that movement is our primary mode of being in the world because before we learn to gather information and express ourselves through language, we are already skilled at interrogating the world and expressing ourselves through our bodily movements. She writes, "We come into the world moving . . . we're not still born, and our conception of infants as pre-linguistic is really off the mark because infants aren't pre-linguistic; language is post-kinetic."[3]

It is this kind of pre-linguistic exploration that Alan Kay noted when he observed that computers seem to be developing in a way that is opposite to infants—away from textual modes of communication and toward a more physical way of exploring the digital world as a baby would do.[4]

In interactions with the environment, our movements are neither random nor rigidly repetitive. There are constraints on how we accomplish a task, such as lifting a box. Constraints in this case are imposed by the shape and weight of the box and the arrangement of our muscles and joints. But there are many ways to lift that box: quickly, thoughtfully, gently. Each way has the practical result of changing the temporal arc of our movement and how we coordinate the interaction of our muscles and joints with respect to the reaction forces we encounter. To a motion-capture system collecting data from cameras or body-worn sensors, the movements associated with lifting the box in different ways appear to belong to different "classes." Yet we, as observers, have no problem understanding that they are all shaped by the same end goal—lifting the box—and that they all depend on the way we coordinate our movements to accommodate the constraints imposed on our bodies by the weight and shape of the box.

To use movement effectively within the interactions we design, we must first acquaint ourselves with how our own bodies move, rediscovering what it feels like to move and interact with the world of our devices as naive users do. Part of this process is understanding how we produce and perceive our own movements. But we must also learn to observe how others move because we are also designing for bodies that are not our own and for use contexts we may never encounter. This process involves understanding how we perceive the movements of other people, animals, or objects. It also involves understanding how motion-capture systems track and interpret movement. For example, these systems might interpret movements as trajectories of body parts observed in space and time like camera-based systems might do, or as changes in the inertia of moving body parts, as accelerometers affixed to the body might do. The discussion that follows introduces the challenges of bridging the gap between the current understanding of, on the one hand, how movements are produced and how bodily skill is acquired and, on the other hand, how motion-capture algorithms extract patterns from movement trajectory data.

As we have already noted, movement is the very basis of being in the world. It is only when we move, or when something else in the environment moves, that it becomes accessible to us via our visual, auditory, or tactile senses. And the way we move, in turn, is determined by the goals of our actions. This is why our actions guide perception, which, in turn, guides future action. When people's actions are directed toward the artifacts we design, we determine the information to be "picked up" because we design the ways our systems respond. It is thus important for designers to understand how bodies move and to be aware of the many reasons bodily movements that are not directly related to a given interaction might be detectable in the data captured by sensors embedded in our devices.

3.2.1 The Biomechanics of Movement

Although it is beyond the scope of this book to provide an in-depth review of the biomechanics of movement, it is necessary here to provide a brief overview of the human skeleton and musculature and their role in movement production as context for later discussions in this chapter.[5]

In the broadest sense, bodily movement is triggered either in response to neuromuscular signals initiated to execute planned actions or in reaction to forces and motions arising from interaction with the environment. The body's neuromuscular system is assisted in accomplishing movement by the skeleton, which serves as a rigid scaffold supporting the internal organs and musculature and maintaining the body's shape. The skeleton thus provides the support and leverage necessary for the body to accomplish movement. It is divided into two parts: the axial skeleton (the spine, rib cage, and bones of the skull and inner ears, all located near or along the central axis of the body) and the appendicular skeleton (approximately, the appendages: the shoulders, pelvic girdles, and limbs).

The work of moving the body is carried out by the muscles. They are the body's actuators, flexing and contracting in response to signals from the brain or in response to forces and torques applied to the body through interactions with the environment. Muscular contractions can be voluntary (initiated as motor commands from the central nervous system) or involuntary (arising in response to reflexes terminated in the spinal cord).

The skeleton is also assisted in its movement by articular ligaments, tendons, and cartilage. Tendons are strong bands of tissue that connect muscles to bones. Tendons act not only as connective tissue but also as springs that passively modulate forces during motion and locomotion, stabilizing interaction between muscles and bones.[6] In running and jumping, for example, muscles and tendons work together, with the tendon absorbing energy as the muscle flexes and releasing energy as the muscle contracts. Stability during movement is also helped by articular ligaments—short, tough bands of connective tissue that permit elastic motion between bones while constraining movement enough to maintain structural stability. Their tension cannot be actively modulated, so they continue to stretch in the presence of externally applied forces.

Most skeletal muscles are composed of parallel bundles of fibers attached at one end to the point of origin of the muscle and at the other at its insertion point, the point on a bone to which the muscle can apply force. The skeleton can therefore be described in terms of structures conforming to the laws of lever mechanics. As with any lever, moving the point at which force can be applied farther from the fulcrum—in this case, moving the insertion point of the muscle on the limb bone farther from the joint of rotation—will result in an increase in the force applied when the associated muscles are flexed but at the expense of the movement's velocity. Conversely, moving the insertion point of the muscle closer to the joint will decrease the amount of force generated during flexion but will increase movement velocity. In addition to the position of its insertion point, a muscle's potential to apply force is also determined by its cross-sectional area, as this defines the number of muscle fibers that can operate in parallel. How someone moves is a function of their bodily form and their level of fitness. As we shall see later, factors such as skill and condition will greatly impact how someone moves, and this, in turn, will give us the opportunity to observe, through movement, the acquisition of bodily skills. If such movement is used as input to an interactive system or application, algorithms that track and classify it will need to take into account how, for example, people of similar heights or weights but differing levels of fitness may move differently.

As the rigid structure that gives form to our bodies, the skeleton also serves as the framework for organizing movement. It defines a person's height and, together with their musculature, gives shape to the profile and dynamic signature that describe an individual's movement trajectory in space and time. For this reason, understanding how the body accomplishes movement is central for designers of systems that capture movement, of algorithms for the analysis of movement, and of interactive systems of all kinds.

3.2.2 Sensorimotor Control

We have so far considered the skeleton and musculature as structures independent of the brain. But movement and postural control would not be possible without the brain and nervous system sending commands to and monitoring the status of tendons, muscles, and joints. Our ability to sense and control movement is central to our ability to pick up information from our environment—from controlling the motion of our head to adjust our visual and auditory perspective when observing a scene to controlling our hands for tactile exploration of our environment.

Our bones contain no nerves but are controlled by neural structures that innervate the muscles and tendons. These structures, in turn, are controlled by the motor cortex and are monitored by the somatosensory cortex, where information from skin and muscle receptors is integrated for tactile and haptic proprioception. Most body movement results from nerve signals that travel from the brain to the muscles. The nervous system is made up of the central nervous system (the brain and spinal cord) and the peripheral nervous system, which connects the body and limbs to the spinal cord. The individual nerve fibers (axons) of the peripheral nervous system run together in bundles, which are what we refer to as "nerves." The hand, for example, has three nerves: the median, the ulnar, and the radial (see figure 3.1). Each contains hundreds of axons, and each controls and monitors different areas of skin and different muscles in the hand.

A peripheral nerve cell may be motor or sensory. Motor axons (also called motor neurons) carry motor commands from the spinal cord to the muscles.

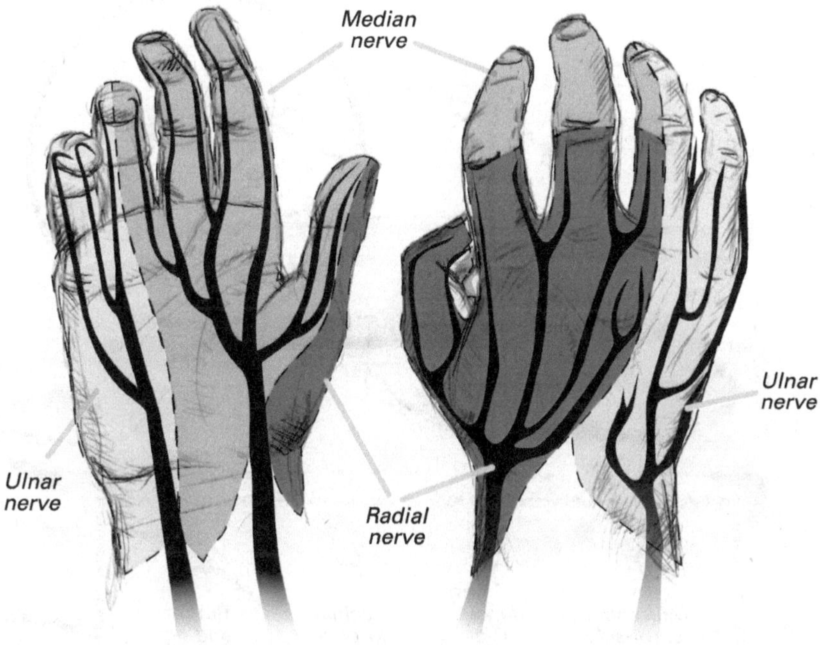

Figure 3.1
The paths of innervation of the ulnar and median nerves, the major pathways for sensory signals to travel from the hands to the somatosensory cortex. *Source:* Original drawing by Hans Anderson.

The firing of a motor neuron activates the muscle cell and thus plays a role in initiating muscular contraction. Motor neurons also control bodily events, such as sweating and bladder function. Sensory nerves convey information in the opposite direction, from the periphery to the central nervous system. They transmit nervous impulses from specialized receptors in the skin, muscles, joints, or internal organs to provide feedback about what is happening to and within our bodies. When you move your arm and your hand encounters an object in your environment such as a table, you understand that the impact you feel is a result of your hand hitting the table and not the table jumping up to hit your hand. Your body is coordinating the signals sent to motor nerves to move your arm with the signals returning from sensory nerves caused by the forces arising from the contact between your hand and the table. If no motor commands were issued to your arm,

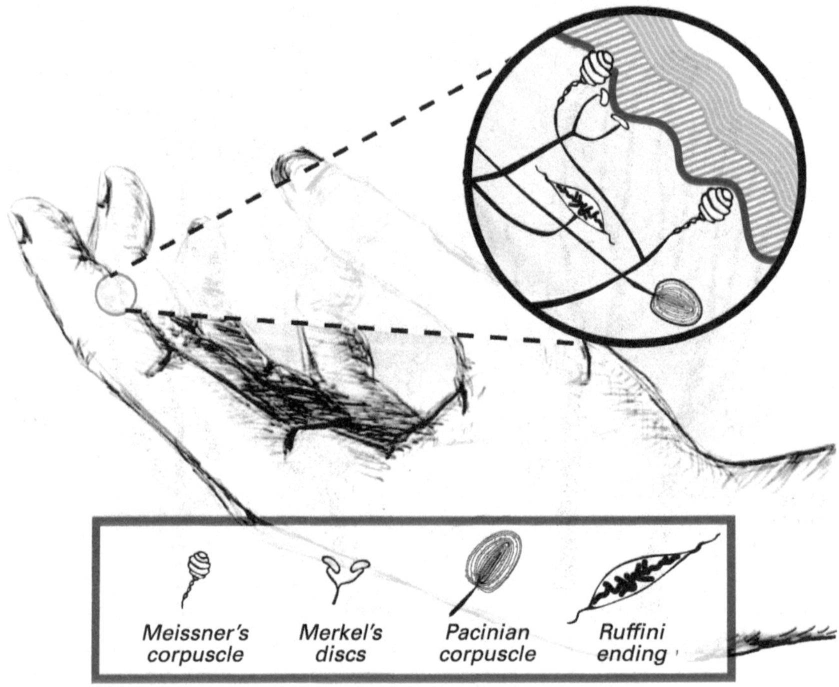

Figure 3.2
The position of different mechanoreceptors in the layers of the skin of the hand. The glabrous or hairless skin contains four types of mechanoreceptors located at different depths within the skin. Each receptor is primarily (but not exclusively) sensitive to a different attribute of tactile sensation. Merkel's disks and Meissner's corpuscles are located closest to the skin's surface, in the papillary dermis, where the epidermal layer projects into the dermal layer, forming a rippled surface. They have small receptive fields, important for both localization and spatial acuity. Ruffini endings and Pacinian corpuscles are deeper in the skin and respond, respectively, to deep pressure and sustained vibration. *Source:* Original drawing by Hans Anderson.

then you infer that something external to your body has caused the table to move.

The position and motion of the body is sensed primarily by "mechanoreceptors" embedded in the skin, muscles, and joints. The skin contains five types of mechanoreceptors, four of which are illustrated in figure 3.2. Together they monitor pressure, vibration, slip, and skin stretch.

In addition, as illustrated in figure 3.3, muscles contain two types of specialized receptors, called the primary and secondary muscle spindle receptors,

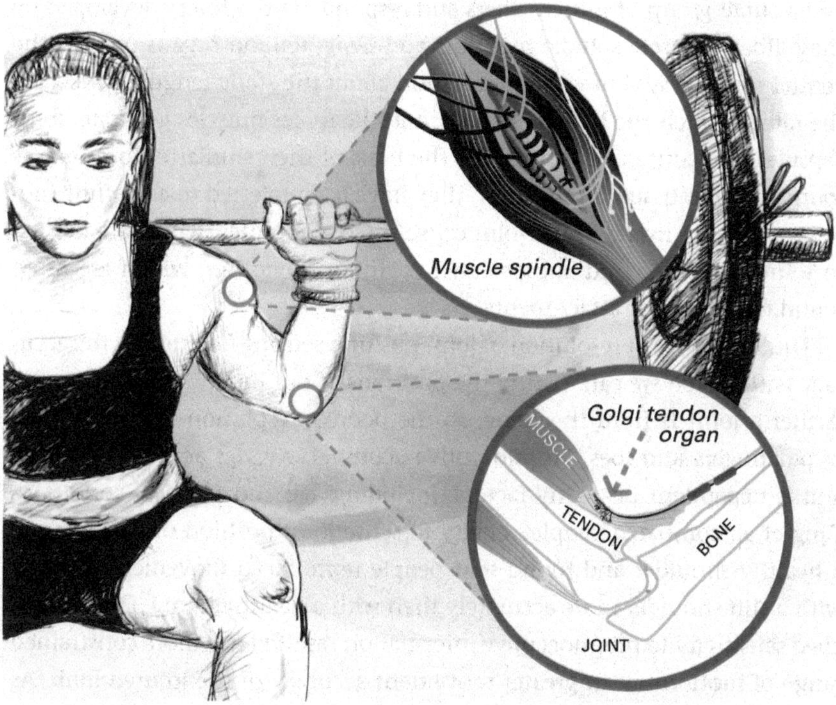

Figure 3.3
Mechanoreceptors in muscles and joints. Illustrates the position and orientation of mechanoreceptors in muscles and joints—muscle spindle and Golgi tendon organ. Motion in the body changes the pressure applied to mechanoreceptors within joints. The compression or stretching of receptor capsules changes the amplitude of their potential discharge, registered as change in position. Changes in joint movement velocity, on the other hand, are registered as changes in discharge frequency. *Source:* Original drawing by Hans Anderson.

which monitor changes in muscle length and may even be responsible for tuning a muscle's response to external forces during motor learning. This is, in part, why movement repetition is such a key component of learning a new skill and also why a teacher or trainer encourages us to be consistent as we learn new movements. By being consistent, we are reinforcing action–response behaviors within our bodies, right down to the level of the nerve signals being transmitted to and received from individual muscle spindles.[7]

A third class of mechanoreceptor found in muscles is the Golgi tendon organ, which is normally found at the junction between the muscle tendon and a small group of muscle fibers and responds to the forces developed by these fibers. Muscle spindle receptors and Golgi tendon organs provide the central nervous system with information about the static length of muscles, the rate at which the length changes, and the forces muscles generate. Joint capsules also contain receptors. On the basis of their similarity to receptors found in other tissues in the body, they have been referred to as Ruffini endings, which are found in the joint capsule; Golgi endings, which are located in joint ligaments; and encapsulated paciniform endings, which are often found near articular attachments.

The sensitivity or resolution of joint position sensing determines the accuracy with which we can control the positions of our limbs. Acuity drops the farther a joint is from the spine, so the poorest resolution is in the joints of our fingers and toes. Proprioceptive acuity also varies across individuals and is dependent on many factors, including age and physical condition. Yang et al. compared people's ability to sense joint position of a stiff versus a healthy shoulder and found that people reproduced movement patterns with a stiff shoulder more accurately than with a healthy joint.[8] This heightened sensitivity to proprioceptive information resulted in a more constrained range of motion but in greater movement accuracy of the injured limb. As we shall see later in discussing the work of Bernstein and others, constraining the motion of the shoulder in this way also constrains the body's ability to respond to perturbations in the environment, but this could be a worthwhile trade-off to prevent further injury and allow for healing to take place.

In addition to receptors in muscles and joints, sensory cues relating to motion and self-motion are also provided by mechanoreceptors embedded in the skin, which provide tactile feedback. However, touch receptors are not uniformly distributed over the body's surface. They are most concentrated in areas concerned with precise movement, such as the lips, fingers, and hands, further highlighting the importance of peripheral receptors in the control of movement. Take, for example, the case of Ian Waterman, the Jersey butcher who, at the age of nineteen, contracted a viral infection that destroyed the peripheral sensory nerves below his neck.[9] Although he could still transmit motor commands to his muscles, he was unable to monitor his own responses to these commands via receptors in his muscles and joints. Nor was he able to feel any sensations of light touch on his skin. He could still feel pain and temperature, as that subset of the peripheral sensory system was unaffected by the virus. Waterman learned to retrain his proprioceptive system so that movements normally monitored by his haptic and proprioceptive subsystems could be controlled through visual and vestibular feedback instead. The only way to ensure that his legs would move in the correct trajectory, rhythm, and sequence to result in walking was to watch their motion to ensure that he shifted his center of mass forward without overbalancing. It was impossible for him to walk in dark or crowded places where he could not see his feet. From this, we deduce that peripheral sensory neurons in muscles, tendons, and joints both regulate self-produced motion and monitor motion imposed by interaction with elements in the environment, such as gravity. Ian Waterman's case strongly suggests that if the pattern of movement that gives rise to success in a given action can be sensed using a second sensory modality, this second modality can, through learning, support the performance of the action.

It is not surprising that such a transfer of responsibility from one sensory modality to another is possible if we consider all actions performed in the environment to be coupled to the global energy array, which provides the individual with information in response to a given action within that environment (see chapter 2). There is increasing evidence to suggest that, even at a cortical level, perception is organized around attunement to task-relevant information picked up via one or more sensory systems and not, as has often been emphasized, around the post hoc integration of coincidental spatiotemporal stimuli from specialized sensory centers in the brain.[10] The implications of this shift are that cross-sensory integration occurs early in neural processing of stimuli and that the sharing of information between modalities can occur even where no direct stimulation is present. For Ian Waterman, therefore, transferring responsibility for the control of walking movements from haptic proprioception to visual proprioception is possible because the visual system was already receiving information relating to postural control from the vestibular system before his loss of haptic sensation below his neck.

Such fluidity in coupling between action and sensed response from the environment would suggest that, in designing interactive systems, there may be much more flexibility in providing feedback and in cuing action than we might once have thought. As the many experiments in sensory substitution have demonstrated, people lacking one sense, such as vision or hearing, may be able to learn to substitute cues from a second sensory modality, provided that there is sufficient opportunity for such individuals to explore the relationship between their actions and the substituted sensory response or to extract meaning from a new sensorimotor coupling between their actions and the remapped response from the environment.[11]

It is important to note here that the contribution of the designer of such systems is not mapping one modality onto another but rather providing the opportunity for the user to make sense of their new experience: to discover, through their own actions, the new sensorimotor coupling that is supported by the designed interaction. In their work, Auvray and colleagues systematically examined conditions under which a blindfolded person could detect the presence of an object on a table using a head-mounted camera and auditory feedback. They found that for a person to attribute changes in the auditory signal to the presence of an object in the external world (distal attribution), three components needed to be in place:

1. The person's movements had to be precisely correlated with changes in the stimulus.
2. The presence of the external object had to be detectable.
3. The external object had to be distinguishable from the perceived background.

In their example, if the camera detected the presence of the object on the table, then, by moving their head, the blind perceiver ought to be able to find the edges of the object and distinguish it from the table. Furthermore, when the perceiver picked the object up, the auditory stimuli relating to the object needed to change in response to how the object was manipulated (turned, raised from the table, etc.).

Sensory substitution, if it is to work at all, is dependent on the active participation of the perceiver to make the connection between their own movements and correlated changes in the new substituted form of sensory information. Sensation alone is not enough. The perceiver must also have control over the camera to learn the coupling between their movements and the sensations that arise. What the above example highlights, for those of us designing interfaces that couple captured movement to real-time feedback from our systems, is that users of our devices will inevitably seek to discover lawful relationships between the movements they make and the feedback we provide. If feedback to the user of the system is designed to promote the discovery of such sensorimotor couplings, then we can build on a user's ability to learn these novel sensorimotor relationships. (See chapter 6 for a further discussion of attunement to properties of virtual objects.) If, however, we fail to take account of the user's need to discover lawfulness via their process of learning, then it should be no surprise that people become at best confused and at worst frustrated with our systems.

3.2.3 General Principles of Movement

The chief problem facing designers of interactive systems, and indeed designers of any movement analysis system, is understanding how a brain and a body choose from many possible movement patterns to achieve an action goal. This challenge was first noted by Lashley and became the basis of Nikolai Bernstein's pioneering work on human motor control.[12] In describing the "motor equivalence problem," Bernstein points out that people can accomplish the same end goal for an action in many different ways.[13] Figure 3.4 illustrates how the same hammer swing can be executed using different combinations of muscles and different joint configurations, demonstrating how much redundancy is built into the anatomical, kinematic, and neurophysiological coordination of movement. Joints and muscles can be organized to achieve infinite combinations of angles, velocities, and movement trajectories. At the neurophysiological level, many motor neurons can synapse onto a single muscle, and many muscles can be controlled by one motor neuron. The result is a highly adaptive system for movement coordination, allowing us to respond flexibly to perturbations in the environment.

Bernstein, whom many consider to be the founder of the field of human motor control, proposed that the human body handles this problem of movement redundancy by "dimensionality reduction." Dimensionality reduction suggests that the movements performed by the body to achieve a movement goal are governed by processes concerned not with the movement of individual joints but rather with the trajectory of the part of the body that must accomplish the goal. For example, to take a sip from a cup, the hand must reach the cup, and the fingers must grip it before the arm is raised to the mouth. In other words, movements are planned with respect to their goals but are organized hierarchically to recruit the muscles and joints required. The trajectory of a movement for the shoulder is determined or constrained by how the hand is moved, not vice versa.

Bernstein formulated this theory after he observed people performing skilled movements, most notably blacksmiths repeatedly striking chisels with hammers.[14] In his observations, Bernstein noted that skilled blacksmiths could choose among many different paths to arrive at the same strike point.

Figure 3.4
Movement traces from a hammer swing. Two dashed lines are used to illustrate how a hammer swing can be completed. *Source:* Adapted by Hans Anderson from an original photograph of Aleksei Gastev produced by the laboratory of Russia's Central Institute of Labor, now in the public domain.

To confirm his observations, he recorded blacksmiths' movements at each of the joints from shoulder to hand and at the point of contact between hammer and anvil. He observed that there was little variation in movement at the point of contact, but there was considerable variation in the movements of the joints along the arm. Notably, the blacksmiths' skills included the ability to reach equivalent end goals through a variety of limb configurations. He points out that the ability to balance the need for consistent movement with the need to incorporate variability is itself a skill because it ensures robustness in motor performance by allowing a blacksmith to select from redundant ways of accomplishing the goal in order to adapt to changing conditions in the task environment. These might include slight differences in the contact force felt through the hammer or tiny differences in the angle of incidence between the hammer and a workpiece or between the workpiece and the anvil. Such redundancy also reduces damage to muscle tissue by distributing strain across different muscle fibers in repetitions of the same action. Latash thus proposes a theory of motor abundance, framing the flexibility of the motor system as an adaptation that allows us to respond to a changing environment.[15] Think of walking on a rocky surface. You plan where to place your feet based on what you see, but the final shape of the movement trajectory is modified by what happens once your foot has met the ground and you transfer your weight onto it. You may need to modify the tension in many muscles to prevent overbalancing if the foot–ground contact is not what you planned.

A central tenet of Bernstein's theory is that any description of movement coordination is meaningless if no account is taken of external factors. In the case of hammers striking chisels, the requirement is that sufficient "reaction" force must be present in the hand's grip to ensure an accurate, forceful hammer stroke. The planning of movement must also accommodate predictable external forces, such as the effect of gravity on hand and hammer and the inertial forces resulting from wielding the hammer. Bernstein suggests that any theory that fails to consider such environmental forces acting on the body would not be a theory of movement but a theory of *miming* movement.[16]

Given that there are multiple ways of achieving an end goal for a movement, what happens when a command to execute a movement is initiated? Consider the following problem described by Alain Berthoz in his book *The Brain's Sense of Movement*: "Write the letter 'a' on a piece of paper in different sizes. Then sketch this same letter 'a' with your finger in space, or with your foot in the sand or on the living room rug."[17] If you could compare the changing directions and speeds of your hand or foot over time (e.g., by recording their tangential velocities), you would notice that, regardless of implement or body part, the trajectory would be approximately the same. When drawing a straight line, the speed is constant, but when drawing a curve, the speed changes in proportion to the radius of the curve, increasing as the radius becomes smaller. In each case, the drawing movements appear continuous, but they are, in fact, composed of segments within which tangential velocity is constant and proportional to the cube root of the curvature of the movement's trajectory.

The relationship between tangential velocity and curvature for such drawing movements is described by the "two-thirds power law."[18] This law holds that a change in one quantity causes a proportional change in a second quantity and that the coefficient that governs this proportional change is a constant. In our example, the cube of the speed of the drawing movement is proportional to the radius of the curvature of the drawn trajectory, and the proportional coefficient relates to the mechanical power required to accomplish the movement. The two-thirds power law is derived from a physical law called the "principle of least action," which states that if a movement between two points obeys the two-thirds power law, the amount of work required to move along a trajectory in a given time period is minimal. In other words, of all the possible routes to accomplishing a given drawing movement, the one selected is the most efficient in terms of the mechanical work to be done.

That this law applies across drawing conditions, regardless of the limb or the end effector, Berthoz suggests, is a further instance of behavioral evidence for Bernstein's theory of a neurophysiological hierarchy in movement coordination.[19] Following Bernstein, other principles for movement coordination have been defined, such as the principle of isochrony, which states that the farther you need to move to reach a point, the faster you will move.[20] Understanding the relationship between motor commands and the movement trajectories they give rise to is important when, for example, we wish to detect patterns in sensor data to observe how movements change over time as people acquire new skills or as the context of an interaction affects how people move.

3.3 The Purposes of Movement

Having briefly reviewed how bodily movement is accomplished, we turn our attention to *why* we move by considering the different circumstances that shape the way we interact with the world around us. With such an understanding, it will be possible to reconstruct meaningful interactions when the world in question is *not* real but rather a virtual world constructed within, for example, an action game.

All animate objects move. Movement is their means of acquiring nutrition, evading predation, and accomplishing the propagation of their species. Humans move for much more than mere survival, however. The reasons we move are many and varied and very rarely occur in isolated contexts. The movements we perform to achieve an action (i.e., goal-directed movements) are often seamlessly interleaved with movements that allow us to explore some aspect of that interaction (i.e., exploratory movements). For example, when we lift a cup of liquid, we simultaneously initiate motor commands to cause the cup to move off the surface on which it is resting, and we acquire sensory information that provides us with some knowledge about the amount of liquid in the cup, perhaps the temperature of that liquid, and the material properties of the cup itself. Even the way we lift the cup—slowly or quickly—can convey to an observer something about our state of mind. For example, are we thoughtful, angry, distracted, or in a hurry?

Some movements are functional but not directly associated with a primary goal. For example, compensatory and recovery movements are adjustments made to restore stability in response to perturbation of a limb or the whole body. Think of lifting a heavy box. Obviously, you must recruit muscles in your back, legs, and arms to help lever the box off the ground, but there are other movements you perform to ensure you stay upright. You subconsciously anticipate the backward shift in your body's center of mass by leaning forward to compensate for the weight of the box and to stabilize your limbs to ensure smooth movement of the load you are carrying. Then, when you have put the box down, there is a series of movements associated with recovering equilibrium and resuming an upright posture. Even maintaining an upright standing posture involves continual and often unnoticed corrective movements to offset natural fluctuations in body sway. These movements, which are ancillary to the primary movement goal, are still functional and, as we will see, will be present in data acquired by most motion-capture systems. Therefore, although the following discussion addresses movement in categories defined by some notion of the purpose of that movement, it is important to keep in mind that such categories are somewhat artificial, structured primarily to focus on aspects of movement that will be of interest when discussing how to extract patterns from data captured by motion-tracking systems.

The goal here is not to provide a means of separating out different features that might place different movement types in different analysis bins; rather, it is to point out potential sources of features, patterns, and variance in captured movement data that can become raw material for use in interaction design. If our job as interaction designers is to embed digital artifacts in the loop that couples the actions of a person with a meaningful response from their environment, then understanding the sources and purposes of movements can guide us in providing contextually relevant feedback in response to how people move.

3.3.1 Goal-Directed Movement

The overarching challenge for those who study movement in the context of motor-control research or interaction design is that, as Bernstein noted, there are an infinite number of ways our bodies can recruit muscles and joints to accomplish one goal. Think again of picking up a cup. You can pick it up by the handle using your fingers, or you can wrap your whole hand around it. You can reach for it quickly or lazily, rest your arm on the table before lifting it, or use your whole arm. All you are thinking about is picking up the cup. But how you pick it up determines the extent to which a given joint or muscle group will be involved. This is what Bernstein means by redundancy in movement coordination. All those possible routes to the same goal allow your body to adapt quickly should one of them become unavailable. From the perspective of the system tracking your movements, the data acquired would look very different. A slow, lazy reach of your whole arm would produce a pattern of change in all joint angles extended in time, whereas picking up the cup with your arm resting on the table would result in little change in the values seen by sensors recording the position of the shoulder, because it would not move.

We in the field of HCI have analyzed for decades how people perform tasks such as typing, manipulating mice, and using touch screens. Often, such analyses adopt existing experimental paradigms to study the effects of manipulating variables in a design space. One of the earliest examples of systematic observations of movement within the context of HCI is found in the work of Card, English, and Burr.[21] Their study compared four methods of selecting text on a screen: a joystick, a mouse, step keys, and regular keyboard entry. In the design of their study, they adopted Fitts's law, an experimental paradigm developed by Paul Fitts as a means of describing human psychomotor behavior.[22] Fitts wanted to find a perceptual motor analogy to Shannon's theorem, which describes the capacity of a system to transmit information.[23] In Fitts's analogy, movements are assigned indices of difficulty in numbers of bits transmitted between the human and the system with which they are interacting. The fastest time required to complete a movement accurately is a logarithmic function describing the relationship between the distance to be traveled and the size of the target to be acquired. Fitts called this relationship the Index of Difficulty (ID) for the task, and he demonstrated that the value of the ID is analogous to the rate at which bits of information can be transmitted between the human and the environment. Fitts further determined that the model predicted a trade-off between the accuracy of movement and the speed of movement, showing that when a task is performed more quickly, accuracy is compromised. Applying this model to HCI, Card and colleagues sought to determine whether they could predict the time needed to position a cursor and select a target for each of the four devices in their sample.[24] They found that Fitts's law predicted around 80 percent of the variance in movement time for the mouse and the joystick, but that the time to acquire a target using either text or step key input was proportional to the number of keystrokes required.

Since then, Fitts's law has been used to predict performance for many human–computer interfaces in a variety of contexts, including determining how quickly people can move a mouse pointer between items on a desktop and how easily people can find and select items within drop-down menus. As the factors involved in the formulation of Fitts's law have been explored in more detail, the relative contributions of the perceptual and motor components of tasks are being disentangled. Bootsma, Fernandez, and Mottet, for example, have determined that the time to complete a movement is a global feature of the continuous exchange between the mechanical or movement-related component of the task and the information relating to the perceptual response that flows from the task environment.[25] In other words, accuracy is predicted by the changing relationship between pointer and target, whereas speed is determined primarily by the biomechanics of the task. As Mottet et al. have shown, this holds true whether the task is to move the pointer toward a stationary target, to move the target to a stationary point in space, or to move both at once.[26] In their work, they demonstrate that as the ID for a task varies, the whole kinematic pattern of a movement changes. By observing such changes, we, as designers, can estimate how well users are able to perform tasks and whether their performance improves over time, allowing us to estimate whether people can become skilled users of our interfaces.[27]

Mottet et al. also demonstrate that moving a pointer to a target can be defined in terms of closing a gap in space and time.[28] Such a formulation of goal-directed movement has been proposed as a fundamental principle for the planning and execution of skilled action in many other domains. If tasks can be described in terms of closing such perceptual motor gaps, then observing how people perform these tasks is a useful way to monitor the acquisition of new motor skills.

The first person to propose that we time goal-directed actions by determining how long it will take to close a "perceptual gap" was David Lee.[29] He proposed that such gaps are defined by variables relating to the state of the acting organism and the state of variables in the environment that are relevant for the task goal. Thus, a baseball player running to catch a ball will follow a curved path that maintains a linear optical trajectory (LOT), meaning that his running speed and direction serve to maintain, within the image reaching his retinas, a rate of increase in the lateral optical angle that is the same as the rate of increase in the vertical optical angle.[30] Using this strategy will result in a position gap between the player and the ball being closed in such a way that they coincide in space and time. This means that the player will get to where the ball will be when it is at approximately head height. Soccer goalies use a similar principle to intercept incoming balls.[31] They track the changing shape that the looming ball casts on their retina. If the shape is symmetrical, the ball is directly approaching the goal with no spin to curve its path. But if the shape on the retina changes, this is a clue to the angle from which the ball is approaching and to whether its trajectory is curved or straight.

Humans are not the only species to latch onto systematic changes in stimuli, such as visual angles, to guide their movements. Shaffer and colleagues have demonstrated that a dog catching a Frisbee uses the same strategy and indeed will alter its path to maintain the LOT if the trajectory of the Frisbee changes midflight.[32] When catching fish, a diving gannet can adjust its trajectory to minimize the time it needs to spend underwater as it closes the gap between itself and its dinner.[33] Its task is made more complex by the fact that, unlike a baseball, the fish may move unpredictably of its own volition, so the bird may have to make unanticipated adjustments mid-flight.

Crucial in all these cases are environmental variables that carry the information relevant to a given goal in a given context. The image of the object on the retina continually changes as pursuer and object move. Gibson was among the first to propose that, in navigating our environment, we rely on tracking the "lawful" way in which such variables change with respect to our own movement in order to plan and execute action.[34] In all the above examples, the perceiver, in planning to close a perceptual gap, is latching onto sources of information within their optic flow that remain constant or, to use Gibson's term, invariant. In the case of the baseball or Frisbee, the invariant is the linearity of the trajectory; in the case of the ball approaching the goalie, it is the angle of the looming image of the ball.

In addition to proposing that we track such invariants to close perceptual gaps for discrete action goals such as catching balls or fish or Frisbees, David Lee proposes that, for repetitive movements, our actions become coordinated with, or "coupled" to, such invariants. In his theory of tau-guided movement, Lee proposes that the time to close a perceptual gap, which he called "tau," is equally valid for gaps that are defined by external variables, such as time-to-contact with an approaching object, and by internal variables, such as the time it takes to place your foot on the next step of a staircase.[35] As I shall propose, becoming proficient in skills that require closing such intrinsic gaps may be no less dependent on becoming a skilled perceiver of one's own movements than on becoming proficient in tasks that require closing extrinsic gaps, such as catching baseballs. Framing a goal-directed movement in terms of closing a perceptual motor gap provides a new tool by which to determine the perceptual cues that carry relevant information for the successful completion of the task. The better our understanding of the relationship between such perceptual cues and the actions they support, the more robust our interactive systems will be.

In summary, goal-directed movements pose very particular challenges for designers of interactive systems because people accomplish even the simplest of tasks in very different ways. This redundancy is a positive adaptation from the perspective of someone seeking to become more skilled because there are multiple strategies they can choose from. But redundancy presents a challenge for a motion-tracking system because it requires a person or an algorithm to "train" the system to recognize many possible paths to the same movement goal. However, the variance observed should not be discarded, as it too contains valuable information for the designer, such as how well people can perform a task, given a number of alternative interfaces,[36] or how people's movements stabilize over time as they acquire a new skill.[37]

3.3.2 Repetitive or Cyclical Movement

At the outset of this discussion of the purposes of movement, I suggested that the division presented here is somewhat artificial because, in carrying out daily activities, our movements do not fall into neatly partitioned categories. One example is repetitive movements, such as the steps we take in walking or running. These can each be thought of as supporting an individual goal—that is, taking the next step—but they can also be thought of as serving a larger goal, such as walking to a particular place for a particular purpose. With the advent of low-cost activity-tracking systems, there is an increased interest both in characterizing aspects of the local goal of such repetitive or cyclical movements and in understanding how the structure of these movements changes over time in service to the larger task goal. The result is that analyzing cyclical movements can be a multilayered process. The features of a sequence of running steps might inform an application that is tracking the progress of a morning run, which, in turn, might feed into a timeline that provides information about improvement in fitness. Similar layers of interpretation exist for activities such as skill acquisition or rehabilitation. It is worth taking time to consider the nature of cyclical movements and some of the particular qualities that might be of interest for an application tracking changes in such movements.

Let us begin by considering the activity of walking. As illustrated in figure 3.5, the sequence of movements by which humans accomplish walking is typically described as starting when the heel strikes the ground and concluding when that same heel strikes the ground again. Leg motion is divided into a "swing phase," when the leg is moving through the air, and a "stance" phase, when the foot is planted on the ground. In biomechanical terms, this sequence is often called a "double pendulum." In the swing phase, the pivot of the pendulum is the hip, and the leg swings below the body. In the stance phase, the pivot is the contact point between the foot and the ground, and the pendulum is inverted with the body passing over the static foot. The swing phase of one foot is coordinated with the stance phase of the other, so that both feet are on the ground for approximately 20 percent of the stride duration.[38]

As a basic description of how people accomplish walking, this is sufficient. But if we were to track the motion of a walker, we might expect considerable variation in how the same person walks in different contexts (e.g., on the beach versus a road) and even in the profile of individual footsteps in the same sequence. Changes in someone's gait can be used to diagnose not only mechanical problems such as joint disease but also motor problems symptomatic of conditions such as Parkinson's disease.[39] A typical analysis of walking would include measuring parameters such as stride length, cadence, speed, dynamic shift in center of mass, and foot angle. These can

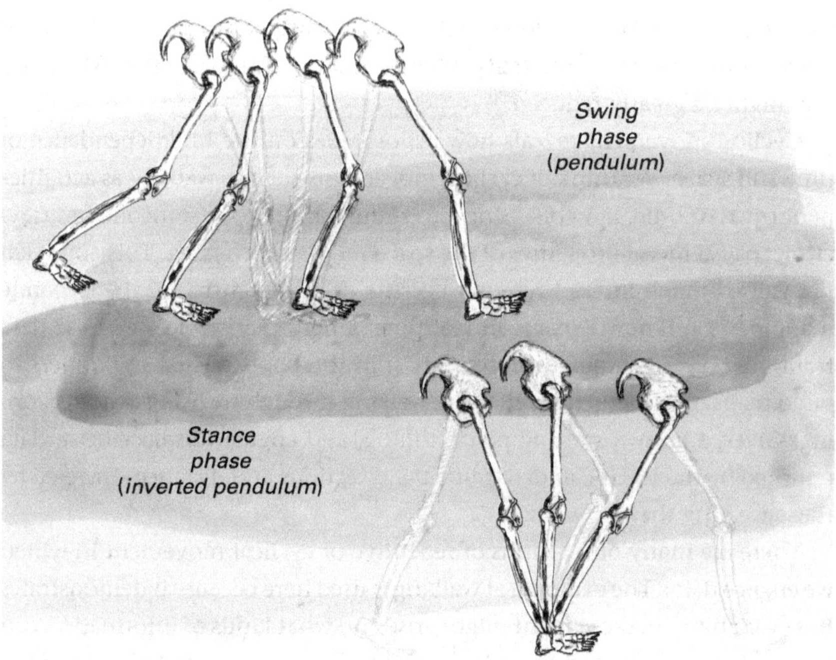

Figure 3.5
Gait cycle of walking person, which can be modeled as a pendulum centered on the hip (the leg in the swing phase) and an inverted pendulum above the stationary foot (the leg in the stance phase). *Source:* Original drawing by Hans Anderson.

all be impacted by external factors such as terrain, clothing, and footwear or by intrinsic factors such as height, weight, personality type, and level of fitness. How someone walks is affected both by their immediate context and by characteristics of their anatomy, physiology, and even their mood.

As with the discrete goal-directed movements described above, useful information can be gleaned about patterns in movement from the way sensor signals vary over time, even for cyclical movements. For example, Hausdorff points out that many gait disturbances in Parkinson's patients are not readily visible during clinical observations but instead become apparent when quantitative measures are applied to sensor data captured from such patients over long periods of time. Issues such as increased left–right asymmetry, diminished bilateral coordination, and a loss of ability to maintain a consistent stride-to-stride rhythm then become apparent. Researchers have suggested that if one could establish a cyclical guide that patients could track using some other modality, such as auditory feedback, the risk of falling might be greatly reduced.[40]

Cyclical movement reveals how action relies on the interdependence of time and space. We think of cyclical movements such as walking as activities performed in time, but collapsing a recording of the path someone has traveled can also reveal properties of the space they are traversing. This approach has gained much interest recently because it has potential for the dynamic mapping of unknown spaces in real time. Emergency services can use data from inertial measurement units (IMUs) in the boots of first responders to build maps of the structures they are walking through by using dead reckoning: Given a known starting point, timing, and angular displacement, data returned by the IMUs can determine the directions and distances traveled by those wearing them.[41]

There are many other forms of repetitive or cyclical movement in which we engage daily. The example of walking is used here because it demonstrates how variability in movement might arise and what kinds of information can be gleaned from analyzing such variability. As we shall see later, rhythmic movement can become a mode of interaction in its own right—for example, in motion-based games and interactive systems for musical performance. Therefore, understanding the potential to observe change and being able to extract information from variance is a useful tool for the designers of such systems.

3.3.3 Exploratory Movement

In examining the purposes of movement, it is important to acknowledge a class of movements whose goal is simply to discover something about the world rather than to act intentionally on it. If you have ever watched a baby playing, you will have noticed that many of their movements are directed toward testing some hypothesis relating to the objects around them: Can I eat this? Can I pull this? Can I crawl on top of this? Such exploratory movements are how the baby learns to interact with their environment. As adults, we no longer need to perform such tests because we have learned how the world works. Occasionally, however, we return to this mode—for example, when we are caught unaware by something in the environment, such as an icy sidewalk, or when we are learning a new skill. Let us return to our example of lifting a box. Most of us have had the experience of trying to lift a box and discovering that it is lighter or heavier than we anticipated. But an experienced lifter will first push on the box to get a sense of its weight. This exploratory movement helps the lifter plan the effort they will need to exert, and it may reveal other information, such as where the center of mass of the box is, if it contains a large solid object that fills it or a small, heavy object that will move.

Exploratory movements are not restricted to the haptic modality. We use vision to plan how we will grasp objects, and we even use sound as a means of learning about our surroundings. But exploratory movements have been studied in haptics because they offer a glimpse into how the brain "seeks" such information—for example, how we build and test hypotheses related to the material properties of objects. In their groundbreaking work on the haptic classification of everyday objects, Susan Lederman and Roberta Klatzky first noted that people are remarkably consistent in the hand movements they use during object exploration.[42] When asked about the texture of an object, people rub their fingers back and forth across its surface, whereas a prompt for information about softness will result in people pushing on or squeezing the object. Unlike goal-directed movements, exploratory movements are designed to selectively focus on a single perceptual attribute.

The reason for mentioning such movements here is that people often combine exploratory and functional movements when performing a task, and such movements can appear within data captured for other purposes, such as tracking the acquisition of a new skill. Hefting a ball before throwing it, for example, can be seen as a mode of exploratory movement, where the thrower is testing the dynamics of the hand–ball system to discover how much force to apply to hit a target.[43] Similarly, a baseball player might heft a bat to test the dynamics of the arm–bat system before attempting to hit a ball. In capturing the motion of the batter, therefore, one would need to predict which bat swings that failed to connect are exploratory movements and which might be "a swing and a miss." Such exploratory movements can reveal the process by which someone is discovering the relevant invariant perceptual cues to track for a given task, and they are therefore a potentially useful resource for the designer of an interactive system.

3.3.4 Ancillary Body Movements

The challenge in designing motion-capture systems is that the system must "see" all the body's movements, including ancillary ones. When you walk or run, the primary movements are those of the legs, but leg movements are usually accompanied by movements of other parts of the body, particularly the arms. One could say that leg movements are the primary functional movements in walking and that arm movements, because they are not so functionally necessary, are ancillary. When capturing a person's gait, therefore, it would be necessary to capture the effective or functional movements of the legs and hips, but it would be extremely informative to also capture the movements of the upper body and upper limbs, as these could be indicators of sudden shifts in a walker's center of mass, which could predict a fall.[44]

One area in which ancillary body movements have been extensively studied is musical performance because it is often difficult to track musicians' functional movements by placing sensors between their fingers or lips and an instrument without interfering with the felt response of the instrument. The observation of ancillary body movements thus provides a useful window into changes that may arise under different performance conditions[45] or as instrumental skill is acquired.[46] Wanderley et al. suggest that ancillary movements in musical performance may be important for a performer's comfort, such as a flourish at the end of a musical phrase, which can release tension in their hands and arms, or they may work to enhance the emotional experience of the performer or their audience.[47] Whatever their purpose, ancillary body movements have been shown to mirror structural elements of performance, such as the articulation of individual notes or the grouping of notes into musical phrases.

In tracking ancillary body movements, the designer of a digital musical instrument may be able to classify movement goals according to various timescales, from local note events to more global articulation of musical structures. Discriminating between different movement categories may also allow for the separation of bodily movements related to maintaining postural stability from those with a communicative or performative intent. In short, one should not discard movement data that does not relate to the accomplishment of a functional goal—in this case, playing a sequence of notes—but instead should search for structure in this data. It might enrich the context within which a given interaction is situated. In distributed musical rehearsal scenarios, for example, where it is difficult to coordinate high-bandwidth video signals, ancillary body movements have been used to control visual and even mechanical avatars to convey body language cues from remote performers, which is critical for coordinating ensemble performance.[48]

3.3.5 Involuntary and Compensatory Movements

A further challenge in designing any system that tracks movement is that sensors pick up bodily movements that arise from processes unrelated to the actions being performed or arise as a by-product of the action of muscles involved in the primary movement of interest. Sensors might also pick up involuntary movements such as muscle tremors or reflexes. Whatever their source, it is useful to understand when and where such movements might appear in sensor data and what their signatures or patterns might look like. Such movements can also often provide useful information, such as how tightly someone is gripping an object or when someone is becoming fatigued.

3.3.5.1 Involuntary movements Some bodily processes give rise to movements that are not under conscious control. Some of these are autonomic movements, which are spontaneous and independent of external stimulation, such as the beating of the heart, the action of muscles in the intestine, or the movement of the lungs to breathe. These movements are controlled by the autonomic nervous system and continue even when we are sleeping. Although most autonomic movements are not observable using typical motion-tracking techniques, they can be modulated by activities that can be sensed. For example, increased physical exertion or heightened emotional arousal will result in increases in both heart rate and breathing rate. Monitoring heart rate can therefore be a useful way of tracking someone's affective state within a game.[49]

3.3.5.2 Reflexes Unlike autonomic movements, reflexes respond to fixed external stimuli—for example, blinking in bright light or the patella response that causes your leg to straighten when you are tapped on the kneecap. Other reflexes occur to stabilize posture, such as the slip reflexes that stop us from falling or losing our grip on a handheld object. Postural movements combine reflex responses with small voluntary adjustments, allowing us to maintain an upright posture for long periods of time. Because these involuntary movements are likely to be captured by sensors embedded in interactive systems, it is important for an interaction designer to be aware of how they may impact data analysis algorithms. On the other hand, the ability to track how postural reflexes change over time has become a central component in studying conditions such as Parkinson's disease, providing researchers not only with a means to observe disease progression but also with the ability to intervene to regulate a patient's gait (e.g., by introducing rhythmic patterns of sound or light to increase the fluency of walking).[50]

3.3.5.3 Involuntary muscle tremor This form of tremor is often observed as an underlying component in movement data. There are two main classes of tremor. Rest tremor occurs when the body is supported against gravity. Action tremor occurs while muscles are performing work. The amplitude of rest tremor increases with stress and is reduced when engaged in a targeted task such as pointing at one hand with the other. Action tremors occur during voluntary muscle contraction and can be categorized as postural tremor, when muscles work to maintain the position of a part of the body against gravity; isometric tremor, when muscles contract against a stationary object, such as squeezing an orange; and kinetic tremor, which arises either as "simple tremor" from the flexion or extension of peripheral joints, such as the wrist, or in the form of "intension tremor," which occurs when guiding the finger toward a target. Detecting hand tremor has also been used to determine hand posture and grip for people holding mobile devices. This, in turn, can be used to discriminate movements arising from active engagement with the device from those that passively arise from carrying it.[51]

3.3.5.4 Compensatory movements Compensatory movements are movements that occur in one part of the body to compensate for movements in another. For example, when lifting a heavy box, it is necessary to make some postural adjustments to maintain balance. When you raise the box, you may move your hips back and bend your legs to lower your center of gravity to prevent yourself from falling backward. Many activities require such compensatory movements. Even an action as delicate as bowing a string on a violin will start with a postural adjustment that can be observed in data captured by a sensor on the foot.[52] This is because most actions require exertion, and when force is applied to move the body or to move something in contact with the body, the equal and opposite force that arises must go somewhere.

Stabilizing the body to absorb that force often requires compensatory movement, which will show up in sensor data if the part of the body involved in this compensation is being tracked. Such movements are not always under conscious control, although, with practice, a skilled performer may be able to control some aspects of the timing and amplitude of these movements. In the violin example above, the shift in weight observable at the right ankle anticipates the start of a downward bowing gesture by about 10 ms; by tracking these movements during live performance, the motion-tracking system can be tuned to anticipate the very start of the downward bow gesture.[53]

3.3.6 Skilled Movement

If you have ever observed someone learning a new skill, such as riding a bicycle, you probably recall how their initial attempts seemed uncoordinated and somewhat chaotic. If you stuck around to observe their progress, though, you probably noticed how, with time, they figured out how to organize their limbs, the rest of their body, and the bike so they could ride without falling. You might also have noticed that their initial process of trial and error was replaced by confident movements they could reproduce each time they got on the bike.

Bernstein, whose work on goal-directed movement was introduced earlier, was one of the first to characterize how movements change as skill is acquired.[54] According to his theory of graded skill acquisition, early stages of skill learning are associated with "freezing" biomechanical degrees of freedom (DOF)—for example, restricting movement of some joints—to reduce the number of ways of accomplishing a movement goal. Conversely, later stages of skilled action are characterized by a more differentiated use of DOF, allowing more free movement of joints to support more efficient and flexible performance. Thus, our novice cyclist might initially focus on a few successful movements, but as she becomes more confident, she will allow more "equivalent" strategies into her repertoire. Initially, perhaps, she will slow down and deliberately orient her body as she turns corners. With practice, she will discover that leaning outward as she approaches a turn causes the bike to turn without having to slow down quite so much. Over time, leaning into turns will be something she does without thinking. Gradually, she will discover just how much she needs to lean to turn, given her current speed and the turn angle. The more strategies she has available, the better her skill set will accommodate perturbations in the environment. In the same way, a novice golfer's game will fall apart in windy conditions, but a skilled golfer will take into account the speed and direction of the wind and adjust their game accordingly.

Many studies have framed the process of acquiring a new motor skill as one of learning how to recruit motor resources to close a perceptual gap. The ability of an organism to close perceptual gaps is a predictor of skilled performance of a task. In observing clarinet players learning new music, for example, Rodger and colleagues were able to determine the rate at which performers arrived at a stable performance of a piece by observing, from the capture of ancillary movement data, how the variability in their ancillary bodily movements reduced over time.[55] Expert performers arrived at a stable and repeatable performance immediately, whereas novice performers often took many days to reach the same goal. Intermediate players arrived at a stable performance within a couple of rehearsals. The research team formulated their analysis of movements in terms of closing a gap between an intrinsic internal guide (the musician's knowledge of where the next beat should be) and the timing of associated ancillary body movements. One important caveat: Not all the musician's ancillary movements stabilized at the same rate. The expert, for example, had sufficient mastery of the piece and the instrument to allow him to move *more* freely as he played. (See section 4.5.2 for further discussion of this study.) It is important in any such analysis to understand what properties of movements are becoming more controlled to anticipate where variability in sensor signals will decrease over time.

Studies such as these demonstrate not only that it is possible to observe the acquisition of new skills through observing changes in movement but also, as Bernstein predicted, that such changes are systematic, converging on performances that are stable and repeatable. Where there is variability in the movement patterns of experts, it arises not from a process of homing in on optimal strategies, as with novices, but from a "freeing up" of possibilities for manipulating the performance in controllable ways. It is important for any interaction design that depends on tracking movement to be able to distinguish between these different sources of variability. Variation in how a user performs a task can arise for a number of reasons. For example, in encountering a motion-tracked game for the first time, a user might try different strategies to find the one that is most effective. As noted in the next section, if players are very skilled, they might introduce elements of performance or flare into their actions for fun or because they are engaging in some form of collaborative play. In each case, there will be considerable variance among data captured for the same task, but the sources of variance will be different: the uncertain movements of the novice versus the performative movements of the expert. This difference provides an opportunity for an interaction designer who might want to provide feedback to guide the novice, on the one hand, or to reward the flare of the expert, on the other.

3.3.7 Affective Movement

Movement visible in captured data may also reflect a person's emotional state. If you observe someone walking on the street, even if you cannot see their facial expression, you can often tell whether they are focused or distracted, sad or joyful. Detecting the emotional state of someone using an interactive system has been a topic of much interest and has given rise to a field of research called "affective computing."[56] As demonstrated by Dittrich et al. for dance performance, one person can understand something of the emotional state of another, or can recognize their own affective state, by observing their movement.[57]

It is important to note, however, that the interpretation of affect in this way is grounded in a shared social and cultural understanding of the expression of emotion through action. Picard and others have suggested that the emotional state of a user can be quantified by analyzing biological signatures such as voice patterns, facial expressions, bodily movement, and skin conductance. Others, such as Kirsten Boehner, have argued that, although these signatures might reflect certain emotional states, affect is perhaps more accurately described as a form of socially and culturally grounded action in which people interactively construct and subjectively experience their emotions.[58] This would mean that the effect of an emotional state on action would be embedded in a broader notion of situated action, of a coupling between the acting person and the responding environment that includes not just physical but also social and cultural responses to action.

It is beyond the scope of this book to address the broader discussion of social and cultural interpretation of action, but it is clear that movement data may convey information bounded by a clearly defined social and cultural context. Musical performance, for example, is interesting from the perspective of how players communicate with audiences through body language. Some performative gestures are congruent with and are also outcomes of functional movements related to sound production—for example, effortful gestures required to produce very fast or very loud sounds or sounds at the limits of an instrument's range. But as we have noted in the earlier discussion of ancillary body movements, other gestures may be used to emphasize the dramatic nature of a moment or to indicate a release in tension at the end of a phrase.[59]

Game designers too are becoming interested in the affective content of movement. In their guidelines for designing movement-based games, Isbister and Mueller suggest that a designer should support what they call "self-expression." In other words, there should be space in the design for players to achieve the same end goal in many different ways, some of which might be more flamboyant than others.[60] They suggest that designers should encourage players to try out more expansive movements and should "celebrate" self-expression by showing players the outcome of this expression. Finally, they recommend that designers offer opportunities for what they call "secondary performances"—in other words, they should offer the possibility for ancillary body movements to contribute to the outcome of an action.

In sum, bodily movements can bear the signature of someone's affective state. Often, the person is unaware that their actions are shaped by how they feel, but sometimes they can choose to use bodily movement to convey an affective response to a situation, such as gesturing to emphasize a point they are making in a conversation. Performers too can choose to shape their actions for dramatic effect. Affective movements may appear in motion-capture data, either shaping functional movements or as ancillary movements alongside goal-directed actions. Finally, affective movements are an outcome of the relationship between the acting person and the responding environment, including physical, as well as social and cultural responses to actions.

3.4 Perceiving Movement

So far, our discussion of movement has focused primarily on self-motion and the perceptual cues associated with the sensorimotor coupling between a moving individual and their environment. It is easy to forget, however, that this environment will include other moving objects and beings, and it is imperative for us to perceive their motion to avoid collision, anticipate action, and even escape danger. There are situations in which a designer might wish to build on this capability, such as creating realistic avatars, constructing immersive environments, or building collaborative interactive games. This section therefore reviews some of the most salient cues about the motion of animate and inanimate things that might be relevant in the analysis and use of movement for interaction design.

In the early 1970s, perceptual psychologist Gunnar Johansson attached light bulbs to the joints of people wearing black clothing and then filmed their movements in a dark room.[61] He showed these films to naive subjects, who easily identified them as representing people walking, just from the motion of the lights on their joints. Johansson thus demonstrated that the relative motion of body joints alone contains sufficient information to support a viewer's perception of human action. He referred to his actors as "point-light walkers," and the recording and replaying of their actions was termed "point-light animation" or "point-light display" (PLD). Although today's motion-capture systems typically use passive reflective markers, which reflect infrared flashes to digital cameras, they are direct descendants of Johansson's apparatus.

Subsequent research has demonstrated that people can identify not only walking humans just from PLDs but also attributes such as the gender of the walkers, their height, their weight, and more. Observers could even correctly identify emotional states in dance performances replayed using PLD.[62] Moreover, point-light techniques have also been used to observe human interactions with objects. Runeson and Frykholm demonstrated that people could estimate the weight of a box being lifted by an actor and could distinguish between real and mimed lifting movements using only PLD.[63] As we shall see, marker-based motion capture and PLD are now widely used across a range of motion-analysis applications, including the study of movement (kinesiology), the study of performance in sport, physical rehabilitation, and motion capture for animation.

For interaction designers, it is important to recognize that the perception of movement is a much more active, dynamic, and malleable process than merely observing the stimulus that arrives at our sense organs (eyes, ears, skin). Moreover, we are continually making decisions about how to move in response to what we perceive, and our movements alter in predictable ways as we become more skilled. The evidence would suggest that such changes reflect increased skill not just in performing actions but also in perceiving when those actions will be successful. Even though we cannot easily observe the acquisition of skill in perceiving, we can still directly observe its effect on action by looking at changes in the movements people make. In short, the designer of a movement-based interactive system must pay attention to how people perceive movement.

3.5 The Role of Context and Skill in Motion Perception

For fans of the Chicago Cubs, October 14, 2003, was a day to remember. The action of one of their fans may have cost them a game against the Florida Marlins. As of the eighth inning, Mark Prior had pitched a three-hit shutout for the Cubs. They led the Marlins 3–0. They were five outs from getting into the World Series. Cubs outfielder Moisés Alou was attempting to catch a foul ball hit by the Marlins' second baseman, Luis Castillo. The ball appeared to Alou to be headed over the left-field corner wall. Steve Bartman, a fan sitting in the front row, reached for the ball. He failed to catch it, but he deflected it away from Alou's glove, and Alou missed the catch. The Cubs lost the game 8–3 and did not go to the World Series.

Why did both Alou and Bartman think they could catch the ball? For Alou, running toward the ball meant controlling his speed and direction of movement to ensure that the velocity of the ball in his field of view remained constant. To accomplish this, Alou may have had to vary his running speed to cancel out any changes in the ball's speed, particularly any change in acceleration as it rose and fell in the arc of its flight. If Alou was attuned to optical acceleration cancellation (OAC) as a source of invariance that could guide his running, then he would intercept the ball before it reached the boundary of the field.[64] For Bartman, sitting in the stands, the basis of his judgment about the catchability of the ball was the shape of the looming image on his retinae. He judged that all he had to do was to reach out his hand to intercept the ball. Alou and Bartman were performing the same task—closing the perceptual gap between their hand and the ball—but based on different information. One was running, the other seated, so what was perceptually relevant to each was different.

As this anecdote demonstrates, the cues salient in closing a perceptual motor gap are shaped by many factors, including awareness of task-relevant perceptual invariants, level of bodily fitness and skill, and the context in which the action is performed. If the goal or the context of an action changes, so do the task-relevant cues that will be tracked.

It is increasingly evident from research on skill acquisition that the way we perceive movement has much to do with the relationship between oneself and the moving object, not just in space and time but also in context. As observers, we stand apart from the action taking place, with very little input from our own proprioceptive systems other than, perhaps, ocular or auditory pursuit of the action. As participants, we observe movement with the intent to act. Playing a soccer game on the beach with friends, we stand ready to move should the ball come in our direction. We are observers not just of the ball but also of the actions of other players. We typically do not move until the ball comes toward us. In faster sports, where players need to run to intercept the ball, their skill lies in being active observers, planning their own motion in relationship to that of the ball to intercept its trajectory.

At the opposite end of this spectrum are sports such as golf, where the player has to act first and then observe, planning the trajectory of their ball before they even hit it. Making the shift from the context of the ball game on the beach to setting up a golf shot in a club match, one might also shift one's perspective from playing an informal game to skillfully performing an action, where awareness of one's own ability to move skillfully becomes increasingly important. A professional golfer, for example, might attempt a shot at which a novice would balk, simply because she would know her own level of skill and have tuned her body and her senses to perceive the potential opportunities for action afforded by a given context—the position of her ball with respect to the hole, a bunker, and so on.

In a study involving soccer players, Dessing and Craig uncovered differences in the information about incoming balls tracked by novice and expert players who were asked to judge where a ball was heading.[65] In some trials, the ball was given a spin that curved its trajectory. The two groups were equally successful at intercepting the ball. However, when experts and novices were both placed in a virtual reality environment and asked to intercept a free kick by moving to block it, experts were far more successful. The experts demonstrated strategies very different from those of novices: They waited until the last possible moment to move. The researchers hypothesize that, by waiting, expert players can pick up more information about the ball's trajectory. Their findings suggest that, in moving and acting upon the world, we can alter the perceptual cues that are available and relevant for us, and that these cues will change over time as we become better at performing a particular skill in a particular context. Returning to the case of Alou and Bartman, if they had swapped places, there is likely no way Bartman would have attempted to run for the ball because he did not possess the training or level of fitness to make the catch. Alou, on the other hand, could easily have taken Bartman's place because catching the ball from the front row required little specialized skill.

The fact that the individual can "tune" perception via skilled action is a topic of much interest, particularly in sport. If a player can hone their ability to predict the path of a ball or the actions of another player, they increase their chances of success. As Jacobs, Michaels, and Runeson have shown, the process of tuning perception appears to involve selecting from multiple potential descriptions of the same anticipated event to find those that provide the most salient perceptual cues for executing the task.[66] Jacobs and colleagues asked people to estimate the relative masses of two colliding balls and discovered that before being given task-specific training, participants tended to use a wide variety of nonspecific cues. But with training, many participants quickly learned to track those cues that were salient to ball mass, increasing their chances of success in estimating the relative masses of the balls. Jacobs, Michaels, and Runeson emphasize how dependent improvements in performance are on context, suggesting that one can become a skilled perceiver if one is trained (or trains oneself) to observe with skill.

If greater skill in moving results in greater skill in perceiving, then the design of an interaction that aims to support users in acquiring skill will need to anticipate changes in the feedback cues the skilled user will seek because users may draw on different sources of information as they become more adept at interacting with a system. Some information may specify aspects of the lawful relationship between action and feedback, and others may not. It is therefore important for a designer to understand how the user perceives which variables are relevant at any given time.

3.6 Movement as a Design Material

Increasingly, designers of interactive systems are addressing the need for better tools and techniques to capture and analyze movement by turning to disciplines with long-established traditions of observing how people move, including animation, product design, and choreography. In this section, some of these approaches are introduced alongside frameworks that are being developed in HCI to integrate them with movement-based design methodologies.

The big question is what practical steps are required to implement an interaction design that builds on movement as a design material. In proposing a path from the messy, noisy world of sensing and analyzing movement data to the clarity of the final interaction design, Benford and colleagues frame the challenge elegantly, suggesting that the first job in implementing such a system is to determine what is expected, what is sensed, and what is desired.[67] Implicit in this process, and illustrated in the Venn diagram in figure 3.6, is the exercise of determining what is constraining the design—namely, what is expected and desired but not sensed, what is not expected or sensed but is desired, and so on for all combinations.

In the upcoming discussion, I demonstrate that no matter how integral a given action might be to a design, if you are not using the correct tracking system, you will not sense that movement, and it will not be available to

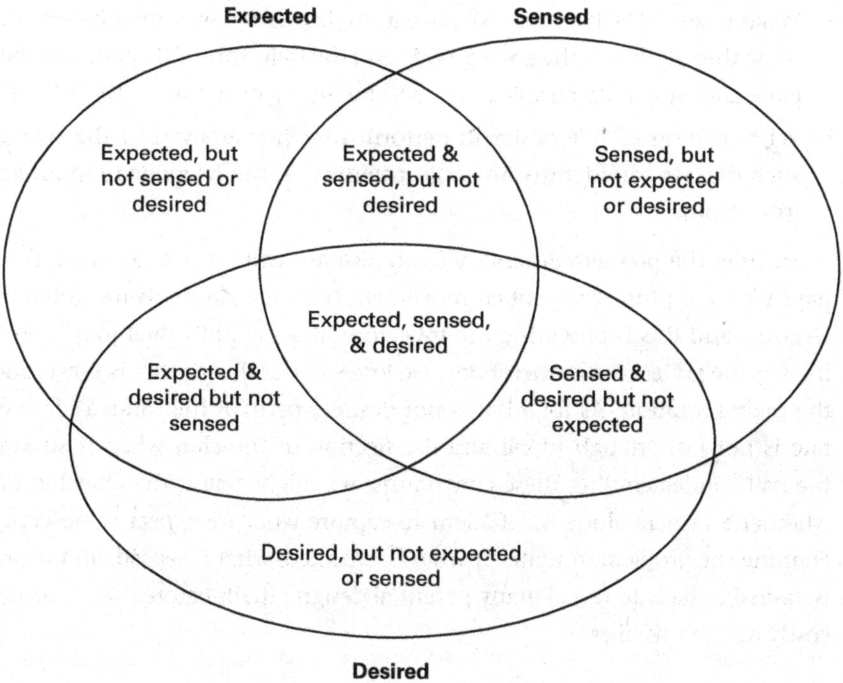

Figure 3.6
Venn diagram: expected, sensed, desired. Movements of people using interfaces can be analyzed in terms of whether they are expected, sensed, and desired. "Expected movements are those that users naturally perform; sensed are those that can be measured by a computer; and desired movements are those that are required by a given application" (Benford et al., "Expected, Sensed, and Desired," 3). *Source:* Used with permission of the Association for Computing Machinery. Permission conveyed through Copyright Clearance Center, Inc.

you as raw design material. Taking time to align the goal of a design with the movements to be tracked and the potential for missing or noisy data is a crucial step in the design and implementation of any interactive system. Let us imagine, for example, that we have decided to observe a golfer while using a camera, with the goal of observing the entire swing up to and including the contact with the ball. We would need to ask:

- What is expected? We would expect to capture the movements of the golfer's body, shoulder, arm, and hand. We would expect to observe the motion and flection of the club and to pinpoint the time when the club contacts the ball.

- What is sensed? Given that we have a single camera, we would be able to sense the rotation of the golfer's body and the trajectory of the arm and the club, and we should be able to extract the timing of contact with the ball.
- What is desired? We desire to perform post hoc analysis of the swing such that we can identify any adjustments that can be made to improve its outcome.

Turning the problem around, we can also ask what is not expected. Perhaps we are capturing too much movement from the garments the golfer is wearing, and this is obscuring the trajectory of some individual joints. Perhaps as the golfer rotates their body, we lose sight of the leg that is powering the body's rotation. As for what is not desired, perhaps the camera's frame rate is not fast enough to capture the flection of the club when it strikes the ball. Understanding these constraints, we might revisit the question of whether a camera alone is sufficient to capture what we expect to observe. Framing the problem in terms of what is expected, what is sensed, and what is desired is likely to reveal many potential design pitfalls before they become costly design mistakes.

3.7 Capturing and Analyzing Movement

As we noted at the outset of this chapter, in designing interactions that use movement, movement should be considered a resource to be used in the same way hardware and software are recruited to serve our interaction design goals. But to use movement in this way, a number of processes must happen first. First, movements must be captured. This is typically accomplished by placing some form of sensing system on or near the body. Sensors can be as simple as the button and track ball on a computer mouse, which together capture the location and timing of mouse clicks, or as complex as a full-body motion-capture suit, which will capture whole-body motion, often using multiple types of sensors to track bending, twisting, and so on. The next step is to extract patterns or features from the raw data and to code them for classification and further analysis. For this purpose, some form of intermediate representation must be defined that can be passed to a person or presented as input for a classification algorithm.

The process of classifying movement means determining which of its features should be attended to in order to arrive at a description of how the person being tracked was moving. Such analyses are not usually open-ended but are carried out with reference to an end goal. A choreographer, for example, may choose to use a system such as Labanotation, with the goal of describing any changes to be made to more closely reflect the artistic goal of a production. In its simplest form (as shown in figure 3.13), Labanotation assigns a unique symbol to each dance position. Laban movement analysis (LMA) describes movement qualities (MQs) with reference to four non-Cartesian dimensions (displayed in figure 3.14): body (what is moving), effort (how an object/body is moving), space (where it is moving), and shape (the relationship between the moving body/object and the environment). In comparison to the needs of a choreographer, the goal for a computational analysis system working with a numeric representation might be to determine whether the movement is close enough to a template movement to trigger a response from the system—for example, did the person move their hand fast enough to intercept a falling barrel in a game of Donkey Kong?

Most of the motion tracking that underpins applications such as fitness trackers, touch-screen interfaces, and navigation systems relies on machine learning—that is, on algorithms that can learn patterns in a set of training data and then recognize these patterns when they arise as a system is being used. As we have noted, the greatest challenge for these systems is how adept human bodies are at accomplishing the same movement end goal in different ways. Ideally, a machine learning system must be flexible enough to allow for alternative strategies but powerful enough to recognize which of all possible strategies are associated with which goals. In this section, I will address some of the challenges directly associated with interaction design, including how to determine the most appropriate system for capturing movement and how to incorporate bodily movement into the process of designing an interactive system.

Whatever the technology used, the goal of all such systems is to track three things:

1. Where you are (context)
2. What you are doing (task)
3. How you are moving (action)

As we shall see below, the task of choosing an appropriate system (or combination of systems) to track and capture movement for a given application will depend on how that movement relates to each of the above goals. A fitness tracker used while running, for example, must track the user's location on a map of their route (context), track their pace (action), and provide them with updated information about the distance to the end of their run (task progress). The choice of tracking method is important because not all methods for tracking movement are created equal. Each has properties that make it more or less appropriate in a given context. The choice of capture and analysis tools will also shape the aspects of movement that are available to a designer because if a given movement has not been captured or extracted, it may as well not have happened as far as the application is concerned.

The following discussion is intended as a guide to help designers navigate the space of some of the most common methods for capturing and analyzing movement. This section will also address the question of annotating movement, for, as we shall see in section 3.8, the ability to observe and annotate how people move is an invaluable skill for someone who designs systems that rely on tracking and using movement.

3.7.1 Capturing Movement

The term "motion capture" (or "mo-cap") refers to any system that records temporal and/or spatial attributes of movement using sensors or body-worn markers. The data thus captured is made available as time-stamped vectors (lists of numbers that describe the movement, over time, of each sensor or marker). This data may either be passed directly to another application for real-time control of, for example, audio/visual effects for a dance performance or stored in a file for later processing.

Many application domains use recordings of movement—from digitally assisted choreographic notation systems for dance to mo-cap for animated movies to movement analysis systems for sport and rehabilitation. Whatever the system, two general principles hold:

1. Every motion-capture system is shaped by the purpose for which it was designed. Whether the observer is a choreographer notating a dance or a 3D camera-based motion-capture system tracking a golfer's swings, the data captured is shaped by a series of decisions that determine the kind of information that is recorded. As illustrated in figure 3.7, two choreographers will notate the same movement in different ways based on what kind of information they are interested in.

   Mo-cap systems based on different underlying technologies will capture and therefore emphasize different qualities of movements. In figure 3.8,

OBSERVATION OF A GIRL SKIPPING	
PURPOSE	**NOTATION CHOICES**
Observation tendencies A notation student examines his own observation tendencies by notating what he notices first in the movement.	A broad description of the movement's aerial action and repetitive rhythm.
Diagnosis A physical therapist analyzes the dysfunctional aspects of the girl's movement.	A detailed description of the awkward sequencing of body parts and inward rotation of the legs.
Style analysis A choreographer watches the girl's movement to get ideas for a dance.	A very broad description of the girl's free flowing quality and curving pathway.

Figure 3.7
Observations of a girl skipping from the perspectives of three different movement practitioners: a notation student, a physical therapist, and a choreographer. Notice how each one focuses on different attributes of the same movement according to the practice-based lens from which they view the action. *Source:* Adapted by Hans Anderson from an anonymous source image.

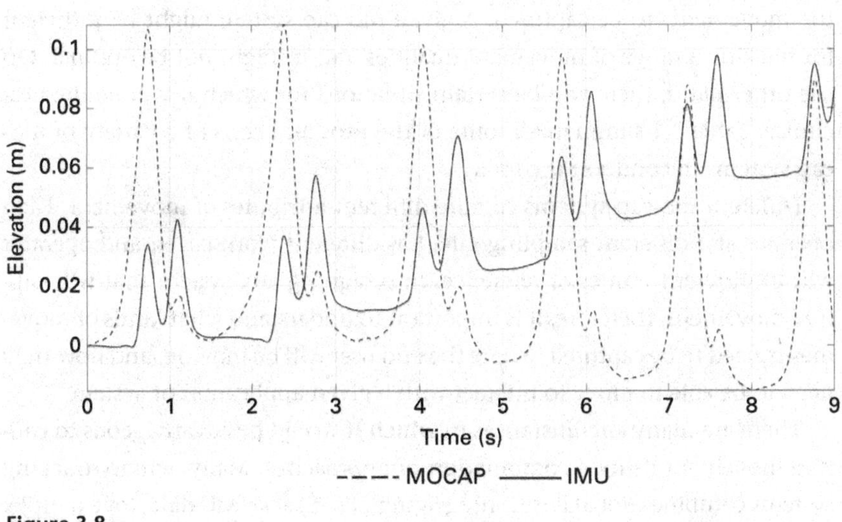

Figure 3.8
Comparison of foot movement traces during walking captured by inertial measurement units (IMUs; solid line) and visual motion capture (mo-cap; dashed line). Foot elevation (m) over time (s) is estimated. The plot shows an 8 s interval, which illustrates five similar periods (steps). The amplitude difference in these two solutions (greater for mo-cap) is the result of using different sensor locations for markers (ankle) and IMU (top of shoe). The IMU solution shows an increasing foot elevation error caused by sensor drift; it drifts about 7 cm in 8 s. *Source:* Original plot of mo-cap footstep data captured by Lauro Ojeda and Sile O'Modhrain.

for example, the trajectories of one movement are shown as captured using a multi-camera mo-cap system and an IMU mounted on the foot.

2. Even the same technology deployed in different spatial arrangements with respect to the moving body will produce different data. Figure 3.9 is a plot of trajectories from markers placed on different parts of the body while capturing one movement sequence.

   The data captured must therefore be understood within the constraints of the system used. The data available at the output of a system will not, in raw form, contain information about features that the system's sensors cannot track. There is no point in expecting an IMU, for example, to provide data about ambient light. More importantly, if a tracking system is only recording movements once every second, the data produced cannot accurately record changes in movement at shorter intervals, although we might perform post hoc interpolation to approximate what happened.

In choosing what kind of mo-cap system to use, it is important to understand the relationship between the capabilities of the system and the qualities of the movements to be captured. A given mo-cap system might be sufficient for tracking a range of movement qualities, but it might not be optimal. On the other hand, there will be certain situations for which it will be the best choice. Table 3.1 summarizes some of the pros and cons of a variety of mo-cap systems in common use today.

Different mo-cap systems capture different attributes of movement. Each operates at a different sampling rate, has different workspaces, and operates within different frames of reference. In designing any system that will capture movement, therefore, it is important to understand what kinds of movements need to be captured, where the end user will be moving, and how they need to be able to move to interact with a given application or system.

There are many circumstances in which it would be advantageous to capture movement using a combination of approaches. Many activity-tracking systems combine Global Positioning System (GPS) and IMU data, for example. IMUs are the small motion-tracking devices embedded in consumer products such as phones and cameras that track the linear and rotational velocity and the acceleration of the objects to which they are attached. Increasingly, this solution is being employed in AR/VR, where IMUs are being added to controllers to overcome problems associated with lack of tracking coverage and/or camera occlusion. Commercial VR tracking systems typically operate in

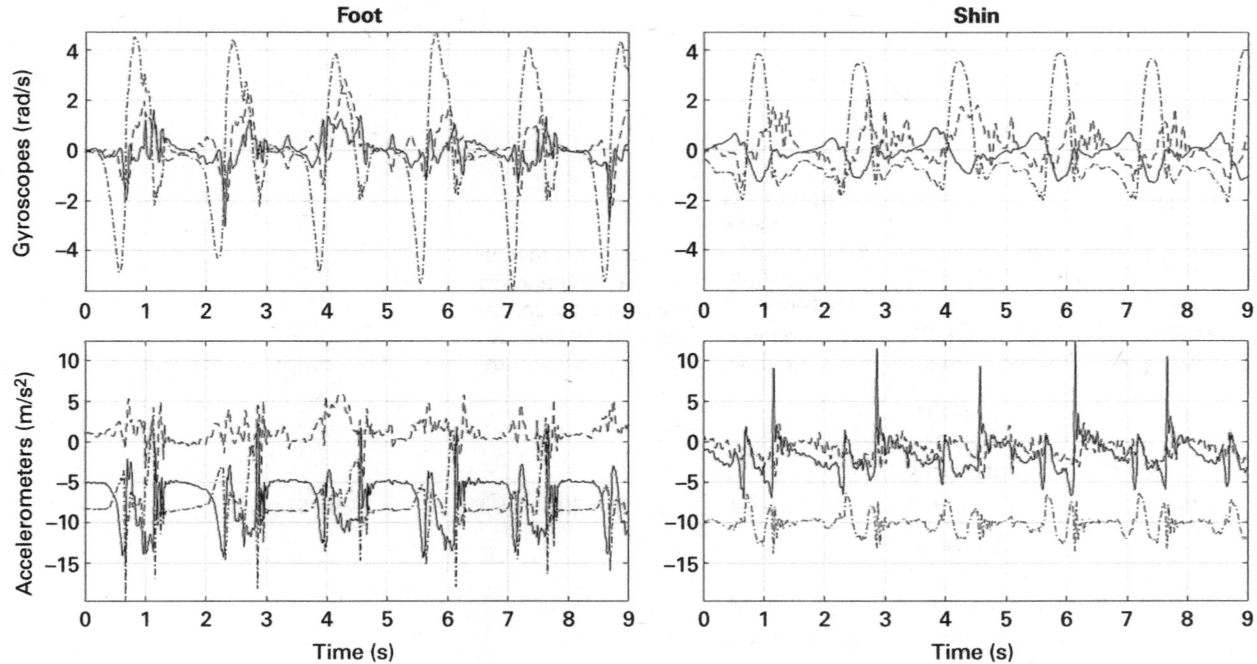

Figure 3.9

Traces of foot movement during walking as captured by a gyroscope and an accelerometer. The left two sub-figures show the raw gyroscope data (rad/s, above) and accelerometer data (m/s^2, below) for the foot over a 9 s interval. The right two sub-figures show the corresponding data for the shin. Each sub-figure plots the *x*-axis as a solid line, the *y*-axis as a dash–dot line, and the *z*-axis as a dashed line. The 9 s captured illustrate six similar periods (steps). The foot and shin data are noticeably different. For example, the *y*-axis gyroscope data for the foot has a much greater amplitude than that for the shin. *Source:* Original plot of motion capture. Footstep data captured by Lauro Ojeda and Sile O'Modhrain.

Table 3.1

Comparisons of methods of motion capture (mo-cap) used in interaction design, including pros and cons

Mo-Cap	Degrees of Freedom	Sampling Rate	Frame of Reference	Workspace	Pros	Cons
Acoustic	1	44.1 kHz	Object/world	Defined by sensitivity and pickup pattern	Audio is ubiquitous Computationally inexpensive High resolution Mics are often directional	Noisy signal
Inertial	9	~150 hZ	Egocentric object	Limitless	Low power Small data Geocentric reference	Latency No absolute position
Visual marker based	N = markers	1 kHz	Absolute, world centered (can also be egocentric if markers contain inertial measurement units)	Setup of cameras 3D workspace defined by mo-cap camera positions (e.g., $15 \times 15 \times 10$)	Absolute position Multiple people and things Better special and temporal resolution	Expensive Not portable Hard to calibrate Occlusion Data cleaning required Rotation data (hard to get)

Video-based motion tracking	Depends on number of cameras, their position, and their field of view	Frame rates between 24 and 2,400, depending on camera and camera settings	Absolute, world centered	Setup of cameras 3D Workspace defined by mo-cap camera positions (e.g., 15×15 × 10)	Cameras are portable and many are auto-calibrating Workspace is only limited by camera positioning and sensitivity Not limited to mo-cap studio Unlike visual mo-cap, whole-body form is also captured	Extracting motion of body parts from video is computationally expensive Can be complicated by motion of other bodies or objects in scene Occlusion of body parts is still an issue Lighting variability, shadowing, and color contrast can also make movement analysis difficult

one of two ways: Either they use head-mounted cameras looking outward to track the movement of usually handheld controllers in an egocentric reference frame (inside-out tracking), or both the headset and handheld controllers are tracked by an external fixed camera located in the same space as the user (referred to as outside-in tracking). Each approach has its advantages and disadvantages, but the possibility of combining a number of body-worn IMUs with either system offers a low-cost, lightweight option both for full-body capture and to avoid occlusion problems for cameras.

Of course, this is just one example of the myriad ways activity-tracking systems might benefit from capturing movement using a combination of approaches. But, significantly, it indicates how VR game designers are actively engaged in solving the problem of full-body, high-fidelity motion capture for consumer-level products in a way that will catalyze further research in human movement tracking and analysis and their use in multiple application domains. This is clearly an important horizon for both designs of these systems and designers who rely on them to create experience-based artifacts.

3.7.2 Analyzing Movement

Methods and mechanisms for analyzing movement are at the heart of any interactive system that tracks movement. Interaction with computers always involves movement, even if just the motion of a trackball or the click of a mouse. As sensing technologies have become smaller and cheaper, the ability to track different aspects of movement in different contexts has increased greatly. But our ability to take advantage of sensing technology for tracking movement, which depends on our ability to find features in a dataset and analyze raw streams of numbers to discover how someone is moving, is less well developed. Designers of systems that use movement need to understand how the movements they expect from their users will coexist with other movements users need to make in a given context.

Designing an interactive system that uses movement therefore requires both the observational skills of a designer—to shape the movement patterns that will be used—and the analytical skills of an engineer to implement the hardware and software infrastructure that will recognize when those movements have occurred. Here, we will address quantitative and qualitative approaches to analyzing movement data that have already been captured before we discuss methods a designer might use to develop an interaction that evokes certain movements and movement qualities.

3.7.3 Quantitative Analysis

Earlier in this chapter, we touched on Fitts's law, a quantitative measure that predicts the Index of Difficulty, which is the trade-off in performance between the size of a target and the distance the user has to move to reach it. The ID is a geometric feature of the task, and it determines measurable changes in performance such as movement duration and acceleration profile. The ID provides a number that can be compared across different devices and users. However, the same movement may be quantified in many different ways, depending on the goal of the analysis. In sports science, it is common to perform quantitative studies that involve both a kinematic and a kinetic analysis of the same movement sequence. A kinematic analysis provides information about the positions, angles, velocities, and accelerations of body segments and joints, whereas a kinetic analysis provides information about the forces and torques involved in producing the movement. In a study comparing the actions of baseball pitchers at four levels of expertise, for example, Fleisig et al. wanted to determine if there were any differences between adult and teenage players.[68] By combining kinematic and kinetic analyses, they discovered that significant differences existed related to movement velocity and movement kinetics but not to position or overall movement timing. They proposed that observed increases in movement velocity (kinematics) and in joint forces and torques (kinetics) were most likely because of increased strength and muscle mass in more physically mature pitchers. If they had chosen to perform just one mode of analysis, they would have had less evidence to suggest that an increase in overall strength was the primary driver in improvement as players matured. Furthermore, if they had chosen to perform a purely qualitative analysis based on, say, baseball coaches' observation of videos, they would not have had the quantitative evidence for the systematic and correlated changes in both velocity and kinematic parameters, such as torque and force. They may, however, have had a more complete description of a player's pitching techniques because the quantitative information that is available is dependent on what parts of the body are marked or tracked using sensors, whereas a qualitative analysis by a coach can often take into account other factors, such as the player's ability to plan their movements with respect to the progress of the game.

For an interaction designer, quantitative analysis provides a way to numerically model how users move. It is the basis of the algorithms behind every gesture-recognition system in use today, from touch screens to VR game controllers. Using quantitative techniques, designers can determine, for a large population of users, the range of ways someone might perform a given gesture that would be considered within range for a given interaction, such as swiping and pinching. With the increasing prominence of e-sports, it is likely that interaction designers will need to account for factors such as increased muscle strength and increased motor skill in analyses of motion-capture data as players spend more time training to interact with game controllers in real and virtual environments.

3.7.4 Qualitative Analysis

But what about those situations where the goal is to analyze aspects of movements that are harder to quantify, such as perceived level of skill or a person's affective state? Data can be collected for both quantitative and qualitative studies using the same methods of motion capture and movement representation. But often, motion-capture data are then used as the basis of qualitative analyses.

Most qualitative analyses involve a human observer watching someone perform an action in real time, from a video recording or from point-light reconstruction of motion-capture data. Qualitative analyses are usually focused on observation of a particular aspect of movement. In sport, for example, several qualitative techniques or observational models have been developed, such as phase analysis, temporal analysis, and critical feature analysis. Phase analysis involves dividing a movement into several segments or phases that can then be attended to independently. In walking, for example, phase analysis might involve looking at the qualities of the stance and swing phases of each stride. Knudson and Morrison suggest that most movements follow a typical pattern that includes three phases: preparation, execution, and follow-through.[69]

More complex skills may be broken down into more phases, depending on the goal of the analysis. Temporal analyses focus more on rhythmic aspects of movement, often using a phase analysis of actions as their starting point. In our walking example, a temporal analysis might include comparison of the timing of heel strike to toe-off portions of a stride for walking on different surfaces or the relative timing of stance and swing phases for both legs. Finally, a critical feature analysis of walking might look at elements such as the amount of pronation of the feet or the alignment of the hips.

Other qualitative or observational methods are more open-ended. In the example cited earlier of audiences observing point-light performances of clarinetists, for instance, no attempt was made to define skilled movements for the audience. We merely asked observers to identify which clarinet players they thought were *more* skilled.[70] Determining which of a player's movements were associated with greater skill would have required closer observation, which might be qualitative (e.g., a phase analysis or Laban movement analysis) or quantitative (e.g., analysis of ancillary body movements). In a 2005 study, Chagnon, Campbell, and Wanderley carried out a qualitative analysis of performances by eleven clarinetists playing the second of Stravinsky's *Three Pieces for Clarinet*.[71] Two listener/observers, who had received some training in LMA (in particular the Laban–Bartenieff method), were asked to notate qualities of the performances in accordance with its four elements: space, shape, effort, and form.

Observers were also asked to comment on the performers' interpretation based only on listening. The observers' comments demonstrated that they considered all the performances to be competent. Many of the performers moved while playing. In some cases, these movements were distracting; in other cases, the movements supported some aspect of the players' interpretations. One player, for example, interpreted the two main musical ideas as characters who appeared to converse, and her body language supported this interpretation. The observers thus tuned into something that would be difficult to extract from a purely quantitative analysis of the clarinetist's movements, although, given that her intent was to tell a story, some form of quantitative technique might have been used to separate her performance gestures into two contrasting groups, one associated with each character.

Indeed, it is sometimes useful to combine quantitative and qualitative analyses of the same movement data. In a study that looked at the importance of the "feel" of a musical instrument to a player's learning process, we used a virtual bowed string to manipulate the feel of the bow–string interaction. Experienced violinists and non-string players were asked to bow virtual strings. In half the trials, the virtual string lacked the friction that would be present with a real bowed instrument, which the experienced player would expect. For such players, the absence of friction should result in poorer performance. For novice players, there would be no such expectation, and performance should not be affected. To assess performance, we combined quantitative and qualitative measures of playability of the bowed string. Quantitative measures were based on how well people could control bow movements—modulate bow position, bow force, and bow velocity—compared with a numerical model. Qualitative measures involved the assessment of tone quality by expert players who listened to the bow strokes.[72]

By combining these quantitative and qualitative measures, we observed that experienced players moved differently and produced a different tone quality when friction was not present, indicating that friction was important to them. Non-players showed little difference in performance based on either qualitative or quantitative results. It would have been difficult to determine precisely how someone was moving just by listening to the bow tones, and it would be equally hard to predict the tone quality of a note just by observing bow movements. The complete picture would emerge only when qualitative and quantitative data were combined.

In sum, the advantage of combining qualitative observations of users' actions with quantitative data gathered from sensors capturing those actions is that a designer can gather information about a user's motivation and the context in which they are performing an action, with a bottom-up representation of how that action unfolded. Quantitative data, such as how users respond when presented with certain visual, auditory, or haptic stimuli, are necessary to determine aspects such as the timing of stimulus presentation. Qualitative data, such as how frustrated a user feels when trying to accomplish a task (often captured using questionnaires), help designers identify and remove elements of a design that are confusing. By combining qualitative and quantitative observations, it is possible to observe exactly when someone became frustrated and what actions they took as they tried to work their way through an interaction. In section 3.9, I will introduce design methodologies that rest on being able to access a combination of qualitative and quantitative data to shape the design of an interactive system.

3.8 Representing and Annotating Movement

As we have noted, all systems for representing movement, whether human or machine readable, exert a shaping influence on the movements they code. In music, for example, one might describe the intended movement of a player's hands by inserting articulation marks into a score. A choreographer would notate the same hand and arm movements differently. In the music notation in figure 3.10, note the crescendo (the hairpin-like marking) that extends through the first measure, indicating that the music should gradually get louder. Then there is an accent and a decrescendo (a hairpin going in the

Figure 3.10
Effort–shape analysis of the score of bars 1–20 of *Marimba Dances, III* by Ross Edwards. Labanotation is by Mary Broughton. (See Broughton and Stevens, "Analyzing Expressive Qualities in Movement and Stillness," music example 1.) *Sources:* Used with permission of the University of California Press. Permission conveyed through Copyright Clearance Center, Inc. Score: *Marimba Dances, III*, by Ross Edwards, bars 1–20 © Copyright Ross Edwards.

opposite direction, indicating that the player should gradually play more quietly). Then the pattern repeats, as you can tell, even if you do not read music.

Now look at the Labanotation above the staff. You will see a similar repetition of words and symbols as the crescendos and decrescendos repeat. The markings, made by a Laban expert watching the performer, say "dab," "rising," and "sinking." They capture not what is in the score but rather how the player moved when performing it. Different players will interpret the same score in different ways, so that while the musical notation will not change (unless the player makes their own markings on the score), the observed movements may be quite different. The music represents ways in which the piece might be performed. The Labanotation captures how the piece *was* performed in this one instance. In other words, the musical score represents a guide for the player, and the Labanotation represents a record of what

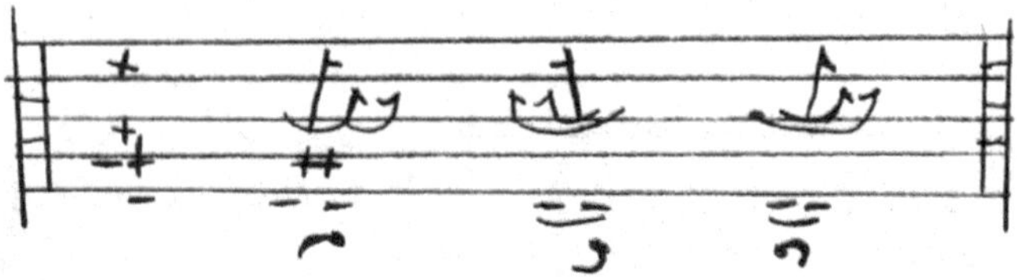

Figure 3.11
Example of Benesh notation. *Source:* Adapted from Julie Kando CC-SA 3.0 https://en.m.wikipedia.org/wiki/File:Benesh.jpg; license: https://creativecommons.org/licenses/by-sa/3.0/deed.en.

happened. But published scores seldom match exactly those used for a piece's first performance because, in rehearsing that performance, the composers and/or performers may have altered details of the score. So, in addition to the score given by the composer, which might be considered a template, there are notations based on observations made by players as they rehearse and, in the case above, a notated record of the performance as it unfolded. Each of these phases of documenting movement has a role in shaping the experience of the player and, in the case of a performance, of the audience as well.

Dance does not have a notational system that is as ubiquitous as the musical score. Where notation systems do exist, they have typically emerged within a particular dance tradition. In addition to differing in their visual syntax, they often differ in the emphases they place on describing spatial, temporal, and spatiotemporal elements of movement. Benesh notation, shown in figure 3.11, focuses on the shape of the body as seen from behind.

Each body part is represented as an element in a matrix on a five-line staff. Positions and lines indicating movement trajectories for individual body parts are indicated on the staff, whereas timing markings and symbols representing more complex movements are written above the staff. This is a body-centric movement representation that prioritizes static poses.

In contrast to Benesh notation, Merce Cunningham's DanceForms (see notation in figure 3.12) places more emphasis on the motion of the body in space and on transitions between poses. This need for representing time as movement within 3D space eventually led to the creation of Life Forms, a computer-based notational system derived from DanceForms, in which poses and transitions are represented in a fully 3D coordinate system.[73]

But the form of dance notation that has most successfully crossed over into the design world is Kinetography Laban, introduced in 1928 by the

Figure 3.12

Merce Cunningham DanceForms. This screenshot is from the first iteration of the DanceForms software, version 1.0, developed by Linda Schiphorst. The system supported the development of choreography using virtual models of dancers. The software was designed for Merce Cunningham. *Source:* https://amt-lab.org/reviews/2020/3/lets-get-digital-visualizing-movement-in-danc.

Austro-Hungarian dance artist and theorist Rudolf Laban (see figures 3.13 and 3.14). Kinetography Laban, also known as Labanotation, is part of a more holistic approach to the analysis and documentation of movement called Laban movement analysis.

Laban believed that of his four movement categories—body, effort, space, and shape—the category of "effort" most faithfully represented the dynamically unfolding and expressive qualities of movement. He also assigned each of his movement qualities four further attributes: time, weight, flow, and space. (Here, space, with reference to effort, is distinct from the MQ of space above because here space refers to the dynamic unfolding of a movement *in* space as opposed to how the body is disposed toward the space around it—whether compressed, expansive, etc.; see figure 3.15.)

Each effort attribute has its own continuum and boundaries (space: direct/indirect; time: sudden/sustained; weight: light/strong; flow: bound/

Figure 3.13
Labanotation symbols and their associated movements, showing different body movements and the Labanotation symbols that might be used to capture them. *Source:* Adapted by Hans Anderson from Rallis et al., 2021, figure 9.

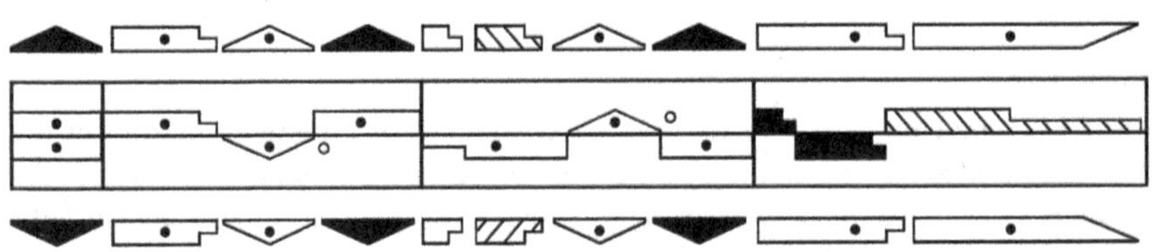

Figure 3.14
An example of a Labanotation manuscript illustrating different symbols. *Source:* Creative Commons Attribution-Share Alike 4.0 International license: https://en.m.wikipedia.org/wiki/File:Labanotation.svg; license: https://creativecommons.org/licenses/by-sa/4.0/deed.en.

free). These attributes are usually represented as orthogonal dimensions of a cube.

LMA's power is its ability to capture through notation the subtle and dynamic nature of expressive actions, whether they are directed at objects or used to communicate with other people. Figure 3.16, for example, records the movement trajectory of someone inserting a light bulb, captured using Labanotation.

Within the MQ of effort, for example, Laban understood time as related to a person's sense of urgency, space to their attentiveness to their surroundings, weight to their impact on their environment, and flow to their sense

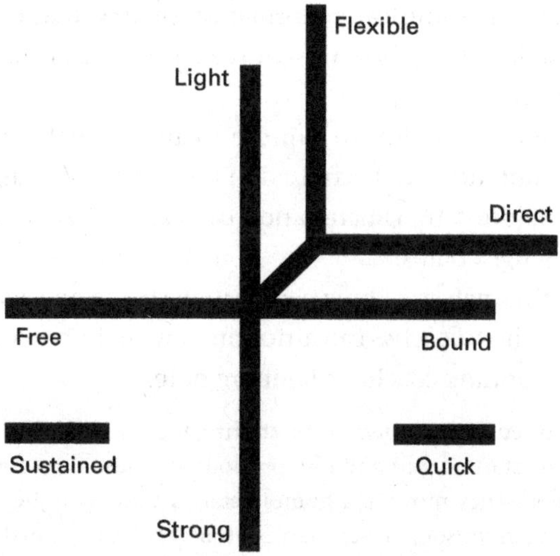

Figure 3.15
Laban's effort–shape notation system illustrated as a three-dimensional cube, with each effort axis labeled. *Source:* Created by Jeffery Scott Longstaff (2012), Creative Commons Attribution-ShareAlike 3.0 Unported https://en.wikipedia.org/wiki/Labanotation#/media/File:Laban-Effort-graph-labeled.jpg; license: https://creativecommons.org/licenses/by-sa/3.0/.

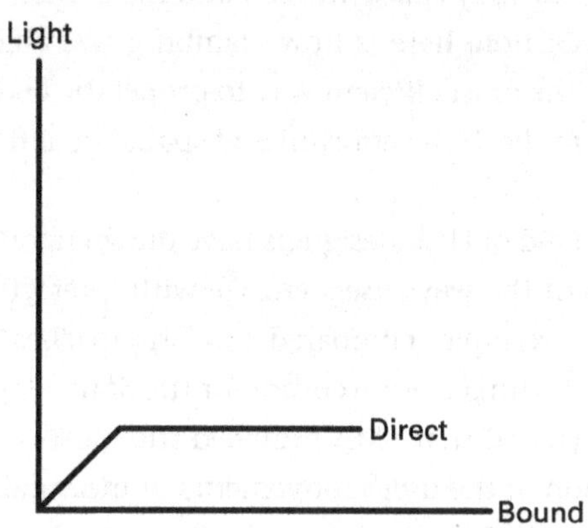

Figure 3.16
An example effort graph annotating the actions of someone inserting a light bulb, using Labanotation. *Source:* Reprinted with the permission of Kristina Höök (khook@kth.se), from Fagerberg et al., 2003, figure 2B, based on Laban and Lawrence, 1974.

of bodily control.[74] In a musical performance, such as that depicted in figure 3.9, it is possible to show how the effort exerted by a player relates to the music being performed.

Partly because of its ability to capture so many qualitative aspects of movement, Labanotation has been used in fields beyond music and dance. As illustrated in figure 3.16, Labanotation can capture sequences of actions such as replacing light bulbs.

In sport, too, the analysis of how people move based on Laban's approach has contributed to how coaches can articulate ways to improve performance. In working with running coaches, Hamburg notes:

> One coach reported how runners at the starting line sometimes rock backward at the sound of the starting gun and lose precious seconds. Their first few steps are highly inefficient. They move in a homolateral pattern, with the same side hand and foot moving in unison, rather than contralaterally. It seemed to me the runners rocked backward in an instinctive attempt to build momentum and propel themselves forward. . . . I coached the runners to move contralaterally (i.e., to reach forward with the hand opposite the lead foot). The body follows the lead hand naturally and quickly into a contralateral run.[75]

This description brings to the fore what Laban viewed as the dynamism of space. Described in LMA terms, reaching forward with the hand and contralateral foot represents a balancing of movement on the sagittal dimension (the plane that divides body space into left and right sides) and helps propel the body forward. Of note here is how Hamburg was able to observe and then communicate the most efficient way to propel the body off the starting blocks, accounting for both the effortful and spatial qualities of the required movement.

Similarly, in the field of HCI, designers have drawn upon LMA as a tool to analyze and represent the ways users engage with interactive systems. Loke and colleagues, for example, compared four approaches to documenting how users interacted with a color webcam for the Sony PlayStation, the EyeToy, and they determined that LMA provided the most complete and most structured description of the user's movements in measurable terms.[76] These would include motion of the body and its limbs in time; the body's relative motion in space (direction, distance, etc.); its degree of motion (timing, duration, and meter); and dynamic qualities of its movements (e.g., heavy, strong, bound). Other designers, such as Jensen, have used Laban's methods to shape the vocabulary of movements that a system will use before building any devices.[77] In an exercise called the Metaphor Lab, Jensen developed a process based on LMA that involved designers first observing movements to identify interaction qualities, then developing metaphors around those qualities, and finally creating physical sculptures that would evoke those qualities in movement when users interacted with the objects.

Designers of digital musical instruments (DMIs), such as Nicholas Ward, have taken this process one stage further, asking users to act out the movements they would associate with controlling a given sound synthesis model before any hardware was designed to control the sound.[78] Ward's goal was to create an instrument whose form and function were dictated not by the constraints of hardware or software but rather by the movements players decided best matched the ways in which they wanted to control the sound and hence play the instrument. This process neatly avoided the design problem of having to decide a priori which subset of the many parameters available in the synthesis model should be made available for control by the user. This simplified the process of mapping the high-dimensional space of the synthesis model onto the lower-dimensional action space reachable by the player.

Because LMA provides a language that can account for properties of movement (such as effort) that are not explicit in other notational forms, many interaction designers have found it useful for describing existing encounters with augmented objects and environments and in developing movement profiles to define new interaction metaphors. LMA combines a well-defined language for analyzing movement with an established notational system that captures functional and expressive MQs at any scale and that accounts for context because, in circumscribed task domains such as music or sport, analyses are always associated with a particular user's skill, bodily knowledge, and experience. Although LMA provides designers with the means to carry out nuanced, context-dependent movement analysis, it still involves third-party observation of those engaged in moving. More and more designers believe that, in designing for movement, we need methodologies that can capture the user's sense of what it feels like to use the things we build. For these designers, the process of creating a product begins with "Designing through movement," i.e., with developing an awareness of how the product will fit with the bodies and within the lives of the people who will use it. This somatic turn in HCI thus answers the need of third-wave HCI to account for people's individual perspectives and circumstances.

3.9 Designing Through Movement: Three Approaches

Our devices shape our movements. Think of the first time you picked up a smartphone or tablet and had to learn how to tap, swipe, pinch, and so on. How natural were those gestures? The set of movements we now associate with such devices is determined as much by their ability to discriminate among the gestures of millions of users as it is by high-level design decisions. A device's gesture set is a compromise between a system that will recognize one user's gestures all the time and one that can recognize most users' gestures most of the time. Device designers are moving toward a model in which a device has a built-in capability to recognize gestures or speech, and through use, we complete the process by continuing to train algorithms to recognize our specific movements and voice patterns. When you take your next new shiny phone out of its box, remember that you are not just using a product, you are beginning a new relationship with the algorithms that form the basis for how that product responds to signals from your body and environment.

Design practitioners now have access to a new material, bodily movement, that can help them interpret users' actions and goals. Any system we build asks users to perform actions to achieve goals. Our interactive artifacts are not privileged in this way over any other object with which we interact in the environment; they merely incorporate computation as one part of their responses. But their form and the ways they respond do shape what we do. Our devices determine our movements.

Behind the phones that train us is a design practice coming under increasing criticism because it assumes that designers know what movements feel good and rewarding to everyone. We have noted that no two bodies are alike and no two people come to their devices with the same sensorimotor skills. By taking a one-size-fits-all approach, interaction design has generated research that treats movement as a homogenous design material.

This homogeneity ranges from theories of embodiment to guidelines derived from the observation of skilled and situated action. I will return to a discussion of the challenges related to the homogeneity assumption in chapter 7. As noted in chapter 2, these approaches place the body at the center of interaction design, taking seriously the proposal that action guides perception, which, in turn, guides future action. These approaches mostly come under the umbrella of third-wave HCI and related design frameworks of the third paradigm.[79]

Placing the body at the center of the design process assumes that the user's bodily actions are a central element to be integrated into a design. Whether the goal is to track the precise, skillful movements of a musician or to design for the playful discovery of possibilities in a game, a movement-oriented approach elevates the capture and use of movement from a cumbersome necessity to a richly featured raw material with which a designer can work. But just because we know how to move and interact with devices, including those with computational capacities, does not mean we know how to think about movement as a material for design. There is a difference between knowing by doing and knowing about doing. Increasingly, designers are turning toward methods for evaluating how their work captures the felt experience of their users. This somatic turn places users' bodies at the center of the design process, building outward toward technologies from bodily experience instead of matching human capabilities to available functionality of devices.

For designers seeking to embrace body-centered design approaches, the gap between understanding movement from a first-person perspective and implementing this understanding in an interactive system is a potential trap because, as Sietske Klooster and colleagues point out, the dynamic nature of interaction design requires that those working with movement be skilled in moving themselves in order to understand these dynamics.[80] More than this, designers need to develop creative skills and methods to transfer this knowledge into their products. In other words, designing for people who will become far more skilled in using your system than even you are requires new tools—tools that let you anticipate what it feels like to use your system and to observe and document the movements of people as part of your design process. You will have to design for the users you are observing, not just for yourself.

For the past twenty years, Lian Loke and Tony Robertson have been developing a methodology for the design of interactive systems that focuses on the role of movement in interaction.[81] It brings together a range of prior approaches, including Suchman's ethnomethodology for observing situated action, Laban's techniques for notating and analyzing movement, and work documenting the role of felt kinesthetic experience in learning new skills.[82] Their methodology, summarized in figure 3.17, considers two aspects of movement. The first is in terms of the three perspectives noted above (user, machine, and observer). The second is in terms of seven key activities that combine to guide a designer through the process of incorporating movement

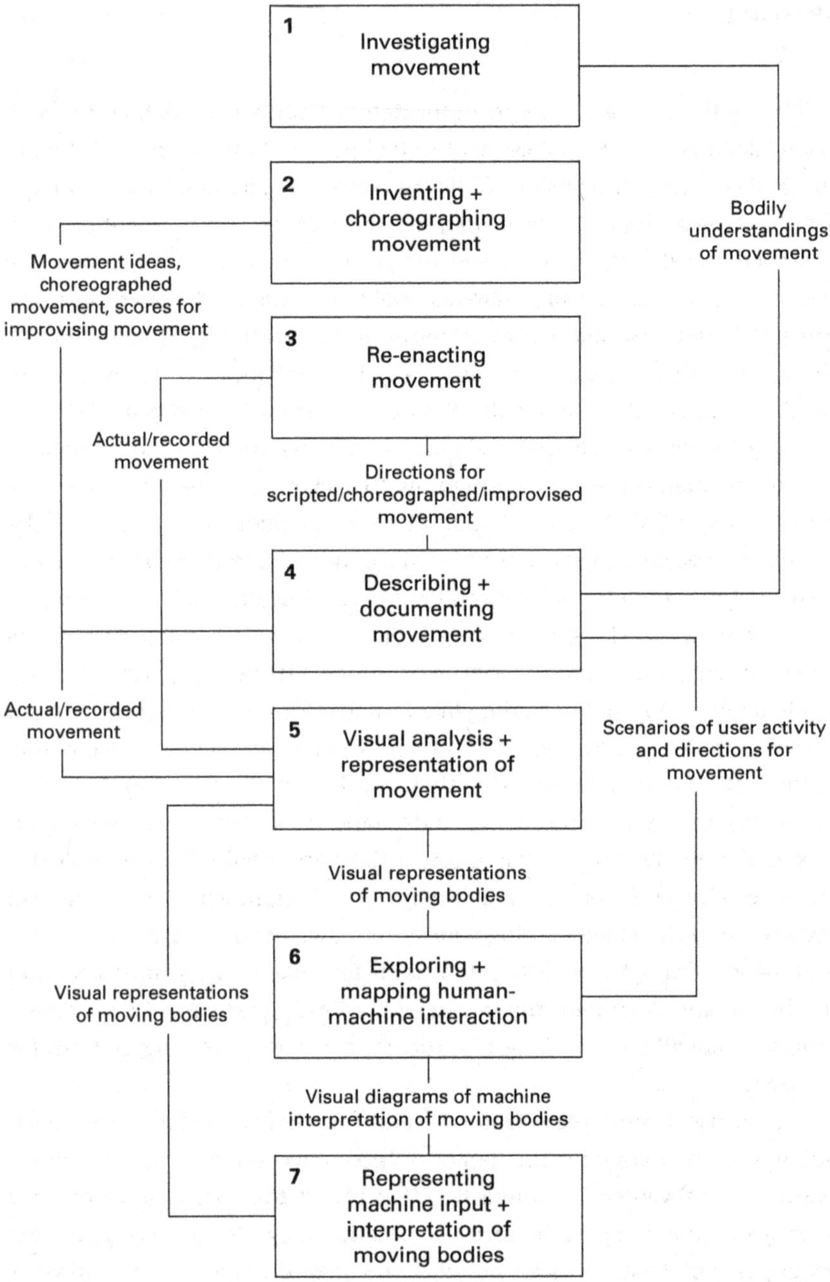

Figure 3.17
Key activities promoted by Loke and Robertson's design methodology. The labels on the arrows indicate the data generated by that activity, and the directions of the arrows indicate the flow of data from one activity to another. *Source:* Loke and Robertson, 2013. Used with permission of the Association for Computing Machinery. Permission conveyed through Copyright Clearance Center, Inc.

into the design of an interactive system. This approach represents a departure from established, technology-centered interaction design methods that view movement from either the perspective of the observer designing the system or the perspective of the machine tracking the user's movements.

In their work, Loke and Robertson suggest that just because accounting for felt experience is challenging does not mean designers are off the hook. It means that we need new tools and methods to account for the diversity of human bodies and experiences and the richness of information present in the movement captured by the systems we build. In the following three sections, I will introduce some existing practices that provide useful points of departure for those interested in developing interactive systems through the lens of design. All three approaches focus on the first-person felt perspective of the end user. They share an emphasis on engaging in an established movement practice to turn one's awareness toward the felt experience of moving. Each practice demonstrates that only when you are aware of how your own body moves can you begin to understand how your products will feel to those who use them.

3.9.1 Approach 1: Design Movement

Design movement as an approach to interaction design was developed by Klooster, Overbeek, and others as part of the product design curriculum at the Eindhoven University of Technology. A central concept of this approach, inherited from the field of product design, is that of designing action before product. Students participating in a product design course were asked to choose a particular context of interaction for which to design, termed a "commotion." They were encouraged to develop a sensitive, empathic understanding of their chosen commotion through observation of their own and others' experiences. From this, they developed a movement vocabulary, called a "choreography of interaction," which they documented using a subset of LMA, developed for the exercise, to build new interaction qualities around the actions associated with their commotion. The students were also encouraged to engage in a process of reflection on how their proposed product would redefine their particular commotion. Only then did they begin to design their product, incorporating the choreography of interaction they had developed.[83]

In "Commit," a commotion that led to the design of a flower vase, Sietske Klooster began not by drawing vases but rather by physically engaging in flower arranging (figure 3.18).

Figure 3.18
Example of a design commotion: choreographed arranging of flowers in a vase. This circular vase, with a slightly concave design, encourages a specific choreographed user interaction with a bouquet. *Source:* Adapted from Hummels et al., 2007, figure 6, by Hans Anderson.

Through this process, four principles for designing the vase emerged, which Hummels, Overbeeke, and Klooster describe:

> First, the centre of gravity of the flower's weight instigated [the designer] to hold the stalk just under the corolla, resulting in an elegant, stable balance and caring lift. Second, the both careful and greedy admiration for flowers instigated her to slide them between her fingers, enabling her to hold more flowers and protectively lift them in the palm of her hand. Third, as a result of the greed of admiration she discovered a whirl of both arms, turning in relation to each other while bringing in and composing a plane of an increasing amount of flowers. While doing so she felt a fourth, not yet possible motivation, to embrace more flowers than she was able to.[84]

The text above demonstrates how, by physically engaging in arranging the flowers and creating a choreography of this interaction, the creator developed a final vase design (shown in figure 3.18) that arose as if the designer's movements, which reflected her emotional responses to the flowers as she encountered them, were frozen in the form of the vase.

An important outcome of design movement is that a choreography of interaction encompasses not only the design of a product but also the context of use in which the product will be embedded. This includes the lived, felt experience of the person who will integrate the product into their everyday environment. Design movement acknowledges that the person, product, and context are all part of an ecology, a lifeworld that includes artifacts that may or may not have computational capabilities. More recently, Hummels and van Dijk have incorporated many of the observational and reflective processes developed for design movement into a practice that considers product design not only as the outcome of a choreography of interaction but also as an outcome of a deep study of the lifeworlds of future users, people who will live with and through interactive objects in the course of going about their everyday lives.[85]

3.9.2 Approach 2: Soma Design

Soma design is a holistic approach to the aesthetics of design that focuses on raising awareness of the connection between sensation, emotion, feeling, and values, "improving on our somas."[86] Soma design builds on Richard Shusterman's somaesthetics theory and requires training one's body to increase one's awareness of movement. The practice asks us to come to an aesthetic appreciation of sensations, imagine through our bodily senses, and put ourselves in the felt place of those for whom we are designing. In her 2018 book *Designing with the Body*, Kristina Höök points to four elements central to soma design: accessing lived experience, slowing down the design process, testing and retesting proposed designs against the desired aesthetic quality of an interaction, and considering interaction as engagement with socio-digital material.[87]

3.9.2.1 Accessing lived experience To imagine what the products you build will feel like in use, Höök argues that you must connect with your own bodily experience. She suggests yoga or Pilates, or any movement-based practice that encourages you to look inward and reflect on how your body feels. In developing soma design, Höök and the members of her lab received two years of training in the Feldenkrais Method[88] and regularly used the skills they acquired to access their felt experiences as they developed new interactive objects and environments.

3.9.2.2 Slowing down the design process Höök argues that developing the skill to attend to what movements feel like requires taking time to experience the sensation of moving. Höök documented her experience of learning dressage, particularly learning how to coordinate with her horse in a "rising trot," where the rider rises and returns to the saddle on alternate strides by the horse. Mastering this transition requires the ability to sense what the horse is doing and to ensure that one's movements are in phase with those of the horse. By slowing down to reflect on and document her experience, she came to a felt understanding of the coordinated dynamics required to accomplish smooth, graceful transitions, which, in turn, resulted in more fluid forward movement for the horse. Taking time to experience and document what it feels like to interact with the systems you are designing, she argues, will improve your ability to articulate bodily experiences and raise your awareness of the interaction dynamics you are developing.

3.9.2.3 Testing and retesting proposed designs against the desired aesthetic quality of an interaction Although we regularly ask users to test our designs, and we observe how successful users are in accomplishing task goals, we rarely take time to consider the aesthetic quality of the experience we have designed. Usually, users cannot articulate why an interaction feels good or frustrating or playful. It is important to assess the degree to which such feelings are evoked when users encounter our products. Only through repeated, reflective engagement with the artifacts as they are being created is it possible to develop a "feel" for the experiences that we are designing. This too is a slow process, antithetical to the existing practices of rapid sketching and prototyping that pervade today's HCI design culture. The work of Hummels and van Dijk, within a practice they call "designing for embodied being-in-the-world" (D4EB), discussed in chapter 6 (section 6.6.7), represents a deeply reflective design process that illustrates the importance of attending to aesthetic properties of designs that will be embedded in people's lived environments.[89]

3.9.2.4 Considering interaction as engagement with socio-digital material Höök emphasizes that traditional approaches to interactive system design rarely consider the felt experience of interacting with particular algorithms, data visualizations, or on-screen widgets—that is, the visceral experience of interacting with digital content. Unlike physical objects, digital materials may change over time in response to a user's actions. Designers therefore have an important role to play in shaping the dynamic arc of an interaction. Höök points out that this arc is not a property of the design of the physical system or even its digital substrate but rather what she terms the "experiential identity of the interaction as it unfolds between the user and the design."[90]

In sum, employing somaesthetic theory as a framework for interaction design requires much more than simply adding on another layer to existing design methods. Höök states that somaesthetics "speaks of changing ourselves, our appreciation capability, our creative practice. It speaks to the fundamentals of what we are, what it means to be human."[91]

3.9.3 Approach 3: Moving and Making Strange

Both approaches to designing through movement described so far suggest that awareness of one's own bodily sensations and capacities and of the physical and digital material of the artifact are not sufficient. The designer still needs to step into the shoes of the end user to understand what it will feel like to invite the object into their lives. Soma design suggests that there are many ways to reposition yourself as a designer to experience your work from the perspective of a user. You might approach your artifact from different heights, place it into unusual contexts, and so on. As we shall see below, the field of product design has a well-developed practice of supporting designers in understanding how their products will fit into people's lives. Loke and Robertson also point out that in becoming accountable to the first-person perspective, designers will come to acknowledge the body as the source of knowledge about movement and as the place where bodily skills are developed.[92] (Their methodology is shown in figure 3.17.) This perspective is supported by a set of techniques for reenacting and experiencing movement. It encompasses investigating movement, inventing and choreographing movement, and reenacting movement (shown as activities 1–3 in figure 3.17). As with soma design, investigating movement involves accessing the experiential qualities of movement through one's own body, using approaches such as movement improvisation, yoga, and dance.

To support their second activity, inventing and choreographing movement, Loke and Robertson identify a number of strategies,[93] including the concept of "making strange"—that is, approaching familiar actions in non-habitual ways so that we experience them anew.[94] This practice of defamiliarization can take many forms. In painting, for example, artists are often encouraged to turn a picture of a familiar object upside down, so that it appears to them as something they have never seen before, and they encounter it as a naive observer.[95] In interaction design, defamiliarization can reposition designers as naive users. The next time you open the door to your house or your car, try to do so with the hand you normally do not use. Then consider why the movement feels strange. What aspects of your movement do you now notice? How does this raise your awareness of the shape of the door handle or the weight of the door? This is how it felt to open this door the first time you encountered it.

Such literal repositioning allows a designer to revisit the experience that causes familiar objects to become, as Heidegger says, present at hand, once more tools to be mastered. Djajadiningrat and colleagues developed a conceptual framework and a design methodology of making strange based on relabeling familiar physical objects with functions associated with digital task environments.[96] They mapped the opportunities for action afforded by mechanical devices such as handles and switches to the functions of a future electronic device such that there was no obvious relationship between the two—a switch might require repeated presses to dim a light, whereas a handle might rotate to turn a light on or off. The resulting "strangeness" of the mapping forced designers to deconstruct the role and function of actions with the proposed electronic device to question the relationship between function and action.

The concept of reenacting movement, the third of Loke and Robertson's activities concerned with the felt experience of users, involves literally acting out various potential interaction scenarios, putting your body in the place of the person who will use the system you are designing. This might seem obvious but is essential if you are building tools for someone whose bodily abilities are dissimilar to your own. Perhaps you are designing for children or for people whose mobility is limited. Designing for people whose abilities are not like yours cannot involve just replicating the movement constraints imposed by a particular disability. The ability to step into the shoes of someone whose lived experience is different from your own involves a deep engagement with their lifeworld and their ways of coping with their environment. I will return to this topic in discussing the work of van Dijk and Hummels in chapter 6.

3.9.4 Reflections

Understanding the process of designing movement from a user's first-person felt perspective begins with turning your awareness toward your own felt experience of moving. In each approach introduced here, the first step is to engage deeply with an established movement practice, to feel what it is like to skillfully move one's own body, to inhabit the dynamically evolving sensation of being physically engaged in moving through the world. Beyond this, these practices serve as scaffolds for turning felt experience into designed products or artifacts. Design movement provides the concept of choreographies of interaction, observations of movements, and their qualities in real-world contexts that can shape the qualities of interactions with digital artifacts. Choreographies of interaction reflect not only which parts of the body are moving but also aspects of how a person approaches an interaction—joyfully, greedily, carefully, and so on. Soma design is a more holistic practice whose goal is to deepen one's engagement with one's own sensory and motor experiences. Soma designers are designers of movement aesthetics as well as interactions because they acknowledge the role of their products in shaping how future users will move. Soma design shares with the work of Loke and Robertson the idea that, in order to truly inhabit the shoes of future users, you have to reposition yourself to experience your designs as they will encounter them. Both approaches encourage the practice of making strange as a way of raising a designer's awareness of the different bodies and life experiences people will bring to encounters with their products.

In the end, design movement, somaesthetics, and the work of Loke and Robertson all serve to enrich interaction designers' understanding of bodily movement, with the shared goal of deepening their appreciation for the richness of movement as a design material. Most importantly, they all focus on deeply engaging with the first-person felt perspectives of those who will use the products.

3.10 Conclusion

No two bodies are created equal, nor do any two bodies move in the same way. Variability in movement among people can arise from many sources, including physiology, level of fitness, mood, level of skill in performing an action, and broader contexts in which the movement occurs. And no individual will perform the same movement in precisely the same way twice. Variance may arise as a function of many factors, including one's level of skill in performing a task or a response to changes in the environment that affect the relationship between what one perceives and how one moves. Any variance in how movements are performed presents both a challenge and an opportunity for systems that track movement. The challenge is to determine the source of the variability. The opportunity lies in using the variability in movement signatures to discover how people are responding to feedback from a computational device or within an augmented or virtual environment.

In this chapter, I have reviewed many of the reasons why people move in order to draw attention to the potential sources of variability in data captured by motion-tracking systems. I have also drawn attention to the information that guides how people move, information that must be available as feedback in interactions where users rely on cues that designers provide. The designer of an interactive system cannot *construct* the coupling between a user's actions and what the user perceives in response to their actions. The designer merely creates the conditions under which that sensorimotor coupling will arise. One goal of this chapter, therefore, has been to synthesize current work that can inform and guide designers in determining how to provide adequate and appropriate perceptual cues for users of their systems. I have also noted that perception of appropriate cues to guide action is itself malleable, so that different properties of the perceived environment become more or less relevant in different contexts or with different levels of skill on the part of the user.

Movement is increasingly a material resource in interaction design, and so I have provided practical guidance for designers of interactive systems. In particular, I have addressed how movement is tracked and analyzed, and I have introduced three methodologies that place the first-person bodily experience of users at the center of the design process. The result of centering the moving, acting body should not only be a system responsive to movements arising in response to auditory, visual, or haptic feedback but also one in which the user's actions reveal properties of the constructed interaction space.

The greatest challenge is that even the simplest action can be accomplished in many ways. Such variability, although it might seem to introduce noise into motion-capture data, is potentially a rich source of information for a designer because it can reveal how a user figures out the relation between their actions and a system's response. In contexts where a user must commit to learning a new, specific, and complex motor skill, the variability in movements captured can inform the designer of the user's stage of skill acquisition. It is helpful to keep in mind the framework proposed by Benford et al.: Be cognizant of the relationship between the movements you expect to observe, the movements that a motion-capture system can actually sense, and the movements that are required in order to produce the desired interaction between the user and your system.[97]

The foregoing discussion of movement and the role it plays in bringing about our experience of the world is the foundation of the rest of this book. I will address time, space, and interactions with objects, in real or simulated environments, from the perspective of an individual who is physically engaged in discovering the opportunities for action their environment affords. As Sheets-Johnstone reminds us, movement is at the very center of our sense of agency, giving rise to our ability to sense time and space and the properties of the objects with which we interact.[98] This is no less true of the computationally supported spaces and objects we build. Ultimately, as the very word "interaction" suggests, everything rests on the actions users perform.

4 Time

4.1 Introduction

Why should we care about time in interaction design? Although it is not something that comes immediately to mind as we design interactive systems, everything we ask of those who use our systems takes some portion of their time. Our artifacts go out into the world of lived experience, and consequently, our users' experience of our devices is situated in their world. The environments we design and build can bring joy, but they can also frustrate. They can be part of creating memories, acquiring skill, and facilitating recreation, but they can also get in the way. Their unpredictable updates and interrupted streaming behaviors mean that, to use Heidegger's language, they are continually moving between states of *zuhanden* and *vorhanden* (i.e., from ready to hand and transparent in use to present at hand, in which case the systems themselves become the object of attention and focus).[1]

This chapter proposes that understanding how our devices fit into lived time will help us to be better designers and better custodians of our users' time. Every book on time points out that clock time is not lived time, that a rigid temporal scaffold of our lives is necessary to allow society to function, and that clock time has also become a means of defining social order, from the factory whistle to the online calendar. Many people's workdays start with literal "clocking in" and end with "clocking out." Our days are broken into slices of time by our calendars. Wages are often calculated on an hourly rate, and people's value in the economy is often determined by what they can charge per hour for their services. Productivity and related bonus structures within many organizations are determined by a person's output per day, month, or year. If we are designing tools that people will use in their daily work, then we, as designers, play a role in people's economic success, whether we acknowledge it or not. If your computer keeps crashing, or if it decides to update in the middle of a task, then it becomes an obstacle to your ability to complete a task in a timely manner. As long as software and hardware manufacturers take people's time for granted by marketing products that are still works in progress, the economies of the world will continue to throw away uncountable hours of life and labor. In the United States and Europe, a person may currently be expected to live for approximately 30,000 days. As designers, we have some obligation to ensure this time is not squandered on our watch. It is time, so to speak, that we take responsibility for wasted time. We can do this only if we understand how the things we build become part of the experience of our users' lived time.

Just as clocks and time cards were introduced in the nineteenth century to structure time for factory workers, today's scheduling and timekeeping tools continue to structure our workdays. As Winograd and Flores have noted, the things we design create new ways of being in the world.[2] And nowhere is this more apparent than in the way our always-on, omnipresent connectivity has collapsed our concept of space and fragmented our perception of time. In this chapter, I will draw on a moment of disjunction, when clocks were introduced to rural Wales in the nineteenth century, to generate a discussion of our current moment of disjunction, as we come to terms with the fragmented "instaworld" we now live in. I will further argue that because our experience of time is brought forth by our actions, the interactive systems we design fundamentally shape the time of the lived experience of those who use them.

4.2 Clock Time and Lived Time

A number of years ago, I happened upon a radio play by Alan Plater called *Only a Matter of Time*.[3] Plater's drama captured the moment in the mid-nineteenth century when the railway companies in England expanded into Wales. Emissaries were dispatched to install public clocks and instruct villagers in reading time. The issue for the railway company was that time, for the farmers and villagers of Wales, was defined by the arc of the sun, so that noon in the most westerly parts of the country arrived eleven minutes after noon in London.

One emissary, a man called Fanshaw, arrives on his horse with a piece of paper in his hat that is inscribed, in Welsh, with the name of the village where he is to place his clock. He approaches a farmer for help. Proffering his paper, he asks the farmer to direct him to the village, whereupon they reach an impasse: The farmer can point him to a village, but he cannot read the paper; Fanshaw can read, but he cannot speak Welsh. And so begins a gentle but probing inquiry by the farmer to discover what this man on horseback wants.

What unfolds, over the course of a whole day, is a dialogue concerning the value of time. It is clear that the farmer is not interested in Fanshaw's time or his trains or his notions of prosperity. When Fanshaw falls asleep in the hot sun, the farmer contemplates murdering him to stop the coming of London time and the railways, but he realizes resistance is futile. They will only send another Fanshaw, and he, the farmer, will end up hanged. He realizes that the train will bring both visible and invisible change—the culture, stories, and ways from London. But he also knows that the Welsh, with their stories and their deep understanding of the land and its time, will ultimately win. He makes Fanshaw wait until he is done turning his hay, until the sun goes down, then walks with him to the village. He makes Fanshaw live according to his time. In the end, the railway is built and, almost 160 years later, a train passes the village—eleven minutes late!

This narrative is of interest here because of its many layers of understanding of lived time. First is the question of whose time controls how we live. For Fanshaw's employers, the only time that should govern people's lives going forward is that of their clocks. But the farmers' hay and the animals it fed would continue to grow and be nourished according to the time of the soil and the sun. And the earth would continue to travel around the sun according to laws far more persistent than the mechanism of Fanshaw's clock. Then there is the question of task time. The farmer was practiced at turning hay and could accurately say how long it would take him to turn a whole field. And there is the question of context: Living in Wales, the farmer knew that a dry, sunny day was not a given and that he had to make hay while the sun shone. Fanshaw might have realized that hay would ultimately feed his horse, or one like it, so it would pay to allow the farmer to take his time.

Somewhere beneath the story of the coming of the railway, then, is a story about a fundamental shift in the governance of time from the sun to the railway owners in London and, with that, a separation of time from life itself. However much effort has been expended in developing algorithms to minimize distraction, there is an equally active community of algorithm designers working to discover how to do the opposite: to distract us and to capture and hold our attention for as long as possible. The most infamous example is infinite scroll, which probably underpins every social media platform you have used. Developed by Aza Raskin, the algorithm allows you to continually scroll through posts and images that are, in turn, continually updated according to what you click on as you go.[4] According to Zahrai et al., the average person spends approximately two and a half hours per day on social media—more time than they spend either eating or exercising—adding up to almost one thousand hours per year.[5] Once captured, our time and attention become commodities that are bought and sold in the increasingly personalized world of online advertising. Advertisers spend billions of dollars each year trying to determine the point in our workflow where they can insert an advertisement that will capture our attention. In 2021, *Forbes* estimated that the global online advertising market (combining digital media, the Internet, and social media) exceeded $7 billion.[6]

Behind the scenes, the competition for our time, our clicks, and our dollars has been developed into a marketplace as sophisticated as the stock exchange—an infrastructure completely invisible to us, its end users.[7] On Google's platform, for example, companies can pay from $1 to $99 per click to have their ads included in a bidding auction that starts when you begin typing words into a search engine, taking advantage of predictive text while you are still typing. Depending on a company's Google-determined ranking, it might not even get into the auction at all. Companies develop advertising campaigns based on objectives they define, and then the Google algorithms take over. Companies ranking highest after the bidding have their paid advertisements displayed on the first page of options returned to the user because, according to a study by SISTRIX, people are almost twice as likely to click on the top result of a Google search than on the one immediately below it.[8] The likelihood of clicking falls to 2.5 percent by the tenth result. If the time and money spent waging an all-out battle for our attention and our clicks were directed toward expanding uninterrupted time rather than on fragmenting workflow, how much more productive could we become?

There is an adage that says, "Technology is anything that was invented after you were born." Each generation is attracted to phenomena that come about on its watch. But the lived experience of each generation is also, to some extent, altered by the technologies it develops. For citizens of England in the nineteenth century, the most transforming technology was the railway. It reduced journey times from days to hours and connected communities that had been isolated in space and frozen in time. But, as noted, the railway relied on shared notions of time, so with trains came town clocks. It could be argued that clocks facilitated the free movement of goods and materials, which, in turn, made the Industrial Revolution possible. But the Industrial Revolution also harnessed clocks to apportion time and quantify labor, dividing the day into hours for work, for recreation, and for sleep, thereby creating the scheduling of activities that still governs our lives today. We still feel bound to account for every minute of our working day in terms of time on task, an idea baked into "productivity tools." Indeed, we now consider it necessary to multitask, to track many strands of activity in parallel—a work practice considered by many as evidence of even greater productivity.

But this is not how we accomplish our best work. Some tasks need our undivided attention, whereas others can be broken into smaller subtasks. Some tasks need time for pre-reflection, whereas others do not. As Jönsson argues, switching between tasks requires setup time.[9] Many of today's work environments support seamless transitions between tasks, but this might reduce, not increase, the likelihood of people having productive, fulfilling working experiences. The fragmenting of time that is a by-product of multitasking works against deep engagement and against a sense of task flow, even as it might also facilitate self-regulation, allowing people to temporarily switch tasks to remain stimulated or to give themselves time to mull over hard problems.

The multilayered processes that are part of today's mediated workplaces are different from the manufacturing tasks that shaped how time was apportioned and accounted for in workplaces of the past. We now need to rethink the relationship between time and work. Accomplishing work today requires a deep understanding of the nature of each task and a nuanced approach to managing the time it takes to set up, perform, and recover from each task goal, even when we might be engaged in multiple activities at the same time. We will not leave the clock behind. Rather, we will be freed from the idea that all tasks can be planned solely in terms of clock time. Like the farmer in his field, we can allow our days to be guided by the time of our experience rather than the time of the clock.

4.3 Perceiving Time

Our perception of time is a paradox. We have no senses for time, but we do have a sense *of* time. Our sense organs do not perceive time, but we do perceive how our environment changes over time. We can anticipate future

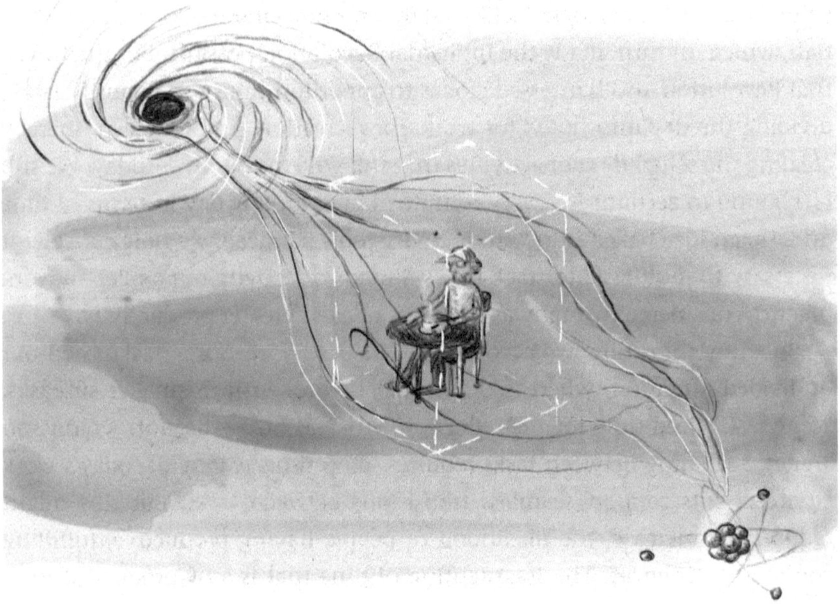

Figure 4.1
Conceptual drawing of Middle World, the timescale between that of atoms and that of the cosmos, as discussed by Richard Dawkins in "Why the Universe Seems so Strange." *Source:* Original drawing by Hans Anderson.

events, and we can often accurately estimate the amount of time that has elapsed as we go about our daily tasks. Humans have not yet evolved the ability to be aware of multiple, simultaneous streams of consciousness. The time we perceive is therefore shaped by the arc of our individual lives, framed by the narrative of our unfolding awareness of the world around us. This unfolding takes place at a pace that we can comprehend, a timescale Richard Dawkins refers to as that of "Middle World," somewhere between the subatomic scale of moving particles and the astronomical scale of the galaxies, on a continuum like that shown in figure 4.1.[10]

The range of speeds of time that we can comprehend is therefore bounded, in the same way as spectra of visible light and audible sound are. Furthermore, because we evolved to live in Middle World, we are attuned to the laws of physics at this scale. We are not aware that solid objects are made of constantly moving atoms with space between them or that objects in outer space move at speeds and under the influence of gravitational forces that we have no bodily frame of reference to comprehend. We are, in short, comforted by the timescale of our biological beings, the presence of matter and the gravitational and atmospheric conditions relevant for our bodies in Middle World. Like the illusion that the solid objects around us are made of solid matter, we are also fooled into believing that our bodies are, to some extent, with us throughout our lives. Yet, the cells in our bodies (with the exception of neurons) are replaced periodically, the longest lived being those in our bones, which renew every seven to ten years.[11] Our bodies, Dawkins claims, are therefore more akin to waves of moving matter that pass through us than they are to permanent biological structures that we own.

Given that we are continually reconstituted through processes that unfold at different biological timescales, how do we account for our lived experience of time? How do we perceive and understand the unfolding of the time of our lives? If we are constantly changing, what parts of us remain stable enough to allow us to perceive the passing of time? One answer might lie in recasting our understanding of time as an active process in which the movements of our evolving bodies bring forth the lived experience of our present, shaped by prior actions and directing our attention and action toward future goals. In this framing, temporal awareness does not arise because we have a persistent physiology or chemistry but rather is continually reframed by the current body–environment coupling that brings forth lived experience.

4.3.1 The Structure of Time Consciousness

The question of how we are aware of our present moment, even as we are also aware that it too is passing, continues to perplex philosophers, psychologists, and neuroscientists because it lies at the very intersection of our understanding of ourselves as biological entities and as conscious beings. Phenomenologists seek to solve the question of our awareness of the present—that which has just preceded it and that which is about to follow—whereas neurobiologists seek to discover how our brains and bodies keep track of the timing and ordering of events and actions that comprise our doings.

Our stream of consciousness is an ensemble of experiences that are unified both synchronically (at any moment in time) and diachronically (across time). We can thus determine whether events happen at the same time or whether one precedes or follows the other. If we allow that we can perceive the order in which things happen (succession) and how long they last (duration), then, Gallagher and Zahavi argue, we have to acknowledge that our consciousness can encompass more than what is happening right now.[12] But how can we simultaneously be aware of what is happening now *and* what has just happened? How can we also anticipate what will happen next? If I get up from my desk and walk across the room to close the door, how do I anticipate what I need to do and how do I plan the sequence of actions that will result in accomplishing each step toward my goal? In his work on time consciousness, Edmund Husserl states this problem in terms of listening to a piece of music. We hear each note not as an individual event but rather in the context of the note that has just passed. He writes, "When the new note sounds, the one just preceding it does not disappear without a trace; otherwise, we should be incapable of observing the relations between the notes which follow one another. We should have a note at every instant, and possibly in the interval between the sounding of the next an empty phase, but never the idea of a melody."[13]

In hearing a melody, therefore, it is not sufficient to account simply for awareness of a note in a given instant, or to account for a succession of phases of awareness of notes and their immediate precursors. As Zahavi puts it: "A mere succession of such conscious phases will not as such provide us with a consciousness of succession."[14] Zahavi points to the paradox at the heart of our subjective experience of time: How can we be aware of the present moment, the now of conscious awareness, while also being aware that this present will soon be part of the past? Husserl accounts for this temporally bounded "now" of subjective experience by proposing a three-part structure of time consciousness:

1. A primal impression, defined as the current or now instant of awareness
2. A retention, an awareness of the instant that has just passed, which provides the context for the now
3. A protension, a context provided by the now for what is about to occur in the immediate future[15]

This is why, when we are engaged in a conversation, we can hold in mind the part of a sentence that has just passed to provide context for what is being said right now. Our familiarity with the structure of language also allows us to predict, to some extent, what is likely to come next.

It is our ability to retain the meaning of the first part of a sentence, even when the words are no longer sounding, that allows us to infer its meaning or intent in the present and anticipate how it might end. And, as Husserl notes, it is protention that opens up the opportunity for surprise. Indeed, it is our ability to anticipate how events taking place in the current moment may be resolved that underpins all temporal forms of art, music, film, storytelling, and so forth—because it provides creators with the temporal machinery to surprise us by delaying resolution of a melody or plot line or by denying resolution altogether. This is how Haydn, in the Andante of his Symphony No. 94, was able to "surprise" his audience by concluding the quiet opening violin melody with a loud chord from the whole orchestra. The audience might anticipate that the sixteen-measure violin melody will end quietly, as it began. The element of surprise works because a loud chord in this context is not something audiences would expect based on their prior experience of such symphonies.

But protention or anticipation is not solely dependent on past experience. There are many ways in which we can anticipate that something will happen, even if we are not sure exactly what it is. In formal terms, the content of protention does not have to be completely determined by the observer. Its function is to provide a forward-looking temporal context informed by what is happening now (i.e., informed by the primal impression). The example of Haydn's audience might lead one to believe that our awareness of the present moment is passive, and all we have to do is to be present. But, as Gallagher and Zahavi argue, time consciousness requires our active engagement to bring forth the lived present, "constituting its meaning in the shadow of what has just been experienced and in the light of what it anticipates."[16] In Husserl's model of time consciousness, retention, primal impression, and protention are not separate but instead are bound together into a temporal, horizontal gestalt, a present in which retention represents the periphery of temporal awareness, much as the edge of our field of view is the periphery of our visual awareness. Husserl provides both a structure to understand extended temporal awareness that affords the perception of things such as melodies and a means to understand how consciousness itself is unified over time. If you want to see a dog from all sides, you walk around it, observing it from every angle. You do not perceive the dog as a collection of disjointed snapshots; it is integrated in your experience of a whole, your current view

Figure 4.2
Present and recollected views of a dog. When viewing an object, we see the view in front of us, but we can often recall the object viewed from different perspectives, particularly if it is very familiar. The shaded views of the dog in this image are intended to represent non-visible perspectives that can nevertheless be brought to mind. *Source:* Original drawing by Hans Anderson.

being given context by your memory of past views (the shaded images of other views of the dog in figure 4.2).

Gallagher and Zahavi argue that this is possible because retention, primal impression, and protention are invariant structures in time that bind the sequence of "now" views of the dog, making possible the temporal flow of visual stimuli from which the invariant structures in light can be picked up by your visual system as you move around the dog.[17] If temporal consciousness had no retentional component, it would be impossible to achieve any synthesis to experience the form of the dog. Therefore, time consciousness is a necessary condition for the possibility of perceiving objects. This, Husserl argues, is why the retentional component of temporal awareness should be thought of as perception, not memory, because the object of awareness never fades or has to be recalled.[18]

Further evidence of the active nature of time consciousness stems from the fact that not all time is experienced in the same way. If you are in pain, every moment seems to last forever, whereas if you are happily engaged in an activity you enjoy, time flies. In other words, the content of experience can influence the structure of time consciousness. In a now infamous demonstration of the impact of experience on the perception of time, Stetson and colleagues conducted a study in which they asked twenty-three volunteers to free-fall backward off a tall building.[19] Each was given a wristwatch programmed to display pairs of rapidly alternating random numbers. Under normal viewing, the alternation was so fast that the numbers could not be identified. The authors' hypothesis was that, if sensory perception speeded up in response to the terrifying experience of falling backward through midair, time would be perceived as running more slowly and the alternating numbers on the watches would be readable. Although all participants reported that they felt as if they were falling for a long time—in their subjective experience, time dilated—none of them was ever able to read the numbers on the watches. Their processing of sensory stimuli did not speed up, nor did time slow down. Only the perception of time changed. The authors concluded that heightened emotional responses, such as fear, may be encoded in memory in more than one way and that, in recollection, such emotionally salient events may be remembered as having lasted longer than they did. In other words, the experience of dilated time is in reference to a memory of the event rather than the event itself.

It is for this reason, Husserl argues, that the structure of time consciousness defined by retention, primal impression, and protention is independent of the flow, or stream, of consciousness because it is this structure that constitutes the flow of consciousness.

4.3.2 The Malleability of Time Consciousness

In an essay on embodied time, Mark Wittman notes that phenomenal consciousness is ultimately the generation of an island of presence in the continuous flow of time.[20] This island of presence represents imminent experience (where the object of awareness can be described without need for recall), when current events are not yet memories and anticipated events have not yet begun. As we shall see, the neural processes that bring forth the now of our experience take time, so that what we experience is not the same as the absolute time of an event's occurrence; the now as we experience it is what William James termed the "specious present."[21] It is this definition of the now of lived experience that I will consider here.

To understand how Husserl's model of time consciousness might impact how we think about time in the design of interactive systems, we must address two questions: How long, in terms of real time, is the specious present? And how does the structure of the specious present relate to that which is beyond its boundaries (i.e., to memory of past events and anticipation of future ones)?

Let us first address the question of the duration of the specious present. As Susan Pockett points out, if the now is too short, experience is no longer imminent; it cannot be described before it becomes a memory.[22] Drawing on evidence from prior work concerning (1) the objective duration of sensations in different modalities, (2) the measured duration required for sensations to enter conscious awareness, and (3) the variability of the subjective sense of time passing, Pockett argues that the duration of "now" can vary from 10 μs up to tens of seconds, even under conditions that are not subject to the emotional extremes of, say, the falling study described above. Pockett also points out that time appears to slow down when we are engaged in externally oriented activities involving frequent or novel stimuli, whereas time speeds up when we are internally focused, in the flow of a task. Pockett's hypothesis is that the length of the specious present is a function of how frequently we sample the external world. Under extreme conditions, such as when avoiding a driving accident, our attention is hyper-focused and, Pockett argues, the rate at which the world is sampled increases: Time appears to slow down, and we seem to have more time to react. But when we focus on writing or on solving a puzzle, our attention is focused inward, our rate of world sampling slows down, and time appears to fly. Our rate of world sampling is inversely related to the perceived duration of the specious present.[23] Whether Pockett's hypothesis is ultimately supported by future experimental findings, it points toward a mechanism that might explain why the subjective length of the specious present is independent of any measurement of duration associated with clock time.

The second question concerns how the structure of the specious present relates to memory of the past and anticipation of the future. As Gallagher and Zahavi point out, the function of retention/protention in binding together the notes of a melody is not the same as the function of memory in supporting the recall of a past vacation or the anticipation of a future holiday: "Whereas recollection and expectation presuppose the work of retention and protention, protention and retention are intrinsic components of any occurrent experience (be it a perception, a recollection, a fantasy, etc.) one might have."[24] As Pockett also notes, the content of awareness in the specious present does not have to be external. It can be a memory, a flight of fancy, or a hallucination.[25]

It should be noted here that unlike memories of objects, virtually rendered objects constitute content that resides in the external world because they are still "picked up" via the same sense organs that relay information about physical objects in our surroundings. In other words, a 3D rendering of a box in a virtual environment is still perceived visually because it has attributes such as color and size; its 3D rendering still structures light and shadow in the same way as material boxes do in non-virtual worlds.

Retention is contiguous with the now, an impression of something that was *just* now, whereas recollection involves bringing into awareness an event that has been completed and so is being brought back from the past, even if it is the recent past. Without this sense of distance, past and present would merge, the flow of time consciousness would collapse, and we would be hallucinating, not remembering.[26] But recollection of the past is not entirely independent of the present, for if an event is to be recalled from the past, something in the present must bring it to mind. Furthermore, without memory, we would not have the capacity to draw on experience to make informed decisions in the present. This is what distinguishes us, on the one hand, from organisms compelled to respond to a single stimulus in the environment in a completely determined manner and, on the other hand, from a completely indeterminate world, which we would confront anew at every moment. As Pöppel states, "Memory is to be found, therefore, only in a world somewhere in between total indeterminacy and total determinacy."[27]

4.3.3 An Enactive Account of Time Consciousness

So far, we have considered the structure of time consciousness in terms of a phenomenological account of experience. But such an account does not explain how our brains and our bodies manifest our awareness of time. Unlike sound or light or mechanical stimulation, there are no organs for sensing time and, as far as we can determine, no parts of the brain solely responsible for timekeeping. But that does not mean that our bodies are not aware of time. Many of the things we do—breathing, eating, sleeping, waking—happen in regular, cyclical patterns driven by bodily signals that tell us of our need for oxygen, food, rest, and so on. As noted in earlier chapters, other activities are bounded in time by the need to close a perceptual gap—to catch a ball or to place our foot on the next step of a staircase. At some level, therefore, our brains and bodies must be keeping track of what has just happened, what is happening now, and what can or should happen in the immediate future.

In explaining how phenomenological models of time consciousness might account for our ability to close perceptual gaps, theorists such as Varela have proposed that, like many other forms of behavior, perception–action couplings are a result of coordinated responses arising from highly distributed and functionally distinct parts of the brain and the body in a process best modeled as a self-organizing dynamical system.[28] In this model, three distinct scales of duration are integrated: the elementary scale, the integration scale, and the narrative scale. The elementary scale is defined at one end by the 10-ms cellular discharge rhythm of interneurons and at the other end by the duration of excitatory/inhibitory post-synaptic potential sequences in cortical pyramidal neurons. Within this range, stimuli are perceived to be simultaneous, even though the processing times of different sense modalities are not the same. For example, auditory events are registered in the primary auditory cortex in 9–15 ms, whereas visual events take 45–55 ms to be registered in the primary visual cortex.[29] The response times of the elementary scale thus serve to integrate events from sense modalities with different processing times so that they are perceived to be simultaneous. This is why, at a concert, you may perceive the movement of a singer's lips and the sound of their voice as simultaneous, even though sound waves travel to you about one million times more slowly than light waves.

The integration scale is the timescale of the phenomenological lived present. It is also the scale at which basic actions, such as reaching and grasping, are executed, where the conscious initiation of an action is integrated with the associated sensorimotor activities that bring about the action.[30] Individual neurons are integrated within assemblies distributed across regions of the brain but with strong reciprocal connections. These assemblies give rise to conscious experience and have a particular pattern of response, in which cell discharges are followed by periods of relaxation and phase transition.

This corresponds to a temporal arc in which experiences arise, are present, and then fade so that the cycle can repeat.[31] To this scale belong reactions to stimuli in our environment such as turning away from a bright light or catching a falling object. Actions are responsive to stimuli in the environment that are perceived, but actions are not planned in advance of perception of the stimulus.

The narrative scale supports the planning and execution of intentional action. Since the time frame of intentional action is dependent on the nature of a given action, it can range from very short (e.g., planning your movements to catch an approaching ball) to very long (e.g., planning a vacation). For this reason, the neural processes involved, although they can still be thought of as dynamical systems, are much more task and context dependent. The self-organization of neural activity is not an abstract process but instead directly relates to what an individual is doing or has just done. This self-organization reflects the embodied nature of perception, action, and cognition. The duration of an event on the elementary scale is not fixed but rather adapts in accordance with a number of factors, including the response times of sensory modalities salient for a given task, an individual's age, their level of fatigue in the moment, and so on.[32]

In terms of Husserl's model,[33] the elementary scale represents the threshold between the perception of simultaneous versus successive events within the now of the primal impression, whereas the integration scale corresponds to the integrated temporal awareness of the specious present as a whole, including retention, primal impression, and protention. Husserl's model can be used to describe the integration of scales of neural activity associated with physical action as well as the structure of conscious awareness. This is because, as Gallagher points out, our experiences and our actions are both characterized by an intrinsic temporal arc.[34] When we walk, we do not stop after each step to assess our posture or to take a conscious snapshot in order to plan our next step. If we are aware of anything about our movement, it is the flow, the way in which one footstep sets up the anticipated trajectory of the next. Action, like consciousness, has a flow structure and an intentional direction: "What we have as the basic datum of experience is a process, through which the primal enaction is already collapsing into the retentional stream even as it is directed forward in protention."[35]

As noted earlier, the time course of neural processing means that there is always a delay or temporal distortion between an event occurring in our environment and our awareness of the stimulus arising from that event, so that experience of the present is always through a small slice of the past. This is why phenomenologists refer to our experienced present as the "specious present": Our experience of the world is more akin to a secondary impression than an immediate snapshot of a knife-edge now. Conscious experience is not simply the sum of a sequence of events at the elementary scale. If it were, there would be no way to accommodate the differences in response times for different sensory systems. As noted above, sounds register in the primary auditory cortex within 9–15 ms, whereas visual stimuli take at least 45 ms to reach the primary visual cortex.[36] Between the elementary and integration scales, stimuli arriving at different times from different sense organs and their associated nonlinear patterns of neural activity are dynamically organized so that they give rise to a stable experience of the living present.

This nonlinearity is also evident in people's judgments of the timing of their own voluntary movements. When asked to estimate the temporal interval between two events a second apart, people's estimates are generally accurate. But if the first event is their action and the second is a result of that action, people greatly underestimate the interval between the two.[37] This effect, called "intentional binding," has been proposed as a way to measure the pre-reflective sense of agency associated with control of motor commands (efferent motor control). Further, people's sense of agency for a given action is also impacted by timing. Farrer, Valentin, and Hupé demonstrated that when actions and their effects fall within the elementary scale (i.e., within 334 ms), people have a full sense of agency; they believe that their actions cause the associated response from the environment.[38] As the time interval shades into the integration scale (intervals between 334 and 707 ms), their sense of agency decreases until, beyond 707 ms, they report loss of a sense of agency altogether. The authors conclude that our sense of agency depends on that window of time during which internal and external agency cues are integrated, and that this is true only for internal agency cues that are time locked to the onset of actions.

Gallagher and Zahavi argue that the unstable dynamics inherent in the relationship between embodied and environmental cues result in the continuous generation of new assemblies of neurons.[39] In the example above, the changing dynamics of the body–environment coupling that result in an altered sense of agency illustrate how, in practice, such reorganization is impacted by the flexibility of the temporal window at the integration scale.

In turn, this indicates the presence of a structure that can adaptively cope with variability in the coupling between the acting body and the responses to action arising from the environment. At the neural level, this provides evidence for the involvement of the active agent within the body–environment coupling that is at the heart of the enactive approach to perception: "The enactivist view holds that perception and cognition are action oriented; that is, they are characterized in most cases by a structural coupling between the agentive body and the environment, which generates action-oriented meaning."[40] Building on the assertion that the structure of time consciousness embodies the coupling of perception and action right down to the level of neural responses, Gallagher and Zahavi propose that the self-constituting nature of retention-primal (enactive) impression-protention provides evidence for a mechanism that underpins an enactive engagement with the world: "Just as I perceive the hammer as affording the possibility of grasping it, or in a different circumstance, as affording the possibility of propping open my window, I likewise perceive the melody as affording the possibility of dancing or of sitting in peaceful enjoyment, etc. Such action-oriented perceptions reflect the temporality of affordances and enactive engagements."[41]

The nub of this argument is that if one perceived the present only as a knife-edge of time, there would be no way to understand whether the object of perception afforded any possibilities for action. "When, for example, I perceive something, I perceive it as actionable. That is, I perceive it as something I can reach, or not; something I can pick up, or not; something I can hammer with, or not, and so forth. If I perceive an event rather than an artifact, I perceive it as something I can intervene in, or not. Such affordances . . . for potential actions (even if I am not planning to take action) shape the way that I actually perceive the world."[42] Just as it would make no sense to perceive the world as a series of knife-edge moments, it also makes no sense to perceive the world without retentions or protensions. The very definition of affordance is the perception of opportunities for future action based on information picked up from the environment in the present. Without protention, we could not plan future actions. Similarly, if all we could perceive were retentions, everything would seem to have happened already, and we would simply be witnesses to a world that had passed us by.

Gallagher and Zahavi propose that experience, as it unfolds in time, is fundamentally enactive, from the highest level of sensorimotor engagement with the environment through action, down to the self-organizing, self-constituting level of the temporal microstructure of each moment of engagement with the world.[43] The philosopher Maxine Sheets-Johnstone, whose work we encountered in chapter 3, goes as far as to ask whether our very sense of time might have its origin in the movement of our bodies.[44] Sheets-Johnstone proposes that, in the case of self-movement, we do not adhere to ordinal time measured in intervals. Rather, she proposes that bodily movements unfold in cardinal time, a qualitatively experienced temporality, a felt time in which movements are experienced as sudden, creeping, prolonged, and so on.[45] "We experience precisely the cardinality of time, not its ordinality. We do not experience kinetic befores, nows, and afters. This tripartite ordinal ordering of time is a sophisticated, reflective attainment that, in terms of the originary temporal structures of self-movement, imposes divisions where none exist, divisions that if present would in fact disrupt what is experienced as a global qualitative dynamic."[46] In this view, our bodily movements bring forth the time of our lived experience, and it is through the temporality of our actions that we shape our encounters with the world.

4.4 Interaction Time

Time consciousness is not an awareness of a single snapshot of time but rather an ensemble of experiences bound together both in a moment of time and across time, both synchronically and diachronically. We are therefore aware of whether events happen at the same time or in succession, and we can estimate the duration of each. In other words, our consciousness can take into account events that are happening right now, how they relate to past events, and how they might shape what will happen in the future. Husserl's phenomenological account of time consciousness accounts for our need to represent this ensemble of experiences.[47]

For designers of interactive systems, the open question concerns the extent to which a nuanced account of time consciousness, brought forth through our active engagement with the environment, will shape and be shaped by the interactive experiences and technical artifacts we create. As we have noted, the interrelated timescales of neural responses to temporal events at elementary, integration, and narrative scales reflect, respectively, the organization of low-level stimuli, the temporal window within which actions are coordinated, and the temporal arc associated with long-term intentional planning. These categories of events arise within the interaction between people and interactive systems, from the detection of simultaneous versus successive stimuli within and between display modalities at the elementary scale to the presentation of coherent experience and the perception–action coupling that creates a sense of agency at the integration scale to the interpretation and planning of actions and events at the narrative scale. It is therefore important to understand the consequences of design choices that impact the timing of events at every level of an interactive system—from its lowest-level system clock to the processing of user actions and system responses to management of media streams and document-handling applications. Here and in the following sections, I will address some of the ways in which time is managed in interactive systems at all these levels, drawing on particular scenarios to frame each aspect of this discussion.

From the reflexes that help us catch falling objects to the poised actions of dance or yoga, the movements we perform unfold at a pace governed by what is possible and determined by what we want to accomplish. When computational devices such as phones and laptops are involved, we typically refer to our actions as interactions; we assume that our devices will respond in some meaningful way to what we are doing. I have suggested that actions upon the world bring forth reactions from the world, and computational devices are part of that world, albeit a part that can be programmed to respond in particular ways. The challenge faced by designers of interactive systems is that people's actions are not temporally consistent, unfolding at a pace that is governed by unpredictable factors such as emotional state, level of attention, and even level of fatigue. Computers, on the other hand, carry out their operations against the backdrop of system clocks that rigidly apportion time in response to competing demands (from input sensors, displays, central processing, power management, and so on).

At every level of the interaction between people and machines, time matters. From the clock speed of the processers at the backbone of gaming PCs to the clocking of audio and video streams delivered to our phones to the speed at which web pages load, there is an expectation on our part that there is a "right time" for things to happen. The challenge for hardware and software designers is to fulfill our expectations by creating a bridge between the time of our machines and the lived time of our bodies. In this section, I take on the temporal aspect of this design challenge in more detail, unpacking how timing and pacing for an interactive system arise from decisions made during every phase of a system's design and how these decisions, in turn, can determine the perceived quality of an interaction from the user's perspective.

We are all familiar with the feeling of frustration when a device is not responding as we expect. Perhaps a keyboard or mouse feels sluggish or a task is interrupted or the computer has stopped responding altogether. At these moments, the device becomes, in Heidegger's terms, *vorhanden*, the object of foreground awareness, taking our attention away from the task at hand[48] (see chapter 2, section 2.3.1, for further discussion). As Seow eloquently points out in *Designing and Engineering Time*, the conversational nature of interactions between people and computers has its own etiquette, or as he puts it, "time matters."[49] If the system responds too quickly, it can give the spooky sense that it is anticipating our thought processes. But if it is too slow, we assume it is broken or that we need a faster device. Either way, there is a sweet spot within which a computer responds as we expect. But our expectations for device performance are not consistent. There are applications and situations where we cannot tolerate any delay or deviation in responsiveness, and there are situations in which we expect the computer to take its time. Managing time in interaction design is therefore a balancing act. It requires balancing the expectations of the user, the demands of the application domain, and the constraints of the underlying system (hardware and software).

Moreover, while computers' operational speeds have increased exponentially, the reaction times of their human users have not. In the 1960s, anything like real-time performance even for the simplest tasks was, at best, aspirational. The requirements of the system dominated (except in safety-critical situations). As computational speed increased and more cycles could be allocated to human–machine interaction, the needs of users became more integrated into the design process. Designers began to talk about real-time interaction; a system should not make a user wait for things to happen. Now that our phones have more computational resources than the computer aboard Apollo 11, the design of the human–computer interface is foregrounded, and manufacturers compete to deliver the highest-fidelity image, the best sound, and the most compelling user experiences.

As previously discussed, the different scales at which we process temporal information reflect different aspects of our awareness of and interaction with the environment. In contrast, the history of the development of computational systems and the way we interact with them has more to do with their context of use, from the earliest safety-critical environments, such as spacecraft and nuclear power plants—where human reaction and system response times were prioritized—to today's immersive environments, with their delivery of real-time, networked, coordinated experiences for multiple users interacting in virtual worlds. For every interaction paradigm, the underlying challenge is the same: How can a designer bridge the gap between time as operationalized by the machines, networks, and devices they use and the way in which humans perceive, understand, and act in the time of their lived experience? The topics presented below address this gap at several timescales and point to some of the open challenges that are still topics of active research.

4.4.1 Response Time

Anyone who has worked in HCI over the past twenty years is likely to say that the appropriate response time for an interactive system should be 200 ms. This is also the most often reported value for "simple reaction time," which is the average time someone takes to press a button in response to a stimulus such as a buzzer or a light.[50] It is the fastest possible time in which a person could respond to the alarm in a safety-critical situation, such as an emergency in a nuclear power plant. As Brebner and Welford note, however, even this simple measure is inaccurate because reaction times are faster for auditory alarms than for flashing lights. For most interaction tasks, response time incorporates more than button presses because the person using the device may have to choose between multiple options displayed on a screen or process information displayed as text.

The most basic definition of response time is the time taken from the start of an interaction to its conclusion. An interaction can be initiated by either a system (e.g., a pop-up on a website) or a user (e.g., a key press). The initiating event starts a clock that runs until some perceivable response is observed—the dialogue box goes away or a letter appears on the screen. Some events, such as the loading of a web page, take time, so the event clock continues to run and the interaction is not concluded until the process is complete and the page is ready to be used.[51]

If, as Seow suggests, we consider the interaction between a human and a computer as a silicon–carbon conversation, then, as with any conversation, we might expect the pace of the interaction to reflect the time taken by each party to process what has happened and to formulate a response. In the case of HCI, the interpretation of events is somewhat asymmetrical, as the processing constraints and modes of interaction for each party are radically different. Any discussion of response time therefore needs to account for both the needs and requirements of the human (user response time) and the constraints of the machine (system response time).

4.4.1.1 User response time User response time measures the period from the start of a user-initiated event to its conclusion. This is hard to measure because it might involve complex user thought processes, such as choosing from various options on a menu. The time taken to make such decisions is referred to as the Hick–Hyman law: The time increases logarithmically with the number of choices presented. Hick demonstrated that as the number of user choices increases from two to ten, reaction time increases, proportionally, from about 200 to about 500 ms.[52] But this assumes that the options are clearly presented and no further cognitive work has to be done by the user to make a selection.

But there is also no point in a system responding faster than a user can process, particularly when reading is required. A recent meta-analysis of reading studies showed that the average silent reading rate for English is 238 words per minute (four words per second) for nonfiction and 260 words per minute (4.3 words per second) for fiction.[53] There is therefore no point in presenting users with material that updates before it can be fully read. Other user-initiated events might be culminations of decision-making processes that cannot be easily captured, such as the decision to stay in and watch a movie, which culminates in directing a device or platform to start showing that movie. User response times are impacted by the contexts within which a device or application is being used. When people use mobile devices in crowded or noisy environments, such as on public transport, their attention is fragmented and their response times are overall slower but vary widely over the course of an interaction.[54] Measures of user response time must therefore take into account a broad range of contextual variables and can, at best, only be specified for a given application within a given context of use.

4.4.1.2 System response time On the other side of the carbon–silicon conversation is the system, which has its own response time, governed by its own design balancing act. As Seow puts it, "System design is driven by hardware, influenced by software and optimized by design."[55]

Ben Shneiderman, a thought leader in the field of user-centered design, defines system response time as "the number of seconds that it takes from the moment a user initiates an activity (usually by pressing an Enter or Return key) until the computer begins to present results on the screen or printer."[56] Like user response time, system response time is governed by a notional timer that starts when the user initiates a command and ends when the system produces a response. Here, however, the time being measured is solely under the control of the computer.

It is interesting to note here how the changing understanding of system response time in HCI reflects the changing priorities of the field. Recalling the discussion of the three paradigms of HCI from chapter 2, the earliest computer interface designs were very much concerned with human factors engineering, characterizing HCI as a form of man–machine coupling, with the goal of optimizing the fit between humans and machines. Guidelines for acceptable response times characterized the needs of the user in terms of basic physiological response times and the requirements of the system in terms of low-level metrics, such as CPU processing time, display refresh rates, and sensor performance of keyboards and mice. The machine-plus-operator model was centralized in the CPU, and time was apportioned accordingly. Design engineers in this era often focused on pragmatic qualities of HCI, such as system safety, and were concerned with functional measures such as simple response time, the time taken by an operator to react to the sound of an alarm or to a flashing light.[57]

In the second paradigm, where interaction between people and computers was characterized as a process of information exchange, response time measures were concerned with how quickly people could move between and click on items on a screen, select options from menus, or respond to items refreshing on a website. Measures here include various formulations of tasks based on Fitts's law (discussed in chapter 3), with its ID measured in terms of the location, size, and number of items with which participants were presented.[58] This era (the 1980s and 1990s) included studies of the utility of the graphical user interface (GUI), measuring its ability to support information exchange between a user and computer for a variety of interface configurations (mouse, trackpad, touch screen, etc.). Here, the ID identified by Fitts is reframed by study designers as the number of bits that can be exchanged between user and computer according to a formula derived from Shannon's

Theory of Information Exchange, with higher bit rates indicating more efficient communication of information.

As the field of HCI has moved from designs motivated by the concept of information exchange that characterized the second paradigm toward design philosophies associated with the third paradigm that emphasize the embodied, situated nature of people's use of technology, there is an increasing emphasis on qualitative approaches to evaluating designs that, in Fallman and Waterworth's words, prioritize "experiences rather than performance; fun and playability rather than error rate; and sociability and affective qualities rather than learnability."[59] Consequently, in assessing the effectiveness of such interactive systems, Fallman argues that it is necessary to reconsider the whole process of evaluation, to reconsider what we mean by "good" interaction design.[60] It is necessary to find new approaches to assessing user experience of software and artifact design, but it is also necessary to quantify the way in which time is managed by system designers behind the scenes, within hardware and software and across networks, in order to deliver rewarding interactions that support users in feeling and being present in the environments we create.

4.4.2 Computational Time

In section 4.1, we discussed the emergence of the measured, objective time associated with clocks. At the human scale (in Dawkins's Middle World), time functions to synchronize events within a common frame so that we can communicate, travel, coordinate meetings, and so on. At the microscale (faster than the time of Middle World), time serves to facilitate traffic on the Internet and ensures that, for example, digital audio and video signals are streamed without glitches. The primary feature of this low-level computational time is that it is isochronous (i.e., whatever the interval of measurement—hour, minute, or microsecond—the "ticks" between measurement points are always the same). Clock time at the microscale is the backbone of all computational systems, but it is a separate time frame from that of "user" or "system" response times, and as we have noted, it is not the time of our lived experience.

As the scaffold on which all interactive applications are built, time at the lowest computational level is yet another factor to be considered by system and application designers. Somehow, the speed at which a given machine or network runs must be mapped onto the time taken for humans to interact with them and with each other. And when multiple devices work together to "serve up" an interactive user experience, such as a distributed VR multiplayer game, individual device clock speeds must be coordinated with each other and with the clocked time of any intervening networks. Thus, the following discussions of real time, coordinated time, and consistent time represent the most challenging issues designers must address in mapping between the operating speed of machines and the experiential time of their users.

4.4.3 Real Time

So far, we have assumed that a computer will take some time to compute a response to a user's actions and that, from the user's perspective, this time is somewhat flexible, as long as it conforms to their expectations of the carbon–silicon conversation. But in some application domains, the computer must respond in real time for an ongoing process to be sustained. Real-time performance is a measure of a computational system's ability to perform correctly, as measured by logical accuracy and speed. If the system takes too long to respond, it fails, even if the resulting output is correct, because it could not deliver on schedule.[61]

Because not all failures are equally critical, they are usually categorized as follows:

1. Hard failures, where failure to meet a schedule is catastrophic, such as in the control system for an aircraft. Results delivered after the deadline are of no value to the system.
2. Firm failures, where occasional failures are tolerable but may result in degradation of system performance, as with media streaming applications. As with hard deadlines, results received after the deadline are of no value.
3. Soft failures, where missing a deadline degrades the usefulness of a result, but it is still considered to be of some value. The main impact for the user is that the performance of the system or service degrades over time, such as a decline in network performance resulting from increased traffic or an equipment fault.

Computer systems prioritize tasks according to the category to which they belong. Processes that must adhere to clock time, such as low-level computations, audio and video streaming, and network communication, are prioritized above the scanning of input devices such as keyboards and mice and above the delivery of screen updates for non-streaming applications. As a researcher in digital musical instrument design and in haptics, I grew up with the specter of real-time performance hovering at my shoulder. I recall having to abandon libraries of convenient functions in C++ in favor of more simple C code, saving a couple of extra cycles so that my audio–haptic violin string model would sound realistic and feel smooth. Those same challenges still exist for interaction design, only now they hover over designers of VR games, in which the movements of one or more players must be tracked in human time and fed into virtual 3D worlds that are rendered in high-fidelity visual, audio, and sometimes haptic modalities. Although these applications are far more complex and resource hungry than my audio–haptic violin string, their goal is the same: to establish and maintain an environment where user actions result in timely reactions from the system and bring forth stable and believable experiences within an augmented or virtual world.

When we talk about real-time performance for interactive systems, we are most often referring to real-time digital signal processing (DSP). The signals concerned are streams of data generated by sensors in input devices such as motion sensors in handheld controllers or VR headsets. The output signals are streams of data encoding the system's response as audio, video, or haptic feedback. The term "signal processing" refers to the stages of computation that happen between the gathering of data from input devices and the generation of feedback to be displayed to users. These stages might include:

1. The analysis of motion data from sensors designed to detect patterns such as walking speeds, hand and head gestures, and so on as discussed in chapter 3
2. The mapping of information from input devices into the application space, which might take the form of registering the change in the user's location, a change in the status of a button, and so on
3. The computation of a system's responses to a user's intended goal within the application, such as recognizing that they have selected some text, picked up an object, or turned to look in another direction
4. The generation of signals to be presented to the user as video, audio, or haptic feedback that represent the system's response to their actions

To be considered "real time," a system must manage all parts of this process continuously, so that no samples are lost or dropped from either the input or output streams. This means that any computation to be performed must be carried out in the period between the reading of input samples and the writing of output samples. Because many processes take more than one sample period to complete, most applications employ buffers to ensure consistent signal throughput. Incoming signal samples can be read into buffers so that they are available when the application is ready to handle them. Similarly, at the output of a process, samples are buffered in readiness to be written to the appropriate output device or display. The processing that happens between these stages is then said to be "bounded" or constrained by the length of the shortest sampling period and the lengths of input and output buffers.

Another complication is that each signal to be processed has its own sampling rate: Signals from different classes of devices run at predetermined speeds (specified in cycles per second, also called "hertz"). Sampling rates for different data streams are determined by how time sensitive the information they carry will be. For example, sampling rates for input devices such as mice are much slower than those for handheld gaming controllers because gaming environments typically expect more rapid and fine-grain hand movements than word processing applications. A further consideration is that output data streams such as those for video screens and audio playback devices (headphones or speakers) have different sampling rates. Consequently, the interval between samples for a video stream will not be the same as for an audio stream. Common video refresh rates (also called frame rates) range from 24 Hz for NTSC (US TV), to 60, 90, or 120 Hz for high-definition video streams. Audio sampling rates are typically 44.1 kHz (CD quality) or 48 kHz (professional audio quality), with higher-definition sampling rates at multiples of both of these base values. Clearly, audio sampling rates are not divisible by rates for video, which complicates the management of digital media streams. Also, because video frame rates are at least an order of magnitude slower than audio sampling rates, audio streams will require higher priority if their associated signal processing and sample handling are to be completed within a time slice that is much shorter than for video. For audio sampled at 44.1 kHz, this slice will be 1/44,100 of a second for each audio channel (i.e., 1/88,200 of a second for stereo audio compared with 1/120 of a second for a video stream with a 120 Hz refresh rate).

As the number and complexity of signals to be handled increases, management of processing and sampling for each stream increasingly burdens the system's central processor. In many high-performance gaming systems, each element requiring heavy DSP is supported by independent hardware, while input and output streaming processes are handled centrally. This minimizes processing delays for each channel or stream of input and output data and minimizes any variability or jitter, both of which can cause a breakdown in a user's perception of reliability and consistency. When this happens, the system no longer appears to be responding in real time and, for VR, the rendered environment becomes unstable.

In his book *Time and Space in Video Games*, Federico Alvarez Igarzábal suggests that for game controllers to feel responsive, game designers should aim to ensure the system responds to user input within a time frame that is as close as possible to the window of simultaneity—the 1/10 scale described in section 4.3.[62] In practice, this means that if the system does not respond within 1/10 of a second, the player will sense that something about the system is not functioning as they expected.

Of course, not every factor determining the responsiveness of a system will be under a developer's control. As already noted, different gaming systems may have different refresh rates, and this, in turn, will impact the speed at which a game loop can run. Alvarez Igarzábal points out that, for a game running at thirty video frames per second, a screen will refresh every 33.3 ms, which is very close to the upper limit of 30 ms for perceiving input action and output response as occurring simultaneously.[63] Citing the work of Mick West, the co-owner of Neversoft Entertainment, who developed *Guitar Hero III: Legends of Rock* (2007), Igarzábal describes the challenge of programming a game so its response is perceived as being simultaneous.[64] He notes that, even for games requiring rapid player responses, such as first-person shooters, a minimum of three frames will elapse between pressing a button and seeing a system respond. This is because the player will press a button during the first frame that will not be registered until the second frame, when the system state will be updated and the central processing unit will generate the information to be passed to the graphical rendering engine. In the third frame, the GPU will render the updated game state, and in the fourth frame, the new state will be displayed to the player (i.e., the player will see the character shoot). In practical terms, it is this inner game loop that ultimately governs what it feels like to act within the world of a video game. In the words of Igarzábal, "The game loop is what makes the clock in virtual worlds tick."[65]

4.4.4 Coordinated Time

The discussion of response time in HCI is centered on the idea that the human and the machine engage in a dialogue, each taking their turn. Although the human user might think ahead, or the computer might carry out work in the background while it waits for the user to take some action, human and computer processes are relatively independent in terms of their time courses. But there are applications where the user and computer are bound by a common clock, when the timing of user actions is critical to the timing of the computer's response and both are framed by an external event structure. Such is the case for music performances that integrate computational processing, either because performance is distributed across several spaces and mediated by a network or because performers are playing instruments that integrate computation (Digital Musical Instruments or DMIs). For the player of a DMI, performance gestures are captured as digital signals by an instrument controller, such as a MIDI keyboard or foot pedal, and are processed to shape the output of a synthesis algorithm that generates the instrument's sound. The external reference frame is the event structure of the unfolding musical performance, which may or may not be defined by a musical score. In either case, the musician needs to be able to predict, for a given action, when a response from the computer will occur. For acoustic instruments, where the actions of a player cause a series of reactions in the instrument that are governed by its physical mechanism, such predictability arises from the physical structure of the instrument and the skill and experience of the performer. The mechanism of the instrument produces sound immediately, but occasionally a musician has to compensate for some delay in response—for example, when a low note on a brass instrument takes some time to sound. Church organists, in particular, must learn to manage such delays because if a rank of pipes is placed far from them, they must play ahead of what they are hearing in order for listeners to experience a coherent performance. As long as this delay is consistent, a performer will learn to adjust their playing accordingly. Problems arise when such delay is variable and the timing of the instrument's response cannot be predicted.

For DMIs that have no direct mechanical couplings between the player and the instrument's sound-producing mechanism, such delay (or latency) and any variability in latency (jitter) can arise from multiple sources, such as sluggishness in sensing input gestures, processing inconsistencies at the operating system level, and latency in audio drivers, to name a few. For this reason, designers of DMIs place emphasis on ensuring that the incoming signals from the player and the output sound signals from the computer are synchronized within a common processing environment.[66] These environments prioritize timely note production over other processes such as on-screen status updates and background file handling. Real-time musical performance requires not only real-time DSP to generate sound in response to input gestures but also bounded signal throughput, so that the sound reaching the audience is synchronized with the movements of the player. Unlike the organist hidden in the loft of the church, who can play ahead of the sound the audience is hearing, the visible stage performer's gestures must be coordinated with what the audience hears.

As studies of music performances have shown, acceptable limits for delays between the time of arrival of visual and audio cues are between 6 and 20 ms.[67] However, as the size of a venue increases, delay tolerance also increases because audience members far from the stage will expect sound to take longer to arrive. When sound in large venues is reinforced by amplification and multi-speaker arrays, a DSP layer is often used to ensure that the sound from the speakers farthest away from the stage is delayed to coincide with the natural arrival time of visual and audio cues from the stage. When delay handling is not employed, people at the back may experience a time warp. The sound of a singer arrives earlier than expected and often comes from behind them, sometimes getting to them before they see the singer's lips move.

When musicians play together, they rely on some means of coordinating their individual performances, so their combined actions yield music that is synchronized in time regardless of their relative positions in space. For live performance, studies have shown that if separated by more than 8.5 m, where the delay because of sound propagation in the air is around 25 ms, musicians experience difficulty maintaining synchronous performance of rhythmic patterns.[68] In large ensemble performances, where distances between players in different sections of an orchestra are often much greater, conductors are tasked with coordinating performances, including timekeeping, because there is likely to be a significant delay between one side of the stage and the other.

Considering that players of acoustic instruments in large ensembles cannot coordinate their playing when delays exceed 25 ms without the help of an external timekeeper, it is no surprise that participants in networked music performances (NMPs) also find it challenging to coordinate their playing when network delays exceed this threshold. This is increasingly relevant for musicians because the advent of high-bandwidth networks capable of supporting high-fidelity, real-time audio and video streams has resulted in a growth in interest around NMP. NMP has the potential to support remote rehearsals, online instrumental instruction, and concerts with performers distributed across multiple venues. Distributed performance spaces lack the acoustic coherence of a physical concert hall. Performers therefore often lose much perceptually salient information, such as reflections from floors and walls, that assist in coordinating timing. In addition, visual communication between musicians in a distributed ensemble is often constrained by the technical resources available to support appropriate video streams. As a result, musicians have to overcome several sources of latency. Often, such latency is a function of the performance of the underlying distribution network, whereas at other times, it arises from limitations that the design of the communication software places on the channels musicians normally use to communicate with each other during ensemble performance. In a comprehensive study of networked ensemble performance, Chew et al. found that musicians' tolerance for latency in NMPs was dependent on two key factors: the tempo of a piece, and the associated types of onset synchronization required to maintain its tempo, and the timbre of the instruments in the ensemble.[69]

In the study, which involved a distributed performance of Hindemith's Sonata op. 11, no. 4, the researchers found that latency tolerance was much greater for the piece's slow first movement than for the final movement, whose texture is characterized by sudden sharp attacks. Similarly, in a performance of Piazzolla's *Grand Tango*, latency tolerance increased when the keyboard player changed their sound selection from an accordion to a more percussive piano timbre. But Chew et al.'s main finding was that if players are experiencing different amounts of delay among their individual locations, they will opt for a shared, common delay to provide a coherent experience for their audience.[70]

Because performers in a networked musical ensemble are dependent on and attuned to multiple timescales—from the stability of audio streams at the sample level to the coordination and timing of events at the level of musical notes—the infrastructure that supports NMP must manage every aspect of the network's technical performance in order to deliver the fidelity that performers and audiences expect. Carôt and Werner have proposed a model that makes explicit the different layers and scales of time that must be concurrently managed to support stable, coordinated NMP conditions, such as those enjoyed by the players and audience of the Hindemith sonata described above.[71] The seven layers of this model, shown in figure 4.3, represent the way in which network processes are managed in order to support two-way communication at each level of functionality during NMPs. The model encompasses a wide range: from the management of timescales that coordinate the passing of bits across physical infrastructure, at the lowest level, to timescales for managing communication between performers. Although developed in the context of music performance, it serves as an example that captures the complexity of interactive, networked virtual spaces that are now the backbone of genres such as e-sports.

Although most research on NMP has focused on minimizing network delays and overcoming sources of instability, some studies have asked what might happen if networks eliminated delay altogether. Would events seem to happen faster in a networked performance than they would in real life? To test this, Chafe and colleagues had pairs of musicians seated in separate networked audio spaces clap a rhythmic pattern together.[72] Microphones were placed as close as possible to their hands to avoid any audio delay caused by the acoustics of the rooms they were seated in, and participants listened to each other's clapping over headphones. The goals of the study were to characterize the effect of time delay on performers' rhythmic accuracy and to identify the range of delay most conducive to maintaining coherence in ensemble performance. As might have been predicted, time delays greater than 14 ms slowed down performances, much as people slow or slur their speech when they hear their own delayed voice fed back to them over headphones. Somewhat surprisingly, delays of less than 11.5 ms resulted in faster performances. In order to maintain tempo stability in networked performance, the authors concluded that delays should be kept in the range of 11–14 ms.

Jitter, or variability in delay, can also affect networked performances. Few studies have considered the impact of jitter in these time-critical contexts. A current study by Cheston and colleagues is beginning to examine how jazz ensembles coordinate improvisations over networks in which varying amounts of jitter are introduced.[73] In line with the work of Chafe et al., Cheston's initial findings suggest that longer network delays cause players to

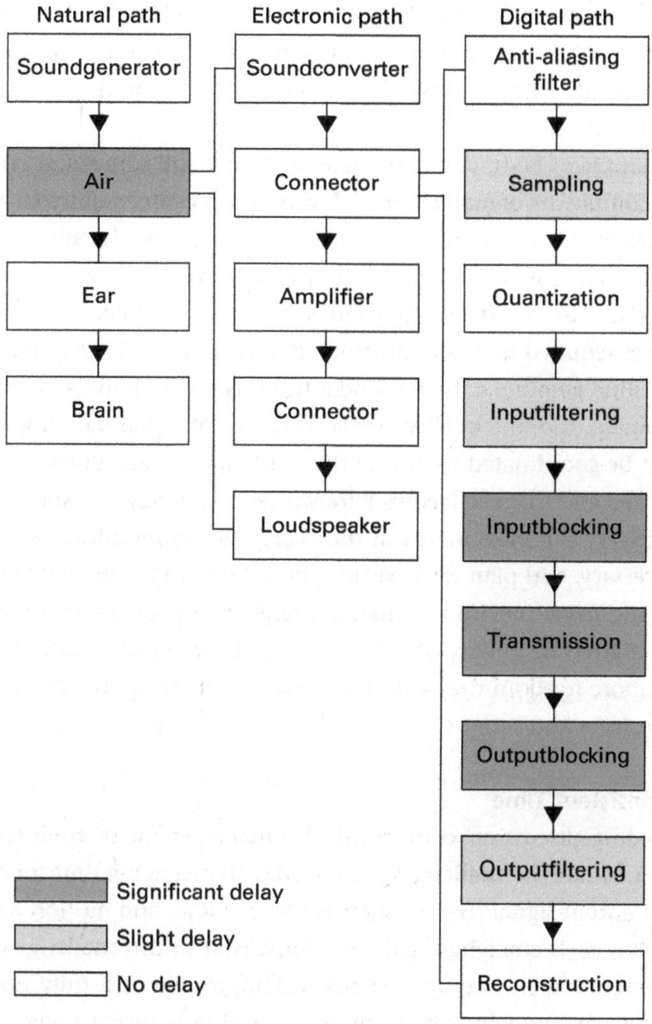

Figure 4.3
Carôt and Werner's seven-layer model of time management in networked music performances, showing the paths of communication between musicians situated in different cities. The model captures the latency present in each of three signal paths: the natural signal path, the electronic signal path, and the digital signal path. The amount of latency introduced at each stage is indicated using white for low latency, black for high latency, and gray for moderate latency. *Source:* Carôt and Werner, 2009, figure 01, Creative Commons Attribution 4.0 International License: https://creativecommons.org/licenses/by/4.0/.

slow down.[74] If the amount of delay is varied, moment to moment, performers appear to adjust their tempo to follow the arc of the jitter curve through these local tempo adjustments, eliminating the impact on the overall tempo of the performance.

The example of NMP described here, although still somewhat rare, illustrates the complexity of managing the layers of processing required to support people engaged in time-critical activities within a geographically distributed virtual environment. Increasingly, the rapidly growing field of e-sports is looking to NMP as a model for managing the layers of visual, audio, and haptic resources required to render distributed environments for a range of competitive online gaming activities where time-critical responses are required. In this domain, too, the quality and consistency of signal handling at every level must be coordinated to ensure that participants' movements are accurately tracked and that the feedback they receive is timely and stable so players can observe the movements of other players, communicate with others when necessary, and plan and execute the actions that will enable them to succeed. The more precise the management of time at the computational level, the higher the fidelity of the experience for the players and, paradoxically, the more freedom they will have to shape the temporal, expressive arc of their gestures and actions.

4.4.5 Consistent Time

The preceding discussion concerning the management of time for NMPs begins to illustrate the challenges associated with managing time for multiple input and output signal streams such as video, audio, and motion or gesture tracking. But such challenges pale in comparison to the challenge of creating a coherent experience for systems seeking to deliver a fully immersive environment, where whole-body motion tracking is increasingly common and both video and audio streams are multichannel to support 360-degree video rendering and surround-sound audio. Furthermore, many systems now include haptic feedback rendered through controllers, gloves, and/or bodysuits. Since the rendering of multimodal virtual scenes involves signal processing stages for every input and output stream, the timing and coordination of processes must be carefully managed to avoid latency or jitter, as inconsistencies can contribute to the breakdown of a sense of immersion for those using VR systems. This breakdown is believed to be a primary contributor to cybersickness, where participants experience physical symptoms of discomfort such as headaches and upset stomachs in response to digitally generated sensory stimuli.[75] In an effort to address these issues, researchers have analyzed system performance to provide guidelines for software and hardware developers so that they can deliver a consistent quality of service (QOS) across increasingly complex hardware and software platforms.[76]

Meeting QOS requirements is, as Seow points out, a balancing act between hardware, software, and design optimization.[77] The quality of the perceptual cues the virtual environment delivers and the quality of experience (QOE) the cues create will determine a user's sense of being present in that virtual reality. But, as Cummings and Bailenson point out, the perceived quality of an immersive experience is less dependent on the quality of rendering of visual and/or auditory stimuli than on the level of tracking of users' movements and support for stereoscopic rendering and wider fields of view when moving about in virtual worlds.[78] In other words, it appears that participants are more aware of how appropriately a system responds to their movement in terms of the dynamics of a visually rendered scene than to the static quality of the visual display per se. Therefore, the rendering rates of VR systems have a measurable impact on the level of presence experienced by participants in VR environments because of participants' sensitivity to the dynamics of the coupling of their movements to the rendering of the virtual world.

For this reason, objective QOS standards, such as those that measure the fidelity of visual and auditory displays, need to be balanced by standards that measure users' subjective QOE, measures that capture how reliably and appropriately systems respond to users as they move and act within immersive spaces. Van Damme, Vega, and De Turck point out that understanding requirements for assessing QOE,[79] defined as the degree of annoyance or delight experienced by the user of a service or application,[80] involves understanding the degree of user tolerance for delay and jitter in each of the modalities rendered. Using measured tolerances reported in the literature, they summarize values and ranges for three measures of signal quality—delay, jitter, and packet loss—for five modalities of presentation: audio, video, graphics, haptics, and VR. They note that this delay should be less than 150 ms for audio, less than 400 ms for video, 100–300 ms for graphics, between 3 and 60 ms for haptics (although they do not specify whether this is force feedback, tactile feedback, or vibrotactile feedback), and 15 ms for VR. The very low value for VR is notable because it suggests that tolerance for delay is greater when multiple modalities coexist in desktop or mobile HCI but not when they are designed to support a coherent immersive experience. For jitter, tolerance thresholds are more consistent across modalities, being less than 30 ms for audio, video, and graphics and between 1 and 10 ms for haptics. No value is reported for virtual reality. The lower value for haptics likely relates to the level of stability required to render surfaces that feel smooth and solid using force feedback.[81] Finally, tolerances for percentages of packet loss, assuming no dynamic error correction, are also very low for all modalities, being less than 10 percent for graphics, less than 1 percent for audio and video, and between 0.01 and 10 percent for haptics. No value is reported for VR.[82] In short, although technical specifications are well known for individual modalities required to support an acceptable QOE, those for fully immersive multimodal virtual environments are underspecified.

For fully immersive experiences, many aspects of a user's enjoyment or annoyance may depend on how well a system manages time. For example, when viewing 360-degree video, users report higher levels of discomfort when there is an abrupt change in the rate of vibrational and translational camera motion. This can be handled by using image processing to compensate for vibration and an omnidirectional wearable camera to avoid abrupt changes in viewing angle and perspective.[83]

Although latency between user actions and system responses in VR may eventually be reduced to tens of milliseconds, it will never completely go away because the transmission of signals takes time. This is particularly true for interactive systems and applications, where feedback in one or more modalities is generated in real time in response to the tracked movements of users. As Waltemate et al. note, studies that have looked at QOE for VR have paid little attention to the potential impact of latency on perception and motor performance and how these interact, particularly for virtual environments employing full-body avatars.[84] Prior work has determined that high latencies influence factors such as a user's sense of agency, sense of ownership of body avatars, perceived temporal coherence of the virtual scene, and motor performance. As Waltemate et al. note, users will continue to perceive agency with respect to images of their own moving hand delayed up to values of about 150 ms. Above 150 ms, their sense of agency gradually decreases until, for values of 500 ms and beyond, they no longer associate their hand with the moving image[85] (see also the discussion of perceived agency in section 4.3.3). In contrast, if the task is to associate a button press with a subsequent flashing stimulus, the user will continue to retain a sense of agency for as long as 500 ms.[86] It has been proposed that this disparity may arise from a much narrower temporal window of perceived agency when it concerns ownership of one's body.[87]

In terms of motor performance, Jörg, Normoyle, and Safonova found that delays of more than 150 ms impacted players' abilities to control characters in a computer game.[88] Similarly, for full-body motion seen through a virtual mirror, delays of 200 ms in the mirrored image have been shown to alter users' gait.[89] In their work, Waltemate and colleagues used a full-body avatar and a virtual mirror paradigm to examine the impact of delays ranging from 45 to 300 ms on their four criteria—motor performance, sense of agency, sense of body ownership, and perception of simultaneity—as well as possible interactions between these criteria. In their study, participants were asked to mimic the actions of an animated image while observing their own actions reflected, after a predetermined delay, in the actions of a projected avatar (i.e., a virtual mirror). They found that both motor performance and perception of stimulus simultaneity were negatively impacted by latencies above 75 ms, whereas sense of agency and body ownership began to decline only for values above 125 ms and never completely broke down, even for the highest latencies tested.[90] But their most compelling finding was that users did not appear to perceive response delays directly. Instead, participants acknowledged that when delays were longer, they made more errors performing motor tasks. What they detected therefore was an increase in performance errors rather than an increase in delay. In other words, participants' perceptions of the quality of an interaction impacted by delay in a virtual environment might depend more on their perception of their own performance than on latency arising from the system they are using.

In addition to the impact of perceived latency on task performance, researchers have also examined how the presence of multimodal stimuli impacts response latency for people interacting in fully immersive virtual environments. Marucci and colleagues examined the impact of increased perceptual load (defined as the amount of information associated with processing sensory stimuli) on the latency of neural responses as a function of display modality.[91] Participants drove a virtual car around a track under a variety of conditions, some of which involved hitting see-through targets that appeared on the track. In some situations, the task was made more challenging by the simulation of rain, thunder, and low visibility. Effects were simulated using combinations of vibrotactile and audio stimuli presented alongside video feedback in either high-load (poor weather) or low-load (good weather) conditions. The finding most relevant to the current discussion was that electroencephalogram recordings showed a significant reduction in neural response latency when auditory feedback was present, either combined with visual and tactile stimuli or with visual feedback alone. This would suggest that reactions to stimuli in virtual environments are faster when auditory feedback is included in VR simulations.

Finally, let us consider the role users' tracked movements play in coordinating the temporal arc of their interactions with virtual environments. The increasing complexity of hardware and software places considerable demands on rendering fully immersive multimodal scenes, and as we shall see in the next chapter, such fully featured renderings may not even be required to situate users in simulated spaces. Recognizing this, developers have begun to take advantage of the fact that we can look in only one direction at a given moment and that we can only feel things that are touching our bodies. This means that if the movements of the head and body can be accurately tracked, only those stimuli relevant to a user's current gaze, pose, and actions need to be rendered. This is not true for audio, as we are always alert to sounds in our environment, even when we are facing away from their source or when our eyes are closed.

Because of this, techniques have been developed for rendering stimuli on demand based on attributes such as direction of gaze and hand pose. Haptic devices have always rendered effects on demand, incorporating sensors that track as many of the hands' degrees of freedom as they can and rendering vibrations, forces, and sometimes torques to simulate contact with a virtual object. In the simplest case, that of haptically rendering the boundary of a cube using a robotic force-feedback device (such as a Phantom), forces are applied to the user's hand only when the virtual tip of the handheld tool penetrates the surface of the cube; when the tool tip is outside the cube, the user feels no force opposing their movement.[92]

Increasingly, hardware manufacturers are also including eye tracking to facilitate foveated rendering, where the parts of the visual scene lying within a user's central field of view are rendered in high definition and those in peripheral vision are rendered with less contrast, reflecting the differential sensitivity of the eye.[93] Combined with increasingly accurate motion tracking, this allows for highly dynamic and seamless generation of complex 360-degree visual scenes.

By rendering visual and haptic stimuli on demand based on how a user is moving and acting in a virtual environment, hardware and software designers can reduce the complexity of rendering for virtual environments to maximize performance, thereby avoiding some of the temporal factors that users find distracting and that can cause cybersickness. In a way, designers of virtual worlds are simply building upon the fundamental role that our bodily movements play in structuring the information we pick up from our environments. In so doing, they enable users of virtual reality to bring forth a virtual experience that is increasingly aligned with the ways they negotiate their physical world.

4.5 Designing Time: Two Approaches

Although the foregoing discussion has focused on managing time to deliver coordinated, coherent interaction experiences, there is a parallel world of design considerations relating to the temporal nature of interaction design itself. In the following discussion, I will draw on two contrasting approaches to demonstrate how we understand the structure of information presented in time and how our actions can shape the temporal arc of our experiences with interactive systems. Our interactions with our environments unfold in space and time. Indeed, such actions bring forth the time of our lived experience, whether we are doing dishes or answering email. Although measures of time and accuracy were central to the work of Fitts and others in the early days of HCI,[94] the current focus on artifact/device and application design now tends to distract us from the role that time plays in shaping how we interact with the computationally enabled things around us.

If, as Gallagher and Zahavi propose, action brings forth time, then interaction with computational objects brings forth time too.[95] I propose here that the time brought forth through interaction with the environment, including computationally enabled devices, is not an instant but an arc, where actions are planned in the context of what has just happened, with awareness of the present, and in anticipation of the future. Although there is little theoretical structure to account for this kind of arc in interaction design, there are examples that can serve as case studies to guide the development of such a theoretical framework. These include sonic interaction design (SID) and the analysis and design of tau guides for controlling movement in dynamic environments, including music performance and virtual sports.

4.5.1 Guiding Movement Through Sound: Sonic Interaction Design

SID emphasizes the temporal nature of perception–action coupling. The unfolding nature of the sounds things make when we act upon them carries information that we can use to plan and execute action in the form of structural (material) and transformational (action-dependent) invariant properties. These sounds also reinforce temporal aspects of causality in interaction. This temporality is often lost in traditional HCI. SID demonstrates that our understanding of action, conveyed through sound, draws on more than an instantaneous moment of time. We can infer, from what has just happened and what we are hearing now, what is likely to happen next.

Although we do not often stop to consider the sounds our actions make, there are many cases in which sound guides us in controlling our movements. Think, for example, of the action of filling a glass with water: As the glass fills, the pitch of the sound gets higher. Other actions have sonic signatures that are informative of their progress, such as the changing sound an eggshell makes as you tap it to break it. When coupled with action, sound can therefore be a means to monitor and control movements. Moreover, such control is often based on temporally unfolding events whose invariant properties are not easy to visualize, such as changes in the sounds that objects make in response to increases in the force or velocity of our actions upon them.

In the early 1980s, William Warren and Robert Verbrugge began to study our ability to determine attributes of sounds based on the actions that might have given rise to them.[96] In particular, they were looking for invariant attributes of the sounds of bottles that were dropped and then, either bounced or broke into pieces, which then bounced independently. In either case, the timing, amplitude, and spectral content of a sequence of bouncing events, illustrated in figure 4.4, is a function of the material properties of both the object, including mass and shape, and the force with which it hit the ground.

Warren and Verbrugge's hypothesis was that the mechanical events of breaking and bouncing are specified by single rather than multiple damped sound events that are quasiperiodic (i.e., sounds that are associated with some resonance but not necessarily with a definite pitch). Further, they hypothesized that breaking events begin with initial noise bursts. Warren and Verbrugge suggested that people would distinguish breaking from bouncing based on invariant attributes of these differing temporal and spectral signatures. To test their hypothesis, they independently manipulated

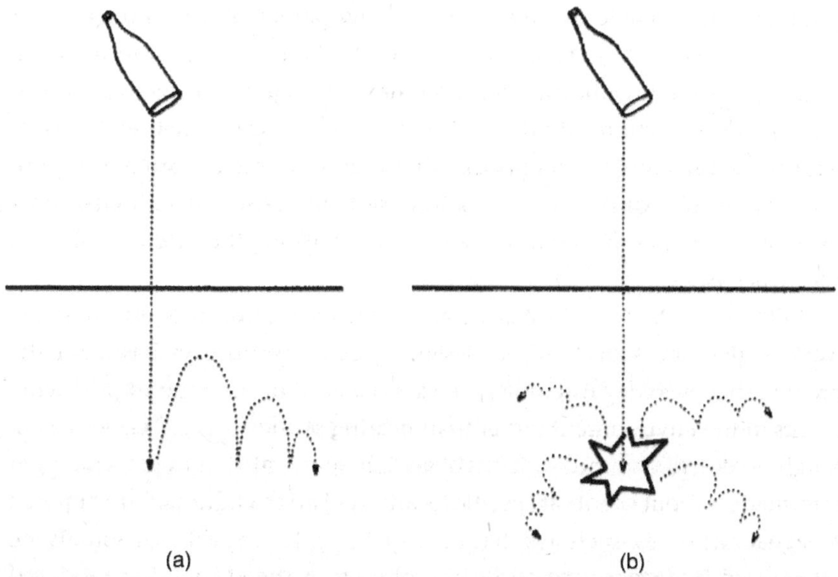

Figure 4.4
Acoustic patterns recorded for breaking and bouncing events for two glass bottles. (a) Bouncing results in a single series of events gradually decaying in amplitude. (b) Breaking results in a series of four events, each of which has its own decay trajectory. *Source:* Warren and Verbrugge, 1984, figure 1. Reproduced with permission from W. Warren and R. R. Verbrugge.

both the temporal unfolding and spectral signature of breaking and bouncing events by inserting breaking sounds into temporal and damping patterns of whole bottle bounces and inserting the sound of unbroken bottles into the temporal and damping pattern of a bottle falling and breaking. Participants compared these two cases with the unaltered recordings of breaking and bouncing bottles. From this, the team determined that the consistency of spectral components across a sequence of damped bounces was more important than their individual spectra and that the noise present at the start of breaking events was not a significant determinant in the categorization of those events.

Warren and Verbrugge thus concluded that breaking and bouncing were distinguished by auditory perception of higher-order temporal invariants associated with the physics of bouncing and breaking. Such invariants, they argued, could be further related to two kinds of information: structural invariants that enable the recognition of properties of a sounding object (e.g., the shape and material composition of a bottle) and transformational invariants providing information about how the object is being transformed through its interaction with the environment (in this case, whether the bottle bounces or breaks when dropped). Warren and Verbrugge's study is significant not just for pointing to the auditory system's sensitivity to higher-order invariant attributes of sounds but also for establishing the field of ecological psychoacoustics.[97]

Building on the work of Warren and Verbrugge, Gaver proposed a framework to describe sounds when designing auditory displays based on the premise that everyday listening is an experience of hearing sound-producing events in the environment rather than hearing sounds per se.[98] Gaver's work sought to describe sounds in terms of audible source attributes that conveyed information about events at specific locations in an environment (as opposed to signal attributes, such as pitch or frequency). He argued that sounds are categorized by listeners based on the physics of the events that produced them. Bottles that are tapped, for example, sound different from bottles that are dropped. Such events might involve solids, gases, or liquids. Within each of those categories, Gaver defined events by the simple interactions that caused them to sound: striking a drum (solid), boiling a pot of water (liquid), or bursting a balloon (gas). Gaber's framework also described more complex events in terms of temporal patterning arising from combinations of simpler events or from the interaction of objects with different material properties. In short, he proposed that any framework for sound design should consider the situated nature of everyday listening by accounting not for the sounds people hear but instead for the information those sounds convey about the behavior of objects in the environment.

The work of Gaver and others on ecological acoustics greatly impacted the design of sound for auditory displays and laid the groundwork for the related field of SID. The basis for SID is that sounds not only provide useful information to guide action but also may bear the temporal signatures of the bodily actions that gave rise to them. In the case of music performance, for example, the temporal unfolding of a note arising from a bowed string is directly shaped by the actions of the performer's hands. As Alexander Jensenius states, such action–sound couplings arise in cases where there is both a mechanical and an acoustical coupling between the musician's action and the sound of their instrument.[99] They represent one instance in a broader class of action–sound relationships that also includes artificially developed mappings between user actions and synthesized sounds. An acoustic piano, for example, represents a direct action–sound coupling by the mechanism that transfers energy from a player's body to the strings of the instrument. In comparison, the finger movements of someone playing an electronic piano are sensed and then mapped to parametric control of a sound synthesis algorithm, resulting in an action–sound relationship that is determined by the digital musical instrument's designer.

SID, as an emerging practice, builds on the second of these approaches, seeking to develop action–sound relationships that are informed by the ecological approach to perception and action. The goal is that the actions of a user should reveal structural and transformational invariants reflecting the modeled behavior of the simulated physical object with which a user is interacting. For example, in a 2005 study, Rath and Rocchesso examined the role that continuous sonic feedback could play in helping users control a virtual ball rolling along a beam, as illustrated in figure 4.5.[100] They point out that any effort to achieve ecologically grounded human–machine interaction needs to take into account the continuous and embodied nature of everyday tasks, such as steering, modulating reaction forces when wielding objects, and maintaining equilibrium. Such tasks require continuous feedback, which we typically receive through some combination of our visual, proprioceptive, and auditory sense modalities.

Rath and Rocchesso hypothesized that by mapping an abstract task that requires a user to control the direction and velocity of a continuous parameter onto an acoustic model of a rolling ball, they could take advantage of our ability to estimate velocity based on microscopic and macroscopic changes in rolling sounds. Such might be the case for an interface for balancing a load on a tilting platform. In other words, by drawing on our ability to decouple structural (material) and transformational invariants for the ball–beam system, the designers provided users with a way to uncover action-relevant acoustic information present in the task environment.

Within the field of HCI, Visell et al. suggested that the temporal nature of auditory and/or vibrotactile feedback (as in the case of the rolling ball) can be a mechanism for supporting the continuous negotiation of intention between a user and a system in cases where there is a bidirectional flow of information during interaction.[101] At the heart of any interactive system is a human–computer interface that interprets user actions in order to carry out

Figure 4.5

Continuous sound feedback from a virtual rolling ball: Synthesized sound is controlled by the tracked position of a glass marbel on a balance beam that is tilted by the user. *Source:* Rath and Rocchesso, 2005. Used with permission of the Institute of Electrical Electronics Engineers, Inc. Permission conveyed through Copyright Clearance Center, Inc.

user intentions. Given that the goal of an interface is to communicate the state of hidden variables that are negotiated between the user and the system using the available modalities of interaction, there is always some level of ambiguity in these transactions. From this perspective, Visell et al. argue, any interaction can be considered a continuous negotiation. This negotiated approach differs from most current thinking, which centers on discrete action accept/reject cycles. These cycles typically depend on exchanges that take the form of "select, confirm, perform" sequences. When interaction is continuous, information can flow in both directions within a temporal sequence that is less structured, where decisions can be made, over time, as new evidence is presented.

Visell et al.'s approach to interaction as continuous negotiation is of interest here because it begins to approach an understanding of interaction as a bodily activity that unfolds across a continuous temporal arc rather than being encapsulated in a sequence of discrete actions. Hallnäs and Redström go as far as to suggest that interaction design involves developing an understanding of what they term "time gestalt," building blocks for HCI that combine space structures (manipulation of buttons, icons, etc.) and the time structures of computation (as described in section 4.4).[102] Such time gestalts are not properties of computational artifacts per se but rather are emergent patterns within given interactions as they unfold in time. Hallnäs and Redström suggest that, in grouping discrete actions into temporal sequences defined by task goals, time gestalts resemble musical phrases, where individual note-producing events are bound together by a temporal arc that builds on what has just happened and projects toward a future goal.

Rethinking interaction design as a process of continuous engagement between user and artifact rather than as a series of discrete interaction events requires a very different theoretical starting point than the kinds of reaction time studies that are typically associated with HCI. In their work concerning the design of continuous auditory interaction, Rocchesso et al. argue that enactive and embodied approaches to HCI offer a fruitful starting point because, as noted in chapter 2, these approaches are based on the premise that action and perception are coupled: That action upon the world guides perception, which, in turn, guides future action.[103] A closed loop exists between a user and their computational device, where control is accomplished through a continuous sensorimotor coupling. Although it is, of course, possible to generate continuous visual, auditory, and haptic feedback in response to any user action, the challenge in doing so is that users can be fooled into inferring causality in the presence of any extraneous, simultaneous system event. In enactive interfaces such as PebbleBox (see chapter 6, section 6.6.6), the manipulation of physical pebbles is coupled to the manipulation of sound grains, and experienced causality tends to more closely resemble physical causality.[104] When one takes advantage of basic interaction gestalts such as pushing, pulling, shaking, and so on, arbitrary bindings between user actions and system responses are less likely to arise because such actions come with their own temporal profiles of energy exertion, sensorimotor coupling, and temporal resolution carried over from the real-world physical counterparts of such actions.

In proposing ways in which designers can build upon such sensorimotor couplings in the design of feedback for continuous interaction, Visell et al.

offer guidelines concerning the kinds of information that can be communicated to users through the temporal patterning of auditory and vibrotactile feedback during interaction with a series of simple mechanical systems over time.[105] Their hypothesis is that people are adept at discovering and learning to use patterns of tactile and acoustic information in order to perform a wide variety of perceptual and motor tasks. Like Rocchesso et al., they suggest that sonic interactions that are analogous to physical, mechanical systems can become models for new ways of interacting with digital information through movement and sound.

As the bottle breaking/bouncing example demonstrated, people are adept at inferring many attributes of the physics of an event from its sound alone. But there are important differences between listening to an event resulting from someone else's actions and listening to an event arising from one's own movement. In the first case, one is passive and has no influence on the sound. In the second case, one's action sounds will bear the signature of one's movements. The implication for interaction design is that the temporal pattern of action sounds must be generated in real time in response to how a user is moving. From the user's perspective, the sounds arising from their motor actions must be predicted by the designer so that users can separate changes in feedback intrinsic to the system from those caused by their movements.

The example Visell et al. provide is that of Shoogle, an application for a mobile phone that uses the metaphor of a collection of balls inside a box to manage incoming text messages.[106] Movements of the phone are transferred to a model of the mechanical system and applied to virtual balls inside a box, determining the velocities with which balls collide with each other and with the walls of the box. Collisions are rendered as both vibrations and sounds. Sounds are rendered using granular synthesis, where samples (or grains) can be swapped, to produce the effect of balls of different materials (wood, glass, etc.). Because the model reacts only to movement, it produces sounds and vibrations only in response to specific user actions, such as shaking. In the context of the message management application, users must shake the phone to determine whether new messages have arrived (as distinct from the more usual notification model that relies on passive haptic presentation through vibration, to attract the user's attention). Moreover, because the feedback to the user arises from the action-driven behavior of the mechanics of the model as it is shaken, mapping to auditory and/or vibrotactile modalities is consistent because the temporal pattern of feedback is derived from the physical behavior of the same underlying model: balls inside a box.

In addition to simulating behaviors that provide continuous feedback based on action-driven disturbances of a mechanical system, some researchers have examined the use of auditory feedback to provide continuous information concerning the tau, or time to closure, of a temporal gap. In HCI, such feedback is useful when monitoring the time required for a process to complete. In chapter 2, I noted that many animals make use of information in their environments to predict the time to closure of spatiotemporal gaps as they avoid obstacles or intercept prey. Auditory displays that build on tau theory allow people to unobtrusively monitor background computational processes, such as the progress of a large file transfer.[107] Whereas such process monitoring primarily involves passive listening, continuous auditory cues can also be used in the regulation of action, such as those informing a user of how much more time and effort it will take to reach a target system state. When asked to estimate the time to completion of a pouring task, for example, people are remarkably accurate at predicting time to completion based on acoustic properties, including changes in resonance as the vessel is filled.[108] If combined with a tilt interaction, such a sensorimotor coupling could inform a user of the remaining time/effort required to actively move data from one location to another. Visell et al. conclude that, in the same way that the design of interactions with data objects in graphical user interfaces is informed by representations of trash cans and file folders, the design of artifacts and interactions that provide continuous feedback in the form of sound and/or vibration could use metaphors based on pouring, rolling, and so on.

In short, if the goal of an interaction is to be responsive to the changing nature of the information being manipulated, then continuous, perceptually coupled feedback derived from a common model will best serve users in shaping and guiding the temporal arc of their movements. When one does so, interactive sonifications will, as Gallagher and Zahavi propose, bring forth the time of an interactive experience.[109]

4.5.2 Interacting in Time

The temporal arc of an action trajectory is guided by contextually invariant information, whether it is present in the environment as extrinsic sensory stimuli or generated internally in response to known constraints such as musical scores or game rules. When our environment contains interactive technical artifacts such as phones, laptops, and digital musical instruments, our approach to planning and executing action does not change: Our movements are still directed toward our action goals and shaped by our prior actions and the opportunities and constraints perceivable in the current moment. As we move away from the event-driven interfaces that characterized first- and second-wave HCI and move toward embodied, situated interaction design, we face the challenge of determining how to support users in discovering, through their actions, what new affordances our artifacts and devices make available.

As I have noted, interactions with the environment are situated in time just as much as in space. It is because we are continually engaged with our environments in a temporally extended way, through our bodily movements, that we can perceive the opportunities for action afforded by our surroundings. When our environment contains computational artifacts, our interactions with these devices are planned and executed in response to the action-relevant information they convey as auditory, visual, or haptic cues. In chapter 3, I noted that moving one's hand to reach a target could be described in terms of closing a perceptual motor gap in time and space, and that, as David Lee proposed in developing his theory of tau-guided movement, the movement trajectories required to close this gap are guided by information available within the environment.[110]

Using the nuanced account of time consciousness set forth by Gallagher and Zahavi, in which our experience of the present is brought forth by our active engagement with our environment,[111] I argue that Lee's tau theory can provide a means for designers to determine the kinds of information that are most relevant for users when they are planning and executing actions directed toward the artifacts and systems we build. As noted, Gallagher and Zahavi propose that our ability to plan and execute an action in the present moment depends not just on information immediately available in our surroundings but also on our awareness of an extended temporal window where current actions are shaped by the immediate past, and future actions can be planned in the context of the present.

If we assume that users plan and execute actions that involve our artifacts in the same way that they plan to pick up cups or open doors, they should be able to determine what just happened, they should understand what is happening now, and they should be able to predict what will happen next in response to what they are about to do. Only by reorienting our thinking toward the temporal arc of interaction design in this way can we move beyond a world where we interact with our digital stuff and toward a world where our digital artifacts do not operate in parallel worlds but are situated in our environment and contribute to bringing forth the time of our lived experience.

In proposing that our experience of time is an outcome of an active engagement with the present rather than a passive observation of time passing, Gallagher and Zahavi provide a model for the planning and execution of action that directly engages the body in bringing forth time.[112] In practical terms, the temporal arc of a given interaction with the environment is determined by the goal of the task with which it is associated. For example, if your current goal is to pick up your coffee cup, the temporal arc of your action will be determined by the distance between your hand and the cup and the speed at which you move. In Lee's theory, tau is defined as the ratio between the magnitude of the gap and the current rate at which it is being closed.[113] Formally expressed, "The control of movement involves keeping the tau of movement (τ X(t)) coupled to a guide tau (τ Y(t)) at a constant ratio (k)."[114]

A tau guide is therefore an invariant source of information in the environment that is used in the prospective control of movement. Such invariants are task specific—for example, the optical tau guide used by gannets when diving for fish (see chapter 2). Tau guides represent a time-dependent coupling between the actions of an organism and a source of information that is either extrinsic (present in the environment) or intrinsic (generated within the neural system). In the action of reaching for a cup, movement is coupled to an extrinsic tau guide: the time to closure of the gap between your hand and the cup. In the case of playing a solo melody on a musical instrument, movements are coupled to an intrinsic tau guide: The actions required to close the gap between the current body/instrument state and the goal state, in order to play the right note at the right time, are coupled to a tau guide defined by the rhythmic and melodic structure of the music.

Whether the guide is extrinsic or intrinsic, it is possible to measure the degree of control someone has over their movements by determining the strength of coupling between the tau of their movement and the tau of the guide they are drawing upon using a regression analysis. The slope of this regression fit (variable *k* in the formula provided) provides a means to assess how precisely a given movement is being controlled. As people become more skilled at performing a given action, they can more precisely control where this peak will be (i.e., whether an action should be more energetic at its start, in the middle, or at its end). With practice, people's chosen movement patterns stabilize, and the value of *k* becomes remarkably consistent. By observing the degree of stability of *k*, it is therefore possible to observe people learning to control movement as they learn to couple their actions to a task-specific tau guide.

Since the early 1990s, when David Lee first proposed tau theory, this technique has been used to measure movement control across an enormous range of species and activities.[115] For example, both bats and hummingbirds have been shown to use extrinsic sensory tau, in the form of auditory and visual information, respectively, to control their approach and landing behaviors,[116] and long jumpers have been shown to use tau information in optic flow to regulate stride lengths to ensure contact with the takeoff board.[117] In infants, too, the ability to learn to catch a swinging toy has been linked to an increasing ability to couple the tau of their movement to invariant properties of the global sensory array.[118] In addition to evidence for extrinsic tau guides, the presence of intrinsic tau guides has been demonstrated in many cases, including golfers controlling their putting and musicians controlling movements to produce notes.[119]

By observing the degree to which movements are coupled to either intrinsic or extrinsic tau guides, Lee's tau theory provides us with a means to observe people's strategies as they plan and execute context-dependent, goal-directed actions. Thus, I propose that tau theory offers a mechanism by which we can observe the planning and control of movements directed toward the artifacts we design. This proposal suggests that in typing and texting and in clicking through menus, websites, and social media feeds, the actions of our fingers are united by the movement trajectories of our hands and arms, much as note-level events are bound together by the movements of musicians' bodies. You may know the feeling that, with regularly used applications, your hands and fingers follow a familiar path. And yet, most usability studies focus on users' abilities to carry out individual actions, such as navigating between websites or finding items on a menu. An exception exists in the literature relating to interruption reasoning systems, where sequences of task-level events are monitored in order to determine the optimal time to interrupt a user with a system notification. For example, Iqbal and Bailey proposed a model for determining the cost of interruption for tasks across a number of domains, including video editing, route planning, and document editing.[120] Their findings, based on observations of users' actions, demonstrated that participants took longer to resume work if interrupted within the temporal arc of a goal-directed task or subtask rather than at its boundary. Such findings suggest that, even for relatively complex tasks, users plan and execute movement sequences with an overarching task goal in mind.

Even when studies consider the temporal arc of interactions, they often do not consider the role of context-dependent cues in shaping the arc of movement sequences. But this is beginning to change. As gaming platforms begin to track more complex body movements, particularly platforms that support fully immersive virtual environments, greater emphasis is being placed on closing the loop between the perceptual information available to a player and how the player moves in response. In a study situated in an immersive VR rugby game, for example, researchers asked fourteen novice rugby players to judge the passibility affordance of a gap between two approaching defenders.[121] The researchers proposed that optical information available to a player as they run toward the defensive line is important in their decision-making process and that this information is key to shaping their strategy. In this case, the optical information players use is determined by geometric properties of the environment that specify the pass-through-ability of a gap.

Moreover, a VR rugby player's ability to perceive the possibility of a given action is also determined by their ability to quickly discover the relationship between an emergent property of the environment and a measurable dimension of their body based on invariant properties of the optic array. Here, the variables modeled are the time/distance to the closure of two gaps: one between the two defenders, the other between the attacking player and the line of defense. More than this, Watson et al. suggest that players make judgments about future actions based on how they are allowed to behave in a given environment.[122] The rules of rugby, for example, state that players can only pass the ball backward. The information that guides players in controlling their movements to close motion gaps can thus have multiple sources, some extrinsic and others intrinsic (based on the player's knowledge of the rules). The results indicated that a tau-based informational model could account for 82 percent of the variance in their data, suggesting that judgments of gap passibility in this dynamic scenario were, for the attacking player, largely defined by judgments of the time to closure of these combined gaps.

Finally, researchers have demonstrated that, with practice, people become more attuned to task-relevant tau guides and make more accurate decisions.

In their VR-based study of soccer goalies estimating the direction of incoming goals, Dessing and Craig found that expert players were far more accurate in responding to the amount of spin on curve balls than novices and that they waited longer before moving, so they could obtain as much information from available optical cues as possible.[123] In short, the more actively users can participate in acquiring perceptually relevant information when using a given technology, the more accurate they are likely to be in estimating the opportunities for action that technology affords and hence the more successful they will be in planning and executing the movements that will result in achieving their action goals. As Craig states, "It is by preserving the perception/action loop that is normally experienced when we navigate through a real environment . . . that the level of behavioral realism, or functional fidelity, is significantly increased."[124]

As more application domains move toward a landscape of fully immersive, interactive virtual reality, there is an increasing need for designers to attend to this perception–action loop, not just in displaying instantaneous cues in response to current movements but also in supporting the whole temporal arc that represents the planning and execution of movements directed toward a particular action goal. Thus, if the more nuanced account of time consciousness set forth by Gallagher and Zahavi provides a conceptual framing for interaction design in the third wave, tau theory can provide a means for designers to reorient our thinking toward the temporal arc of interaction design and determine the kinds of information that are most relevant for users when planning and executing actions directed toward the artifacts and systems we build.[125]

4.6 Conclusion

Creating a bridge between the time of our machines and the lived time of our bodies is complicated by the fact that different domains of interaction have different expectations in relation to time. In HCI, time unfolds at a pace similar to that of a conversation, and our expectation is that computers will respond no more quickly or slowly than a person with whom we are talking. With interactions that rely on real-time continuous control, such as music performance or immersive gaming, our expectation is that computational resources will have no discernible impact on the temporal coordination of individual or group movement. When we step into fully immersive virtual environments, we expect those worlds to look, sound, and feel real and to react to our actions just as real-world objects and people do.

Although computers have gotten faster, humans have not. We still interact in the time of Middle World. This represents a challenge for bridging multiple scales of time, but it also represents an opportunity. If we can accurately and continuously track how people are moving, we can simulate, on demand, the visual, auditory, and tactile experiences required to support their action goals. In other words, we can marshal time that runs faster than the speed of Middle World to deliver enjoyable and rewarding experiences for people, no matter how computationally demanding a given application might be behind the scenes. To do this, we need to rethink our approach to HCI. We need to place less emphasis on the design of artifacts and devices and more emphasis on the temporal nature of the sensorimotor engagement that enables us to bring forth the experiences that devices are designed to facilitate.

The proposal laid out here is that interaction with computational artifacts—like interaction with any other object in our environment—brings forth time and that the time brought forth is an arc, where actions are planned in the context of what has just happened, in awareness of the present, and in anticipation of what is to come. Since there is little theoretical structure to account for time in interaction in this way, I have proposed that we can adopt the theoretical framing of tau-guided movement, which has been shown to guide the design of perceptual cues in simulating dynamic environments such as virtual rugby and soccer games. Rather than providing a one-size-fits-all approach, tau theory is sensitive both to the context in which movements are executed (extrinsic and intrinsic cues) and to the bodily capabilities that a user brings to that context. In many ways, tau theory encapsulates the values of third-wave HCI because it provides a practical methodology for delivering ecologically grounded, context-specific perceptual information to support users in planning and executing action. For designers of interactive systems, an open question concerns the extent to which this more nuanced account of time consciousness, brought forth through our active engagement with the environment, will shape and be shaped by the interactive experiences and technical artifacts we create. Only time will tell!

5 Space

5.1 Introduction

Imagine a familiar space. What comes to mind? A room in your house, or a familiar pathway through your favorite park? If I asked you to describe the space to me, what would you say about it? I have been asking students these questions for many years, and overwhelmingly their first phrases describe colors, light, and visual forms. Sometimes, they describe how their spaces sound, but very few talk about the physical experience of their body moving and acting in their space. And yet, as we noted in chapter 2 when describing the experience of babies learning to walk, our relationship with the spaces in which we live is grounded in our bodily encounters with our environment. As toddlers learn to stand and walk, they must relearn how to move around their world with a higher center of mass and a new vantage point. For the toddler, what has changed is not the geometry of their lifeworld but rather their relationship to that world and their bodily capabilities to act upon their surroundings. Being taller, they can reach higher; being upright, they can see farther in the direction they are moving; and being bipedal, they can eventually move more quickly. In other words, we encounter spaces not as static beings but rather as active inhabitants with things to do. Through our actions, we shape our environments. But our lived environments also constrain our actions. In real life, we cannot pass through walls, we cannot all climb mountains, and roads determine where we can drive most cars.

Today, we are learning how to inhabit entirely new spaces mediated by the technologies we are creating—virtual worlds in which the laws of physics and even of our bodily forms and skills are mutable. In such worlds, we are often asked to suspend disbelief, to forget what it is like to move through space, and, like the toddler learning to walk, to learn how to negotiate the world anew. Whereas toddlers encounter the same world from a different postural perspective—walking versus crawling—the challenge facing designers and users of augmented and virtual environments is that the world itself can change. And because simulated environments are situated within the confines of real spaces that may or may not align with such worlds, their builders must take care in constructing even the most basic virtual sensorimotor relationships to ensure safe and rewarding experiences for users.

When most of us think of the spaces we live in, we think of structures we can go inside (houses, schools, stores) or perhaps the way these structures are organized (streets, neighborhoods, towns, cities). Our constructed environments are dominated by the laws of Euclidian geometry, with angles and vertices that are organized along familiar axes and oriented with respect to gravity and the ground. Such spaces have fronts, backs, and sides, they have ups and downs, and, most importantly, they divide our world into inside and outside and are often subdivided into "rooms," spaces within spaces, which have historically facilitated the compartmentalizing of activities—education in classrooms, business in boardrooms, cooking in kitchens, and so on.

As our ways of living and working continue to converge, mediated by shared, networked experiences, notions of space and place are collapsing. Our living spaces are often also our working and recreational spaces, taking on multiple identities and functions. But our individual spaces of habitation are merged into one online workplace with a distributed physical embodiment. If, as Winston Churchill claimed, "we shape our buildings and, afterward our buildings shape us,"[1] then it seems we should ask ourselves what kind of buildings, what kind of lived spaces, are we now shaping, and how, in turn, are those spaces shaping us?

In the discussion that follows, I will argue that developing compelling experiences in blended and virtual worlds means we must first understand the multifaceted nature of our encounters with the spaces of our lived experience. The first part of this chapter is therefore concerned with how we perceive and act in lived space. Because we inhabit our world as embodied beings, our experience of space is brought forth by our bodies as they move through the world. Space and time are therefore inextricably bound together through our bodily movements. It follows that the design of any virtual space is fundamentally the design of encounters with visual, auditory, and haptic sensory imagery rendered in time. Thus, in the latter part of this chapter, I will also draw attention to challenges faced by designers in building augmented and virtual spaces to be encountered in time and in managing the expectations of moving bodies within the worlds they are designing.

5.2 Encountering Spaces

In their thought-provoking book *Spaces Speak, Are You Listening?*, Barry Blesser and Linda-Ruth Salter draw attention to the fact that our auditory and visual encounters with spaces are fundamentally different.[2] Whereas objects become apparent to us visually when light is reflected from their surfaces, they only become apparent to our ears if they are caused to make sounds. Furthermore, when we illuminate a space with a light, we become aware of its geometry and its surfaces, but we do not usually pay attention to the light source itself. When we play a sound in a space, however, our attention is drawn to the sound source—a voice, a musical instrument, a radio—and we are only secondarily aware of the way the space responds to sound and shapes what we hear. Architects, too, tend to focus on the visual design of spaces, forgetting that every building has an aural architecture as well.[3]

Perhaps the question to be asked is not whether architects should pay more attention to the aural architecture of buildings but rather how spaces, such as buildings, serve to support the activities of those who inhabit them in the first place. The fact that the spaces we construct persist in time as monuments to the activities we perform in them—church services, concerts, manufacturing—draws attention away from the bodily activities for which they were designed. Concert venues, for example, have small, elevated stages for performers and large areas clear of visual obstruction where audiences sit, usually facing the stage. Factories, however, are designed around workflows, with individual workstations arrayed along benches or close to conveyor belts to facilitate each person's activities as they contribute to the production process. Viewed in this way, the spaces we build are primarily spaces for action.

As we saw in chapter 2, James Gibson proposed that our practical and formal understanding of space should be founded not on abstract constructs of geometry but rather on the layout of the surfaces within the environment that support an individual's actions.[4] For workers in factories, this might mean ensuring that the tools they need are within reach and that the actions they need to perform are within the capabilities of their bodies—that is, that the spatial layout of the environment is organized to support a particular individual's action goals.

If we can conceive of space in terms of its action-relevant properties, such as the layout and structure of surfaces, the feel or sound or look of a space is then relevant only as it pertains to what an individual wants to accomplish in that space. If you want to listen to a string quartet, then an aural architecture that supports clear and resonant music will be more important than the color of the carpet. But if you want to film in that same space, the behavior of surfaces in the presence of light will be of great concern. From the perspective of the performers who are engaging with the hall by making music, the relevant properties of the space will be different again—they may prefer an uncluttered wooden stage that reflects the sound of their instruments back to them so that they can achieve a more blended sound.

Indeed, spaces can be heard as well as seen. We hear a lot about large spaces (the size of rooms), small spaces (the size of containers as we pour water into them), open spaces, and so on. But this is not something we typically think about. Imagine someone brought you into a strange house, blindfolded, and let you into a room. What could you learn by singing or speaking in that space? Although you are rarely called upon to do this, I suspect you could describe many aspects of the space far more accurately than you might think. As you speak, for example, you might notice how surfaces such as windows reflect your voice back to you, whereas surfaces covered in fabric do not. If you turn as you speak, you might also notice how these reflections change as you move.

Many animals, and some people, are adept at using only auditory information to build up a picture of the spaces around them. Bats, for example, emit high-pitched clicks and chirps as they fly through the night sky and then listen for the reflections of these sounds to determine where they are situated, a process called "echolocation." Bats can intercept moving prey while flying very quickly through three-dimensional environments that contain obstacles such as trees, walls, and roofs. Some blind people also draw extensively on echolocation to navigate and learn about their surroundings. They make use of the sounds of their own footsteps or sounds from other people or objects in the environment—sounds that can reveal the shapes of rooms, distances to nearby objects, and material properties of surfaces.

Musicians, sound engineers, and acousticians are also acutely aware of how sounds interact with spaces. Many musicians describe the space they perform in as part of the sound-producing mechanism of their instrument because performing requires them to manage the way in which the sound they produce interacts with the room in which they are playing. The composer Pauline Oliveros, who developed and promoted a practice she called "deep listening," pointed out that, for the player of a musical instrument, "any space is as much a part of the instrument as the instrument itself."[5] In other words, the sound an instrument produces is shaped by the environment in which it is played, and the performer modifies their actions and adapts their playing to account for the effect they anticipate the space having on their instrument's sound.

Acousticians, who design and construct performance spaces, greatly influence the work of both performers and sound engineers because they ultimately create the acoustic canvas upon which these artists paint. Performers and recording engineers will travel long distances to work in particular spaces because their acoustic qualities play such a vital role in shaping what the audience ultimately hears. The iconic sound of recordings made in the early days of Motown, for example, emerged partly because, to add reverberation to performances, engineers annexed the attic space of Hitsville, the recording studio at 2648 West Grand Boulevard in Detroit. They ran cable from the main studio to a speaker at one end of the attic and, through it, played live sound from the studio. A microphone at the opposite end of the attic picked up the sound. This delayed, reverberant sound was then added to the mix. The attic space therefore came to shape the Motown Sound in the same way a concert hall shapes the sound of any performance in the space.

In sum, many attributes of spaces can be revealed by the way in which energy from a sound source reflects from their surfaces. We can make judgments about the size, the shape, and even the material content of a space by tuning into the behavior of reflected sound. For musicians, sound engineers, and acousticians, the auditory "color" of a space (bright, warm, etc.) is determined by which spectral components of an emitted sound are absorbed or reflected by the surfaces in the space and how these patterns of reflection and absorption evolve over time. For a bat, the sounds reflected from surfaces reveal a space's geometric properties and the paths of moving objects. In every case, it is an individual's actions that make the sounds that make the properties of the space come alive.

With the rise of augmented reality systems that superimpose virtual spaces within real rooms, there has been an increasing interest in perceptual room matching, for, as Neidhardt et al. note, even after years of research on listeners' perceptions of room properties, it is still unclear which of the available acoustic cues are used in estimating the size and geometry of rooms.[6] Further, as Blesser and Salter note, matching real and virtual spaces is complicated by the fact that the boundaries of auditory and visual spaces are not always the same.[7] Whereas solid surfaces, such as walls and windows, define the limits of an auditory space, visual space can extend through transparent surfaces, such as clear glass. Similarly, visual spaces can be bounded by curtains or thin walls through which sounds can travel. In other words, the spaces of awareness do not necessarily coincide. In real life, we are often drawn to what is happening outside a window or distracted by sounds from the hallway or another part of the building. Virtual worlds that do not give users experiences of peripheral visual or auditory awareness can therefore seem to lack depth.

If we let go of the idea that our perception of space is driven by an internal three-dimensional snapshot of a world specified by geometry and consider instead that space is defined by surfaces presented to us and their ability to support our actions, we can reconsider what it means to construct augmented and virtual spaces. When we consider the important characteristics of the spaces we inhabit, differences between visual and auditory space are not our primary concern. We wish to know if the information needed to support our action goals is available to us, in an appropriate form, within the global energy array (see chapter 2). Rather than building virtual worlds, we need to think in terms of providing action-relevant information so that a user can discover which affordances or opportunities for action the rendered environment supports. If a participant in a game takes on a different body in the virtual world—such as that of a bat, say—the relevant affordances of that world will shift. But as long as sufficient information is available in the constructed energy array, the user will be able to negotiate the constructed virtual world.

As will become apparent from the case study we will investigate in section 5.5, acoustic information arising from direct contact with the environment can also be used to guide actions, as is the case for a blind person who uses a mobility cane to detect the material properties of surfaces to be walked on. However, to understand how one might apply this approach to the design of augmented and virtual environments, we must understand its implication for our perception of the *physical* worlds we inhabit.

5.3 Framing Action in Space

Our activities are always situated within an environment that shapes how we act and provides context for what we do. Whether driving on a country road or making a cup of tea, we are aware of the properties in the environment that guide our actions—the width of the road, the surfaces on which we can place our cup, and so on. Such surfaces and boundaries comprise a reference frame that provides opportunities to act in particular ways but also constrains our actions.

A reference frame represents the locations of entities in space. Think of a map, where "You Are Here" denotes your current position within the map's frame of reference, so that you can determine where you are relative to other locations on the map. But the map is not the world. Your next task might be to orient the frame of reference of the map with respect to the rest of the world. If the map is affixed to a structure, this is relatively straightforward—you need only translate the two-dimensional depiction on the map into the three-dimensional world around you. However, if you are holding the map, you have one more variable to contend with: your own bodily orientation (and hence that of the map) with respect to the world. A map will indicate compass directions (north, south, and so on), but we do not typically carry compasses with us and, unlike homing pigeons, we are not sensitive to the earth's magnetic field. We have to find a landmark in the world that is also on the map, orient ourselves to that landmark, and then orient the map to ourselves. If the map does not provide a "You Are Here" indicator that is reflected in the three-dimensional world, we need to find at least two points in the world so that we can determine both our location and our heading (the direction in which we are facing). Only then can we be sure that our frame of reference corresponds to that of the map. Although using a map to locate ourselves is perhaps the most familiar example of actively thinking about aligning frames of reference, we do this all the time. Think of how seamlessly you reorient yourself to the objects around you in your bedroom when you get up in the morning. You do not stop and think how to reinterpret the world from a standing position every day.

In the study of the philosophy of mind, frames of reference for perception and action have typically been described as either egocentric or allocentric. In an egocentric reference frame, objects are located with reference to the body of the perceiver. Take a scenario where a cup on my table is to the right of me and a window is in front of me. If I turn my body 180 degrees, these relationships change: The cup is now behind me and to my left. My egocentric space is first person, and the locations of objects within it are always relative to the orientation of my body. Egocentric space supports the planning and execution of actions that involve interactions between the body and the environment, such as locomotion, reaching for objects, and so on. As Campbell notes, "Egocentric space is thus a space one should think of 'as a participant,' as someone plunged into its center, as someone with things to do in that space."[8]

Allocentric spatial reference frames are defined independently of any perceiver. They take many forms but typically specify the position of objects relative to an external reference frame, such as magnetic north or GPS. No matter where I am standing, San Francisco is always north of Los Angeles and west of Chicago at a latitude of 37.8 degrees and a longitude of –122.4 degrees. An allocentric reference frame is a third-person, objective framework that is usually stable but that may or may not be absolute. Although compass headings are currently given relative to magnetic north, the earth's magnetic poles periodically flip so that, at some point in the future, this frame will also be reoriented. As Gallagher notes, allocentric reference frames can be further specified in either absolute coordinates (as in the example above) or in object-relative intrinsic coordinates, as in "my laptop is sitting to the right of the window," where the intrinsic reference frame is that of the window.[9] Gallagher further points out that, in English, we typically use two different frames of reference when describing relationships between nearby objects. When describing the layout of, say, cutlery on a table, we might say, "The spoon is beside the fork (intrinsic frame of reference) or the spoon is to the right of the fork (relative frame of reference); we tend to talk of absolute reference frames only on a larger scale. We say that San Francisco is north of Los Angeles, but not that the spoon is to the east of the fork."[10]

Philosophers from Kant to Merleau-Ponty have noted that dividing the world into allocentric and egocentric spatial reference frames creates a problem for the individual who must move and act in the world. How is one's own body to be represented in space, for one's body is the source and the center of phenomenally experienced spatiality?[11] As Merleau-Ponty states, "Far from my body's being for me no more than a fragment of space, there would be no space at all for me if I had no body."[12] For Merleau-Ponty, my body, although present as an object in the environment that is visible to others, is experientially transparent to me: "I observe external objects with my body, I handle them, examine them, walk around them, but my body itself is a thing which I do not observe [in action or in the act of perception]: in order to be able to do so, I should need the use of a second body which itself would be unobservable."[13] And yet my body must be present to my awareness in some way because my sense organs pick up information from my environment. This information, in turn, guides the way I dispose my bodily resources toward the world.

Merleau-Ponty argues that, in addition to ego and allocentric reference frames, there is a third bodily reference frame, present to our awareness not through outwardly directed sense modalities of sight, hearing, or external touch but rather through proprioception, the body's intrinsic sense of the relative positions and motions of our joints, muscles, and limbs. Gallagher argues that this bodily reference frame is both innate and absolute: It exists before birth and is calibrated to encode the movements of body parts relative to each other and in coordination with external forces acting upon the body, such as gravity. This bodily reference frame is neither egocentric nor allocentric. It is the frame of reference that "applies to the lived body as perceiver and actor, in a first-person perspective. In precise terms, it is the proprioceptive frame of reference."[14]

Gallagher's proposal is that this intrinsic, experiential reference frame is the starting point for our bodily engagement with the egocentric and allocentric spatial reference frames within which we act: "At the experiential center of things, you find an absolute—the intrinsic spatial framework from which one reckons the egocentric frame of reference, and the object-centered frame of reference (in this case . . . simply by putting ourselves in the place of the object and saying what is on its right or left), and even the purely allocentric positioning of things."[15] In short, even our most abstract spatial reference frames, such as GPS or a compass rose, grow outward from an intrinsic bodily understanding of our position in our lived environment. We understand our position in terms of an egocentric reference frame because we plan and execute actions that are directed outward from our bodies toward the objects in our world.

If, as Campbell argues, egocentric space is the space in which we have things to do, we must also be able to understand the ways in which our doings impact and are shaped by the actions of other beings and objects in the space around us.[16] Although egocentric space describes the world with reference to our body and its location within it and allocentric space describes relationships between the locations of objects independent of any reference to our body, neither framing can easily account for the coordination between our body and the environment once we wish to manipulate objects. In egocentric space, the axes relevant for the coordination of action are those aligned with the body or with parts of the body engaged in the goal-directed task. I reach my arm forward from my body to pick up a cup. By contrast, the axes that define relationships between objects in allocentric space are not directly relevant for the coordination of action. It does not matter that the cup is to the west of the lamp. What matters is that it is within my reachable space, to my right.[17]

Beyond awareness of the relative locations of things, object manipulation requires the coordination and guidance of action that engages our bodily resources across multiple sense modalities and within a joint-based coordinate system, all in service of accomplishing a task goal (see section 3.3). Of course, as Klatzky points out, coordinating one's body to walk to a particular destination also involves coordinating action within a spatial reference frame, one that unites your egocentric location and heading with the location of a destination in allocentric space.[18] But the whole-body motion involved in navigating through the environment constitutes a different kind of coordination of spatial reference frames than that required to pick up objects or use tools.

In practice, these two ways of organizing space, for navigation and for manipulation, are not mutually exclusive. We seamlessly transition between different ways of coordinating our bodily movement toward the world without even being aware of it. In the course of a couple of minutes, you can get up from your chair, walk out of your house, and get into a car to drive to the store—a sequence of events that requires you to direct your actions toward the world within continuously varying reference frames.

In short, our understanding of space is actively constituted through our actions upon the world. Our understanding of the spaces of interaction with digital artifacts and environments thus depends on our being able to relate our intrinsic experiential reference frame to our egocentric orientation, with respect to designed spaces of interaction, and, through them, to the allocentric space of an entangled digital/physical world.

5.3.1 The Many Ways of Framing Space

Studies addressing reference frames for perception and action have been distributed across a number of fields of research, each concerned with its own task domain. As a result, multiple ways of organizing the space of perception and action have emerged, each of which is supported by both behavioral and neural findings that point toward its utility in guiding action.

5.3.1.1 Peripersonal and extrapersonal space The neurologist Russell Brain was the first to propose that we perceive the space around us in light of the actions that it affords us.[19] Brain observed stroke patients with damage to their right hemispheres and noted that they often had selective impairments affecting coordination of action in space for either near-body or distant regions, and he determined that action space might be organized, at the neuronal level, into two parts: "grasping distance" and "walking distance." In the early 1980s, Rizzolatti and colleagues, working with macaque monkeys, systematically studied the coordination between visual and somatosensory processing of stimuli within the space around the body.[20] They determined that, within the space that could be reached, which they termed "peripersonal space" (PPS), there was evidence for a coordinated neural response across visual and motor regions of the brain that differed from responses to stimuli beyond arm's reach. Over the past forty years, neurological and behavioral research on peripersonal space has revealed many ways in which such space functions to support us in manipulating objects, particularly in co-opting tools to accomplish tasks beyond our reach.

There are two aspects of PPS of particular interest to designers of interactive systems: the multimodal nature of object properties that are registered at a neural level within PPS and the reshaping of PPS as tools that are picked up and put down. Andrea Serino points out that there is no physical boundary between peripersonal and extrapersonal space, between what one considers to be near or far from one's body.[21] But there is a growing body of neurological evidence suggesting that we are primed to respond, across multiple sense modalities, to objects that come close to our bodies either because we reach for them or because they approach us. When an object is close to one's body, it generates a response in multisensory neurons with a tactile receptive field centered in the closest body part and with visual and auditory receptive fields anchored to the same body region. This multisensory receptor field extends for some distance around the body region in question, and it varies in size according to the nature of the relationship between the body and the object. If a bee approaches your hand, for example, you will be very aware of its position and of the part of your body in immediate danger of being stung. So, you will track the sound of the bee and its position in the air, and you might anticipate the tactile sensation of a sting on the part of your hand closest to the bee.

Increasing evidence from behavioral studies with human participants indicates that the processing of tactile information on the body is influenced by visual and/or auditory stimuli. A person's body responds faster when visual or auditory stimuli are spatially congruent with the site of anticipated tactile stimulation than when they are not.[22] This implies that, for augmented and virtual environments that use haptic or tactile feedback, there is a benefit to ensuring that tactile effects associated with object manipulation are always colocated, in peripersonal space, with the visual and/or auditory events to which they belong. In other words, manipulating, with a tactile glove, a virtual object close to the hand will be more effective than using the same glove to move an object that is beyond arm's reach.

In addition to being richly multisensory, neural representations of PPS have been shown to be malleable in terms of spatial form and temporal persistence. Evidence of spatial malleability comes from a body of work that demonstrates how PPS is restructured when a tool is co-opted to extend one's reach to allow a person to interact with an object. The earliest work in this area was conducted by Iriki, Tanaka, and Iwamura, who worked with macaque monkeys. They trained the monkeys to reach for food with a rake. Then they demonstrated that the visual receptive fields of neurons associated with responses to events near the hand responded when the rake came within reach of the food.[23] In an extension of this work, the researchers trained monkeys to recognize their images on video monitors. They demonstrated that when these monkeys could track the movement of their rakes only on-screen, the receptive fields of visual neurons associated with the hand extended to incorporate the video *image* of the hand. With training, the monkeys could transfer the locus of tool–food interaction to a position in space incongruent with proprioceptive feedback from the arm, whose actual position they could no longer see.[24]

Behavioral studies have also demonstrated that our perception of the extent of PPS is modulated by tool use. Witt et al. carried out a study in which they asked participants to reach for projected targets either with their hands or with batons.[25] Targets were located both within reach and out of reach. For out-of-reach targets, participants had to reach as far as they could. They then had to visually gauge the distance from the target of either their hand or the tip of the baton. People underestimated the distances to targets for which they reached with a baton, believing them to be closer than targets for which they reached with their hand. But the most interesting finding was that when people passively held the baton, as opposed to reaching with it, their judgments of distance to the target were the same across both baton and hand conditions. The restructuring of PPS based on reachability judgments was therefore affected by an intent to use the tool, not simply by holding it.

A second observation emerging from the literature is that the location of the area of PPS that is remapped during tool use is dependent on how and where the tool is functionally coupled to the environment (e.g., by a gripper or other tool head, not by its length or physical form). When asked to use two batons of the same length, one with a gripper in the middle of the shaft and the other with a gripper at the end, participants restructured their PPS to match not the tool's physical extent but rather the location of the gripper, the functional coupling of the tool with the task environment.[26]

Finally, many studies have reported that PPS remapping during tool use happens within seconds of picking up a tool and disappears rapidly once a tool has been put down.[27] Although the temporal dynamics of PPS restructuring for temporary tool use are poorly understood, a recent study by Serino examined the case for modified PPS for extensive functional tool use.[28] Serino compared the performance of blind participants who regularly used long canes with the performance of sighted but blindfolded participants in an audio/tactile cross-modal congruency task. Each participant had to respond as quickly as possible to tactile stimuli presented on their hand while ignoring concurrent sounds presented either close to their hand or 120 cm away. Each participant held either a full-length cane or a short, counterbalanced cane handle (the control condition). Participants in both groups who held the short handle responded faster to near sounds than to far ones. When holding the longer cane, blind participants showed the same faster responses they had for the handle alone, but the responses of sighted but blindfolded participants were slower. The researchers concluded that the blind participants, who regularly used their canes to generate sounds that would reflect from surrounding objects, had established a persistent mapping of their PPS such that their projected interface with their environment could be selectively extended to the tips of their canes.[29]

Understanding how people estimate their capacity to reach for and grasp objects within PPS has implications for the design of virtual environments where the malleability of the environment, and even of the size and shape of one's avatar's body, result in a continual remapping of peripersonal space for the VR participant. Studies now emerging on the mapping of PPS for participants in VR suggest that users are sensitive to changes in the action capabilities of their avatars when a virtual environment contains sufficient specifying information to support their action goals.[30] Further, Serino's result with blind cane users suggests that, for skilled users of a tool in a virtual environment—even if the tool is partly or wholly simulated—PPS for a particular task may be more stably remapped and the accuracy of reaching movements may be more persistent than for intermittent users of the same tool. Not only might this allow designers to account for variability in performance between novices and experts, but tracking how reaching performance changes over time may also provide a method for observing the acquisition of skill in tool use in virtual and augmented environments.

5.3.1.2 Personal, action, and vista space The foregoing discussion of peripersonal and extrapersonal space has been concerned with what Brain called "graspable distance," bounded by what is within reachable space and often extended by tools that create physically continuous links to objects in the environment. Research on human navigation and orientation has given rise to alternative ways of framing space. One of the most widely accepted ways of structuring space at the scale of walkable distance and beyond was proposed by Cutting and Vishton.[31] They framed visual space as a series of three concentric circles surrounding the body and shown in figure 5.1—personal space, action space, and vista space—based primarily on the nature of visual information that is available at different distances from the observer.

Cutting and Vishton's definition of "personal space" is very similar to what I have called PPS, in that it encompasses the space within (and just beyond) arm's reach. This definition may be specific to Western societies. We are comfortable with others entering this space only in intimate situations or if we reflexively suppress our sense of intrusion because, for example, we are in a crowded space. Personal space typically extends 2 m from our viewpoint. It is the space within which we can manipulate things in our

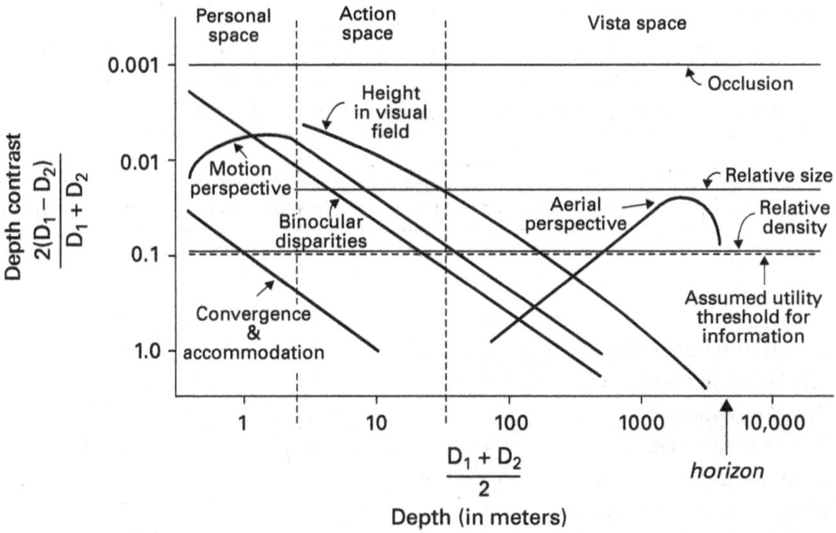

Figure 5.1
Visual cues supporting depth perception, illustrating the effectiveness of different visual depth cues within three defined spaces surrounding the viewer: personal space, action space and, most distant, vista space. *Source:* Cutting and Vishton, 1995, figure 1. Used with the permission of Elsevier Science & Technology Books. Permission conveyed through Copyright Clearance Center, Inc.

environment without the need for whole-body locomotion (and it includes our view of our feet while we are standing). In this space, we take advantage of more precise mapping between hand motion and object motion relative to extrapersonal space and between haptic and visual proprioceptive cues for manipulation. We also take advantage of more robust stereoscopic and motion parallax cues for vision and greater angular precision for visual control of movement.[32]

Action space is defined by Cutting and Vishton as the space in which we engage in actions directed toward our intimate boundary.[33] Action space extends from about 2 to about 20 m from one's current viewpoint. This is the space in which we speak to strangers, in which we have to throw objects instead of handing them over, and where visual information derived from sources such as binocular disparity (BD) and motion parallax can be used to assist us in intercepting objects and avoiding obstacles.

Cutting and Vishton's largest circle is vista space, the space beyond 20 m from the current viewpoint. At this distance, visual depth information is

Table 5.1
Approximate order of importance of visual cues for perceiving egocentric distance in personal space, action space, and vista space, with cues labeled "1" identified as being most important. Source: Adapted from Jerald, 2015, table 9.2.

	Space		
Sources of Information	Vista Space	Action Space	Personal Space
Occlusion	1	1	1
Binocular	9	8	2
Motion	6	7	3
Relative / familiar size	2	2	4
Shadows / shading	7	5	5
Texture gradient	5	6	6
Linear perspective	4	4	7
Oculomotor	10	10	8
Height relative to horizon	3	3	9
Aerial perspective	8	9	10

lacking because BD is no longer an effective cue, and we have to depend on pictorial cues such as shading, texture gradient, and linear perspective to judge the relative position of objects in space (see table 5.1).

Although Cutting and Vishton defined their spatial reference frames based on categories of functional visual cues, the sources of information that differentiate personal, action, and vista space have correlates in other modalities. Haptic space, for example, is mostly confined to the realm of reachable personal space, although, as we will see in section 5.5, the presence of a handheld cane (real or virtual) extends haptic perception of personal space to preview the topography of a path of travel and avoid obstacles. Auditory feedback within action space provides information about the locations of surrounding sources of direct or reflected sound and, in vista space, provides information about more persistent sonic landmarks (e.g., busy roads, the presence of railway lines, etc.). It is also important to note that spatial reference frames are rarely mutually exclusive. As you navigate in your car, you are aware of how your PPS is extended to incorporate the dimensions of the car and of vista space as you negotiate the road and its traffic.

No matter how they are defined in theory, spatial reference frames come into being to support us in accomplishing actions that are goal directed and context dependent. In this view, our understanding of space is actively constituted through our actions upon the world, and our ability to successfully manipulate digital artifacts and to move about in digitally mediated environments depends on us being able to relate our current space of action to the spaces we encounter in our entangled digitally augmented physical world.

5.3.2 Malleable Reference Frames for the Support of Action

When a user's action requires a response simulated by computational means, we must observe the correspondence between the spatial frame of reference of that action and the frame of reference of the system's response. For example, one of the reasons why mobile devices are difficult to use while you are actually mobile is that the activity of walking takes place in a world-centered reference frame, but responding to messages or emails takes place in the device-centric world of the phone. In addition, the activity of walking takes place at the scale of whole-body movement, whereas tapping on a screen is a fine-grain motor activity taking place at the scale of the hand. Trying to perform both activities at the same time involves coordinating not just different scales of movement and disparate reference frames but the continuous cyclical movement of walking with the discrete goal-directed activity of typing a word or scrolling through an email—a modern-day equivalent of simultaneously patting your head and rubbing your stomach.

In the next chapter, we will return to the implications of designing for multiple simultaneous reference frames for users who will interact with real and simulated objects. The current discussion will focus on how we coordinate action situated in a world-centered reference frame but that must be specified in a body-centered, joint-based coordinate system. Through our bodily movement, we are continually mapping actions specified in joint-based coordinates onto a world-centered reference frame oriented by, for example, the direction of gravity or the position of the horizon. When the environment we are moving in is partly or wholly simulated, it becomes the responsibility of the interaction designer to map users' movements, tracked using sensors in the device or the environment, onto the response of their system and to do so with an awareness of the demands of the physical environment in which a user is moving and acting. As noted in the mobile phone example above, failure to account for the demands placed on the user by competing reference frames for movement and action can result in their attention being inappropriately focused.

Think of the car accidents caused by people texting while driving, missing cues relating to a potential collision in the world-based reference frame of the road while they are focusing on the device-centered reference frame of the phone's screen. Car manufacturers have endeavored to solve this problem by remapping the screen-based controls of the phone onto the steering wheel of the car and its screen onto the larger in-dash display and facilitating voice-based interaction, so that there is less of a mismatch between the reference frames for driving and interacting with the mobile device. In prior work, we have also explored the remapping of mobile phone controls into a body-based reference frame to demonstrate that, as in driving, coordinated reference frames can improve people's ability to attend to their immediate surroundings while walking about and interacting with their phones.[34] In short, taking time to design interactions that pay attention to the coordination of reference frames for movement and action will result in interactive systems that are less likely to divide the attentional resources of their users.

The coordination of bodily orientation and movement in space is primarily the responsibility of the proprioceptive system, whose component subsystems include:

1. The vestibular system, composed of the semicircular canals and otoliths of the inner ear, which sense orientation of the head with respect to gravity
2. The haptic proprioceptive system, composed of mechanoreceptors in the skin, muscles, and joints, which provides us information about the position and motion of the body and limbs
3. The visual proprioceptive system, composed of the eyes and their associated musculature, as well as the muscles in the neck that orient the body with respect to the external environment
4. The auditory system, which, for the world beyond the visual field or in darkness, provides information in a world-centered reference frame

Proprioception, often referred to as the "sixth sense," modulates the body's response to multiple sources of stimulation in the environment (light, sound, gravity, contact forces, etc.). In *The Brain's Sense of Movement*, Alain Berthoz shows how the body marshals the components of the proprioceptive system to solve the problem of stabilizing head position and gaze when the body lacks a stable reference to the ground.[35] He uses the example of trying to stay upright in a moving boat, illustrated in figure 5.2:

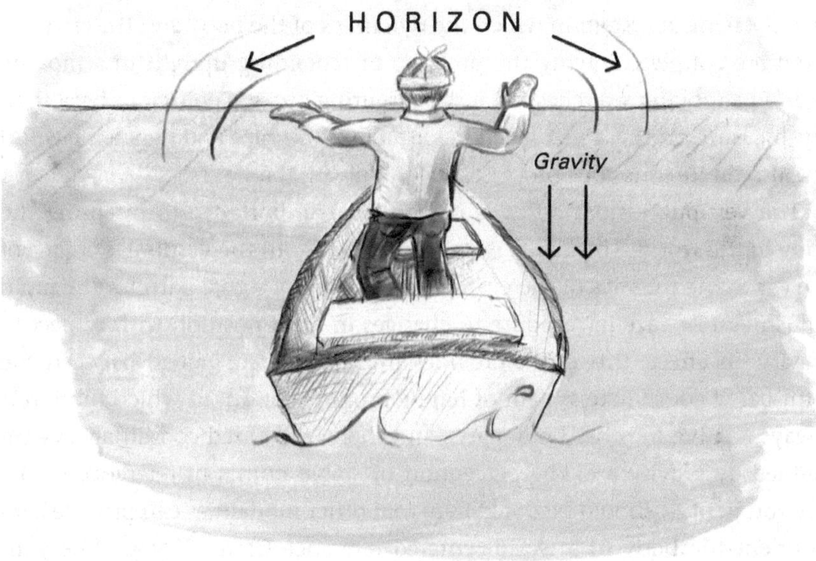

Figure 5.2
Visual and vestibular cues used by a person balancing in a boat. When standing in a boat, the body marshals the components of the proprioceptive system (the visual horizon, gravity as sensed via the vestibular system, and muscle tension to stabilize the legs). Together, these allow a person to solve the problem of stabilizing head position and gaze when the body lacks a stable reference to the ground. *Source:* Original drawing by Hans Anderson.

> Imagine that you are sitting in a boat and a wave comes and rocks the boat; your head will move in rotation with the boat. If you want to stay upright, you will have to make a movement—straightening your body—equal to, and in a contrary sense from, that of the boat. A system of coordinates encodes the movement that the eyes, head, and even the trunk must make to compensate for this rotation because of the motion of the boat. These systems of coordinates are the directions of tension of the muscles of these different parts of the body. The brain thus has to transform the sensory information, encoded in covariant coordinates, into motor commands expressed in the systems of coordinates of the muscles and their planes of action. (p. 48)

Berthoz's hypothesis, based on many years of observing pilots learning to move in zero gravity, says that the strategies animals have developed for accomplishing successful movement outcomes are based on solving mechanical problems in an efficient way. Perception of the world and action upon it are bound together by success in solving problems arising from the behavior of a mechanical system in which the dynamics of the body and the environment are coupled. Solving the problem of remaining upright in a moving boat on a moving sea relies not just on figuring out at a conscious level that, say, focusing on the horizon can help stabilize the gaze and hence the world but also on unconscious processes such as postural reflexes.

The vestibulo-ocular reflex is of particular importance in orienting the body in the world. This reflex is a cortical relay from the vestibular organs of the ear to the muscles of the eye, which provides a signal path for the rapid adjustment of gaze in response to changes in head position with respect to gravity. In effect, this reflex provides the most direct bridge between the joint-based coordinate system of haptic proprioception, in which motion is always relative to other body parts, and the world-based coordinate system defined by gravity and the perception of stable points of reference in the environment. It should be noted here that other modalities can provide cues to orient the body in a world-centered reference frame. If you close your eyes and try to balance on one foot, for example, you will notice how even placing one finger lightly on a wall makes it much easier to balance. This is effectively the haptic equivalent of fixing your gaze on the horizon—you do not need to push hard to restore your balance, you merely need proof of a fixed, world-centered frame of reference. Your touch can be light because your body only requires information about the presence and orientation of the external surface; it is not needed for mechanical support. Here, it is the information about the sway of the body detected through the small motions of the pad of the finger on the surface that generates continuous haptic flow to support the control of posture.

Audition, too, can function to orient bodily movement. In an elegant study, Lewald and Karnath demonstrated that stimulation of the vestibular organs in one ear (by flushing the ear canal with cold water) resulted in a shift in the perceived location of a sound toward the opposite, unstimulated ear.[36] The effect persisted even in the presence of a visual focal point to stabilize the head and lasted for at least five minutes, much longer than it took for the stimulated ear to return to body temperature. The authors suggest that the persistence of the apparent shift of auditory space points to a role for the vestibular system in orienting the body with respect to an external auditory stimulus fixed in a world-centered reference frame.

Everyday activities take place within multiple—often overlapping—spatial reference frames. For example, the egocentric reference frame, in which objects are situated relative to our body (the door is in front of me), is congruent with the allocentric reference frame, in which objects are situated relative to each other (the door is to the left of the table) or relative to the world (the door faces north). The coexistence of reference frames functions to support the planning and execution of actions whose goal may be situated in a world-centered (allocentric) reference frame but whose constituent movements are organized within a body-centered (egocentric) reference frame.

A further point to stress is that frames of reference for the perception and control of movement are not rigidly defined. Think again of Bernstein's blacksmiths, who were introduced in chapter 3. Bernstein claimed they could select from several ways of striking an anvil (illustrated in figure 3.4), implying that the goal of their action—hitting an object fixed in a world-centered reference frame—could be accomplished by the flexible coordination of limb segments specified in a body-centered reference frame. Bernstein's claim suggests that the coordinate system specifying the relationship between body segments must be malleable—that the coordination of movement must be flexible to respond to changes in movement choices or environmental conditions.

Evidence for such malleability has been provided by Turvey and Carello.[37] They systematically explored the mechanisms underlying the capacity of the haptic proprioceptive system to organize resources such as mechanoreceptors and their attendant neural substrates into what Sverker Runeson terms "smart perceptual instruments."[38] A smart perceptual instrument is a configuration of the perceptual motor system, assembled in real time and able to respond to specific, task-relevant invariants. It is like having a toolbox with a tool for every possible task.[39] In the case of the haptic proprioceptive system, Turvey and Carello determined that the physical invariant that guides or "tunes" this organization is the felt direction of the motion of a center of mass—for example, along the length of a limb or across a joint.[40] If you are wielding a hammer, the associated smart perceptual instrument is organized to account for the combined weight of your arm/hand and the hammer, the angle of the hammer, and the location of the nail. The forces and torques supplied by your muscles are applied at the right point in time and space to hit the nail.

Ultimately, though, the coordination of frames of reference for a given action is a function of the goal of that action. In the case of the man in the boat, it is the goal of establishing a stabilizing gaze that governs how bodily resources are recruited to accomplish stabilization of the head. In the case of the blacksmith, it is the control of the velocity and direction of movement that governs where and how hard the hammer will strike the anvil. For an interaction where the goal may be constructed in a coordinate space not congruent with the physical environment in which one is moving, such as in virtual reality, the challenge will be to provide appropriate feedback congruent with the perceptual invariants that must be present for a user to successfully accomplish a task. For a person balancing a virtual boat, one might want to provide a virtual horizon to allow them to stabilize their gaze, and for wielding a virtual hammer, one might provide inertial feedback concerning the location of the hammer's center of mass. The goal, in both cases, is to orient the user with respect to a virtual world-centered reference frame. For example, to simulate the weight of a sword in VR, one designer uses a smaller real object with a modified center of mass to make the sword feel longer.[41]

In fully immersive VR, the real-world referents are subordinate to the structure of the virtual environment, with only gravity persisting as a world-centered referent. In AR, real and simulated frames of reference are aligned to leverage as many world-centered referents as possible. For tangible user interfaces, where real artifacts act as handles to a virtual substrate, the coordination of reference frames depends on the number of degrees of freedom of hand/object or bodily movement that can be tracked.

5.4 Action, Perception, and Getting About in the World

I have emphasized the role our bodies play in uncovering the possibilities for action available to us in our environments. As we have noted in the discussion of virtual soccer and rugby in chapter 4, game developers can specify such information in response to movements of players relative to the objects and surfaces of their simulated worlds. Such information is generated on demand in response to the players' current positions and their movement trajectories. The information players rely on to plan their actions relative to their surroundings is generated, in real time, in response to what they are currently doing.

The ability of VR simulations to support people moving and acting in virtual spaces has rekindled a debate begun by Gibson: the question of whether spaces are specified for individuals by geometric properties or by the nature of the action-relevant information available for a given organism as it seeks to accomplish a task.[42] In his observation of pilots learning to land planes, Gibson came to the conclusion that the kind of spatial understanding required to coordinate action successfully, with the rapidly changing layout of one's surroundings, could not depend on continually updating a fully specified internal representation of the world. Instead, Gibson argued, pilots (and all flying organisms) coordinate their landing behavior using information present in the global energy array specifying relevant information, such as rate and angle of approach. Gibson concluded that our understanding of space is founded on the layout of surfaces present in the environment that can support our actions, not on constructs such as the lines, planes, and points of abstract geometry.

Because we spend so much time within buildings and environments that we have constructed, whose forms are outcomes of abstract geometry, we have come to equate space and its perception with the perception of a constructed Euclidian world. But proponents of Gibson's ecological approach to perception suggest that the abstract geometries of the built environment are a by-product of the architectural design process, not a requirement for supporting us as we negotiate our everyday activities. From the ecological perspective, our perception of space is founded on our perception of surfaces and objects that can "afford" the perceiving organism something: Space is not an abstract thing but rather a place with particular attributes that support the activities we want to accomplish there.

As we noted in discussing Gibson's work in chapter 2, the opportunities for action the environment affords a given organism are not properties of the organism or of the environment. They arise because of the relationship between the capabilities of the organism and the structure of the environment. Through their movements, organisms pick up information that specifies how the environment will support their actions, including information that specifies distance, orientation, and the relative position and motion of other objects. It is therefore not the spatial properties of the environment per se that are the basis for an organism's perception of the environment's affordances. It is the discovery of affordances as opportunities for action that specifies the spatial properties of the environment.[43] Action-dependent perception of affordances is central to our lived experience. It is how we understand our relationship with every aspect of our lived environment, including distances, surface orientations, and many other properties we associate with conventional notions of spatial perception.

The main concern of the field of ecological perception has been figuring out the nature of the information that organisms are attuned to as they "get about" in their environments, whether it is the changing structure in the electric field that tells the eel of the location of its prey or the changing structure in light that informs the soccer goalie of the spin on an incoming ball.[44] The same specifying information can guide multiple actions or behaviors. If the person observing the incoming ball is not a goalie but rather a bystander watching the game, then the looming image of the ball on their retina will most likely cause them to duck. Even before Gibson began to work with pilots in the 1940s, the biologist Jakob Johann von Uexküll (whose work on *Umwelten* I discussed in chapter 2) noted that information for an organism could be specified with reference to different actions, depending on their current states and task goals.[45] Citing the example of the changing roles sea anemones play for hermit crabs—camouflage from octopuses if they are being hunted, shelter while their shells are hardening after molting, and food when they are hungry—von Uexküll notes that the crab is driven not to match a particular image or form of the anemone but rather to satisfy a need: to find shelter or to eat.[46] Von Uexküll referred to this search for a general property of the environment to fulfill an organism's need as "tone," suggesting that the specifying information to which organisms are attuned might not be object specific but quality or tone specific. If I need to reach a high shelf in my house, I might search for something that I can climb onto that is of an appropriate height. It might be a chair (on which I usually sit) or a box (in which I would normally put things). In my current search, though, the "tone" that guides me is attuned to the height I require and its potential to support my weight.

In sum, the opportunities for action that are discoverable for an organism are a function of the relationship between the bodily capabilities of the organism and the structure of the environment. Opportunities for action within an individual's environment can be specified within the global energy array in more than one form—acoustic, mechanical, and visual sources of energy can each specify affordances. The degree to which action-relevant information is perceivable by an organism will depend not only on its individual capabilities but also on the level to which it is attuned to information shaped by the particular environmental properties that will support its current goals. As Fajen and Phillips argue, spatial perception is the discovery of these affordances and the ways in which they specify the spatial properties of the environment, not the abstract perception of its geometry.[47] It is through action-dependent perception of the affordances of our world, they argue, that we can estimate distances, determine surface orientations, and deduce many other properties we typically associate with the perception of space.

In navigating through the world, it is likely that people selectively attend to one or more sources of energy as they attune to invariant patterns that can guide their actions. This poses a challenge for designers of augmented and virtual environments because we cannot predict which source of information will be most relevant for a VR participant at a given moment in time. The proposal I put forth in the following discussion is that designers need to render a sufficiently rich sensory environment to enable participants to discover and attune to sources of task-relevant specifying information within the simulated energy array. In short, in designing spaces for action within interactive systems, we likely do not need to focus on realism, but we do need to pay attention to the ongoing information that individuals use to plan and execute actions with respect to the people, surfaces, and objects around them.

Approaching spatial perception from this action-oriented perspective is of value for designers of augmented and virtual environments because it emphasizes the role that the bodily capabilities of users play in bringing forth the lived spaces of their experience. Such an approach may also cast light on cases in which perception of real and simulated environments does not always align. In accounting for attunement and tone, we can come to understand the relational nature of bodies and things and the co-construction of navigable spaces that together create what users ultimately experience as they "get about" in the augmented and virtual environments we design.

5.4.1 Getting About in Real and Virtual Spaces

The world we inhabit is full of things, some that move, some anchored to the ground, others we can pick up and manipulate. As we navigate this world, we are continually choosing from among many ways of moving and acting on the basis of our bodily capabilities and the things we encounter. If we walk through a partially open door, we determine whether we can continue forward or whether we need to turn our bodies and estimate the "passibility" of the gap framed by the door. Estimating the passibility of a gap as a function of the dimensions of our body is a "body-scaled affordance."[48] If the door in question is the closing door of an elevator, then our estimation of passibility has to take into account the speed at which the gap is closing and the speed at which we can get to it. We need to account for our action capabilities as well as our body dimensions; we need information from our immediate surroundings that will guide locomotion and action in a rapidly changing environment. These are action-based affordances.[49]

Here, I will draw attention to research that has uncovered sources of information relating to both body-scaled and action-scaled affordances that can guide the actions of people as they get about among things in real and virtual environments. People judge their ability to pass through gaps or intercept moving objects based on visual and/auditory stimuli in their environment. Understanding how these stimuli change in response to movement is thus necessary for anyone designing interactions for augmented and virtual environments. If people are using changes in optic flow such as the expansion of an image on their retina as they approach an object, or changes in relative velocity of objects on either side of their car as they drive around a curve, then they will also depend on these cues when these real-world interactions are partly or wholly simulated. Therefore, an understanding of how people perceive these changes, and which cues they use to guide their actions, will result in interaction designs that guide action in ways that preserve the lawful coordination of action and perception upon which participants have come to depend.

5.4.1.1 Perceiving body-scaled affordances: Passibility of gaps Let us return to the example of passing through doorways. A person has to judge whether they can pass through the gap formed by an open door and its frame. We do not mentally calculate the ratio of our body's height or width to that of every door we pass through. Yet, people rarely collide with the edges or tops of doorways. In the 1980s, Warren and Whang carried out a series of studies to determine what information was present in the relationship between people and their environments that allowed them to make visual judgments, such as whether they could fit through gaps. Participants with differing body dimensions were videoed as they walked through gaps of different sizes to determine the ratio of shoulder-to-gap width that people judged to be passable.[50] Warren and Whang propose that an individual's perception of a body-scaled affordance, as determined by the fit between their body and the structure of the environment, is not necessarily defined by a clear boundary between things that are doable and things that are not. They argue for a critical point where a goal could be accomplished if some effort was expended to transform the relationship between one's body and the environment. In their gap-passibility study, the critical point was when participants chose to turn their shoulders to pass through gaps, which occurred when a gap was one-third greater than the participant's shoulder width.

Warren and Whang also determined that participants based their judgments on intrinsic body-scaled information present in optic flow. They demonstrated that the width of a gap is specified, in units of eye height, by both the angle subtended by the inside edges of the gap and the angle of declination of the ground at the base of the gap.[51] Because both angles are fixed (at least in the case of a static gap) and because the relationship between a person's eye height and shoulder width when walking is also fixed, the relationship between both sources of information is stable, or invariant, within the optic flow field as the person walks toward the gap. To confirm their hypothesis, Warren and Whang varied the height of the ground beneath the gap and determined that raising the height of the surface beneath the aperture made openings appear to be more passable. In practical terms, the invariant relationship between the pattern of flow in the optic array generated by the approaching gap and the eye height of the person walking toward the gap demonstrates that opportunities for action in an environment are relational. Affordances are present in coordination between the structure of the environment, the structure of an individual's body, and an individual's action capabilities.

If small alterations in the structure of the environment, such as changing the elevation of the surface beneath an approaching gap, can alter people's judgments of the width of that gap in a real-world situation, then how accurate does the information in optic flow in virtual environments need to be in order for participants to assess the passibility of rendered gaps based on eye height–scaled information? This question was recently posed by Bhargava et al., who compared passibility judgments for gaps formed by real and virtually rendered sliding doorways.[52] Their goal was to determine whether the higher-fidelity optical cues supported by newer devices would result in more accurate distance judgments than those previously made.[53] In Bhargava et al.'s study, participants were asked to judge passibility of gaps from a distance of 4.4 m (14.4 ft). If participants were uncertain, they were permitted to walk toward the gap but not to pass through it. The experimenters hypothesized that, by moving toward the door, participants in both real and virtual environments would be able to access information present in optic flow that might increase the accuracy of their judgments.

The authors found that participants were equally likely to walk toward gaps in real and virtual conditions, indicating that their levels of uncertainty were the same for real and for simulated gaps. The authors point out that the calibration of the virtual environment to participants' eye height allowed for judgments based on eye height–scaled information in the virtual condition, just as in the real world. Somewhat surprisingly, the lack of a virtual avatar did not impact perception of passibility based on information about shoulder width, suggesting that visual information relating to one's shoulder dimensions was not necessary when making passibility judgments in virtual environments.[54]

Although eye height–scaled information is reliable for judging the passibility of gaps between obstacles that rest on the ground, it is not useful in cases where the angle subtended by the inside edges of the obstacles includes an overhang, such as a ledge. Similarly, if the angle of declination at the base of the gap is altered because of the presence of stairs or a slope, then eye height–scaled information is less dependable. In such cases, two further sources of body-scaled information have been identified, both of which are dynamic—that is, relying on the user moving through the environment. These comprise changes in the optic array related to the observer's stride length and the amplitude of their lateral head sway as they approach the gap. As Fajen notes, the relationship between body width and lateral head sway amplitude during normal locomotion is relatively stable, so that head sway–scaled information can be used to calibrate the size of approaching apertures in units of body width.[55]

Similarly, as David Lee determined, the rate of change in optic flow for an observer approaching an object or aperture is directly proportional to tau, the time required to close the intervening spatiotemporal gap, optically specified for the moving observer in units of stride length.[56] To determine whether such information could be picked up by participants navigating

Condition	Information			Screenshots
	EH	HS	SL	
Post	✓	✓	✓	
Tall post		✓	✓	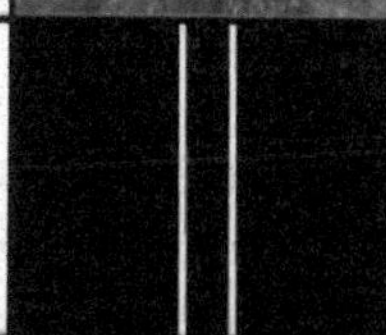
Wall			✓	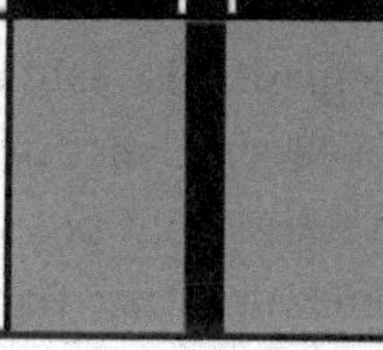

Figure 5.3
Body-scale visual cues available to a walking person. EH: eye height; HS: head sway; SL: stride length. Checkmarks represent sources of information available in each condition. *Source:* Fath and Fajen, 2011, figure 5. Used with permission of SAFE Publications Ltd. Permission conveyed through Copyright Clearance Center, Inc.

virtual environments, Fath and Fajen asked people to walk through rendered gaps formed by virtual obstacles viewed using a head-mounted display.[57] Participants' movements were tracked, and participants were free to rotate their shoulders if they judged that this would allow them to pass through gaps. Each of their three experimental conditions was presented as a different virtual environment, and in each environment, different combinations of eye height–scaled, stride length–scaled and head sway–scaled sources of information were available (see figure 5.3).

Fath and Fajen found that participants in all three study conditions timed their shoulder rotations correctly to pass through gaps, suggesting that dynamic visual information encoded in optic flow scaled by stride length and/or head sway could be used when eye height information was not sufficient to specify the passibility of gaps in virtual environments. For both real and virtual environments, dimensions of the world can be specified in dimensions related to those of the observer's moving body.

It is important to note here that Fath and Fajen's findings are only relevant when participants can engage in full-body locomotion within virtual environments. When, as is often the case, a participant is moving a virtual avatar viewed through a head-mounted display using only their hand and a game controller such as a joystick, the association between the movement of one's virtual body proxy and the resulting optic flow is no longer preserved. In a study comparing full-body locomotion through an immersive virtual environment with motion through the same environment using a joystick controller, Snyder and Cinelli found that the point at which participants judged gaps to be passable without body rotation differed between conditions.[58] When controlling avatars with a joystick, participants rotated their bodies for larger gaps than participants who moved through the virtual environment. Participants also said it would take them longer to close the spatiotemporal gap between their current position and the gap when using a joystick than when they could physically walk within the virtual environment. The authors concluded that participants in the joystick condition, where only visual and hand movement information were available, underestimated the distance traveled toward the aperture and judged wider gaps to be impassable. In situations where information concerning locomotion is incongruent (e.g., based on hand-eye coordination rather than eye height–scaled information arising from full-body locomotion), the authors suggest that game designers need to account for the conflicting sources of information specifying gap passibility. As interaction design increasingly involves tracking and designing whole-body movements such as walking in augmented and virtual environments, knowing how to communicate whether gaps are passable is key to ensuring that participants are successful in negotiating the obstacles they encounter. Understanding the perceptual motor skills that people rely on when navigating gaps in their physical surroundings is the first step that we, as designers, can take toward developing designs that render appropriate invariant properties in the partly or wholly simulated energy array on which our participants depend.

5.4.1.2 Perceiving action-scaled affordances: The case for shrinking gaps In addition to body-scaled affordances, an individual's movement capabilities can also specify opportunities for action within their environment. These action-scaled affordances can also be specified by information present in the global energy array, but the degree to which an individual is attuned to this information depends not just on their body dimensions but also on their perception of their movement capabilities. In the case of the elevator doors discussed earlier, the person walking toward the elevator has to take into account the speed at which the doors are closing as well as their own ability to reach that gap while they can still pass through it.

Estimating the passibility of that shrinking gap is a type of problem we encounter many times daily, such as when we thread through the foot traffic on a crowded sidewalk or time our crossing of a busy road. Of interest here is the nature of the information by which we judge whether we can pass through shrinking gaps. Similarly, when we have to intercept a moving object—to catch a ball or take a cup of coffee passed to us—we draw on information in our environments to help us position our bodies in the right place at the right time. As with body-scaled affordances, action-scaled affordances are specified for the individual by information to which they are attuned—information that arises because of the coordination between their body, the environment, and their action capabilities.

We might anticipate that our ability to avoid moving obstacles while going about in the world depends on a number of factors, including our reaction time, our level of attention, and maybe even our perceived level of fitness. To put it another way, the action-based affordances specified for us by information in our environment, usually in the form of optic flow, allow us to determine when we can and cannot avoid moving obstacles by passing through shrinking gaps. In chapter 4, we noted that information present in optic flow is sufficient for players in a virtual rugby game to judge their abilities to pass between two approaching defenders[59] (see section 4.5.2) Information in the optic array specifies the passibility of a changing gap for the attacking player (assuming that current conditions persist) in what Bootsma calls the player's "current future."[60] Watson et al. demonstrated that players, in determining whether they could run between a pair of approaching defenders, combined or coupled two tau values: one specifying the time to close the gap between the two defenders and the other describing the time to close the gap between their own current position and the line of defense. It is important to note that these research participants did not move through the virtual environment or rotate their bodies. Instead, they made their judgments from a seated position and gave their responses by pressing a button. Their judgments of passibility were based on perception of information within optic flow, not on information arising from the kind of perception–action coupling they would encounter if they could move their whole bodies in either a real environment or a fully immersive virtual space.

The transition between gaps perceived to be passable and those perceived to be impassable was sharply drawn. The third phase—adapting one's body shape—that Warren and Whang found for people approaching gaps in real environments simply did not exist for these participants.[61] It is therefore difficult to determine whether players based their judgments on their perception of body-scaled affordances alone or whether they could access, from prior experience, action-based affordances that might arise in this dynamic game scenario. This ambiguity in determining the source of specifying information for action-based affordances is likely to persist in VR applications as long as there are multiple technical approaches to rendering virtual environments and tracking the movements of participants.

To better understand the sources of information that might specify action-based affordances for passibility of shrinking gaps in virtual reality, Fajen, Matthis, and Fath conducted a series of studies where they asked participants to judge the passibility of virtual shrinking gaps when either standing still[62] or walking toward two converging objects.[63] In each case, the experimenters manipulated both time to closure for the converging objects and the distance between the participants' starting positions and a shrinking gap formed by two converging virtual pillars. They tested the hypothesis that information specifying action-scaled affordances is only available when an observer is moving and so can adjust their actions to be in the right place at the right time to safely pass between the pillars. When the participant is not moving, they have no information about self-motion, so they have to plan their future actions based on a combination of prior experience and the tau of the shrinking gap as they see it. Fagen and Matthis's task required participants to estimate the gap passibility in three conditions: one in which participants stood still but the pillars were only briefly visible to them, a second in which they moved toward pillars that remained visible to them, and a third in which the pillars disappeared before participants were permitted to move. If participants could access information specifying the passibility of the shrinking gap only when they were moving, then, the researchers hypothesized, their judgments in the move condition should be more accurate than in either the standing or the delayed move condition. The study's findings revealed that being able to move while viewing the closing gap had no effect on participants' judgments of passibility. Brett Fajen notes that this should not be surprising because it is often necessary to make judgments of passibility when you are stationary, such as whether you can safely cross a road without a signal light.[64]

Fajen and Matthis further demonstrated that people appeared to make action-scaled judgments of gap passibility based on optic information alone, without drawing on prior knowledge about their own movement capabilities.[65] In formal terms, a shrinking gap is passable if the minimum locomotor speed needed to safely pass through the gap (v_{min}) is less than the actor's maximum possible locomotor speed. In terms of spatial variables, v_{min} is equal to the minimum distance that the observer must travel to pass between the obstacles divided by the amount of time remaining until the size of the gap is equal to the width of the observer's body.[66]

According to Fajen and Matthis, an individual can draw on four sources of information present in optic flow to determine whether a shrinking gap is passable[67]:

1. Their body: The speed of locomotion required to pass through the gap, v_{min}, already takes into account both the size of the converging obstacles and the size (in units of shoulder width) of the observing person's body.
2. Eye height: v_{min} is already specified in body-scaled units of eye height relative to time—in this case, the time needed for the obstacle to cross the observer's path. Also, as we noted in section 5.4.1.1, units of eye height–scaled information are intrinsic to the body, meaning that if the environment is rescaled (as in VR), the participant will recalibrate their movements to account for the altered dimensions of their world.
3. Self-motion: People can judge the passibility of shrinking gaps independently of whether they are moving themselves.
4. Motion relative to other objects: People can distinguish between changes in optic flow caused by their own movement and those caused by the movement of objects in their environment.

Fajen and Matthis argue that since none of these sources of information directly relate to prior knowledge of one's bodily capabilities, prior knowledge may not contribute significantly to estimates of action-scaled affordances.[68]

Nonvisual sources of information might also contribute to decisions about gap passibility, particularly when observers are moving (even if judgments made *without* self-motion are accurate). To test this hypothesis, Fajen and Matthis carried out a modified version of their shrinking gap study.[69] This time, participants walked toward the converging columns in their path, and the experimenters manipulated the relationship between the participants' rate of locomotion and the visual feedback they received. The rate at which participants felt themselves to be moving via proprioception no longer corresponded to the rate at which they saw themselves moving through the virtual environment. Prior work has shown that judgments of self-motion based on proprioception can be influenced by many factors, such as the increase in effort required to climb a steep slope[70] and one's perceived level of fatigue.[71] Fajen and Matthis therefore determined that we must be able to rapidly adapt to alterations in the relationship between changes in optic flow, and that this adaptation may be influenced by changes in what it feels like to move.[72]

One's ability to separate the contributions of self-motion to optic flow from the contributions made by the movement of other objects may depend, in part, on nonvisual sources of information. In other words, when we must exert more effort than usual, such as when we climb a steep hill, we may perceive that we slow our rate of movement to compensate, but the rate of movement of everything else around us does not change. We recalibrate our perception of changes in optic flow caused by our own movements based on changes caused by the movements of other objects.

In their study, Fajen and Matthis took advantage of this ability to flexibly recalibrate visual flow. By amplifying the visual response of a virtual environment to its participants' movements (increasing their apparent walking speed by one and a half times), they determined that nonvisual information (the *feel* of walking at different speeds) contributed to participants' abilities to distinguish self-motion from the motion of external objects. This is because participants began to attribute more rapid optic flow to their own movements. This is important for designers of virtual worlds because it suggests that users can quickly adapt their actions to account both for differing ratios of movement to optic flow (e.g., by donning seven-league boots) and for the action capabilities of body morphologies for which they have no prior lived experience. I will return to the question of calibration and attunement to rapidly changing information in the environment in section 5.4.5.

The issue of the availability of body-scaled and action-scaled information points to a central tenet of the ecological approach to perception—namely, that the important information in an individual's environment is the information relevant to their immediate action goal, specified in terms of their action capabilities. To be relevant to the acting individual, spatial perception must specify properties of space in action-relevant terms—the distance to a gap in units of eye height, the passibility of a shrinking gap in terms of coupled estimates of time-to-contact (for both person-to-gap and gap closure), and so on. In either case, the moving person does not need to perceive the distance to the gap or its width in metric units. They need only determine whether the gap is, for them, passable.

5.4.2 Intercepting and Avoiding Things

In any action-oriented account of spatial perception, time and space are bound together by movement. In the case of passing through gaps, this is as true for simulated environments as it is for the real world. In the following, I will consider sources of information that can guide the spatiotemporal arc of actions in real and virtual environments for two more categories: intercepting and avoiding moving things and reaching for real and virtual objects.

5.4.2.1 Intercepting moving things In walking or running to intercept a moving object such as a ball, people have been shown to move in a remarkably consistent way, adopting a strategy that follows an invariant source of information termed the "constant bearing angle,"[73] as illustrated in figure 5.4.

The bearing angle represents the angle between two imaginary lines, the first between the observer and the moving object and the second of which is a fixed plane, such as the edge of a baseball field. The bearing angle functions

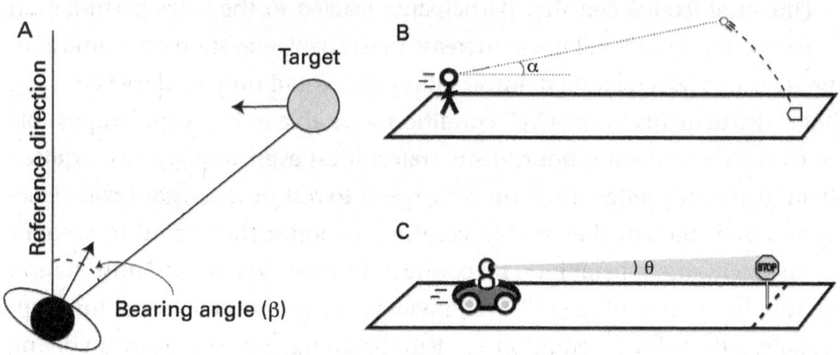

Figure 5.4
Optical variables used in control strategies while (a) intercepting a moving target, (b) running to catch a fly ball, and (c) braking to avoid collision. *Source:* Fajen and Phillips, 2013. Used with permission of the American Psychological Association. Permission conveyed through Copyright Clearance Center, Inc.

as an action-scaled source of information to guide the moving person. As long as they can move in such a way that the bearing angle remains constant, they will intercept the ball. If the bearing angle decreases, the ball will pass in front of them before they reach the point of interception. To correct for this error, they will need to accelerate. If the bearing angle increases, the moving person will need to slow down to intercept the ball.[74]

Strategies such as tracking optical acceleration cancellation are important when the initial distance to the point of interception is more than 30 m, at which point cues specifying absolute distance, such as optical expansion, motion parallax, binocular disparity, and vergence, cannot be detected.[75] The fact that people pay attention to OAC when intercepting fly balls was demonstrated in an elegant study by Fink, Foo, and Warren, who created a virtual baseball outfield and then manipulated the optical information they provided to participants as they ran to catch balls.[76] To determine the optical information participants were tracking, the researchers perturbed the path of a simulated fly ball. They predicted that participants would be able to intercept the ball after perturbation only if the appropriate information specifying its trajectory was available. In some trials, visual feedback to head-mounted displays gave only information specifying the linear relationship between the angle of the ball's trajectory relative to the path to home base (information used to track its LOT). Other trials incorporated information about the rate of change of the ball's vertical angle (information used to track OAC).

Fink et al. concluded that participants reacted to the ball's perturbation by recovering a modified linear increase in the ball's elevation over time after the trajectory was disturbed, information that could only be retrieved using OAC.[77] Participants in the OAC condition were able to intercept "impossible fly balls" (those having nonrealistic trajectories) even though this required them to abruptly adjust their running speed to restore a constant ball velocity and even though they had to continue to adjust their speed to account for the ball's unpredictability. By contrast, in trials where only information relevant for maintaining LOT was provided, participants could not intercept perturbed fly balls. In addition to demonstrating that strategies involving action-scaled information are specific to an action goal, this study also demonstrated that participants in virtual environments are sensitive to the presence of action-specific sources of information when these sources are present in the visual feedback they receive.

If only some of the action-specific information can be retrieved from a simulated environment, participants may still be able to complete a task, but their performance may be diminished. As we shall see in discussing calibration in section 5.4.5, when a simulated environment does not allow participants to retrieve body-scaled or action-scaled information relevant to their task, participants are likely to make inaccurate judgments of spatial attributes such as size, distance, and the reachability of virtual objects.

5.4.2.2 Avoiding things In any task involving coordination between an organism and their environment, the same action-relevant information can result in different outcomes, depending on the objective of the perceiving organism. In the Chicago Cubs example from chapter 3, the fan who caught the fly ball from his seat in the stand might equally well have decided to duck and avoid the ball. He could have used the same information concerning tau (i.e., time-to-contact) to move out of the ball's path. He was standing, so he would only have needed to modify his body posture so that the ball could pass over his head. But there are many situations in which simply changing our body posture will not protect us from colliding with or being struck by an object. To avoid crashing into someone or something while driving a car or riding a bicycle, we both steer away from the object and apply the brakes. We base our reactions on information about what will happen if we do not change our trajectory. We determine how hard to brake (i.e., how quickly to reduce our current speed) based on our prediction of the distance between our vehicle and the object in our path and on the estimated time-to-contact with the obstacle.

In developing tau theory, David Lee demonstrated that as an individual approaches an object in their path, the information relevant to avoiding a collision is optically specified by theta, the angle subtended by the object, which expands as the individual approaches the obstacle.[78] Lee demonstrated that, in braking, tau (the time-to-contact with the obstacle) is specified by the ratio between theta, the rate of expansion, and theta-dot, the acceleration in theta's rate of expansion. Lee further demonstrated that tau-dot, the first temporal derivative of tau, can be used to control braking because it specifies how time-to-contact changes when the moving vehicle either accelerates or decelerates (see figure 5.4, panel C). When a driver is decelerating at a rate that will bring them to a stop just before hitting the obstacle, tau-dot will be equal to –0.5. But if the driver does not brake aggressively enough, tau-dot will be less than –0.5, and they will hit the obstacle. If tau-dot is greater than –0.5, braking is too aggressive. To control the rate of braking so that it is not too sudden but one still avoids a crash, the driver must keep tau-dot at a value of –0.5. A driver is not thinking this as they try to avoid a crash, but they instinctively control their rate of braking by tracking their movement relative to the position of the obstacle.

When the obstacle is also moving (e.g., it is another car), the driver must consider the rate of closure of the gap between them and the obstacle based on two independent values for tau: the first defined by their own speed, the second by the speed of the other vehicle. Remarkably, in navigating the world, people perform these feats of spatiotemporal coordination with moving objects safely and with little effort.

Whether we are intercepting objects or avoiding collisions, the information in the environment that guides our actions is information we can perceive that concerns what will happen in the future, assuming that current conditions prevail.[79] As Fajen and Phillips note, perception of one's "current future" is not typically regarded as a spatial property.[80] And yet, taking the examples above into account, perception of one's current future is as important in negotiating space as the perception of properties such as size and distance. For the designer of augmented and virtual environments that support the active engagement of participants in controlling their own movement, being able to deliver relevant spatiotemporal cues from which the driver of a virtual vehicle or the catcher of a virtual fly ball can derive tau and its derivatives is key to supporting the successful completion of the targeted tasks.

5.4.3 Reaching for Things

In discussing information that guides people as they get about in the world, I have so far focused on actions that involve whole-body locomotion, such as passing through a gap, running to intercept a fly ball, and braking a car to avoid a collision. But many of our daily activities involve reaching for and interacting with objects that are within, or just beyond, our peripersonal space. Designers of interactive systems and applications are concerned with how opportunities for action facilitated by artifacts we build "turn up" within users' reachable space. In the next chapter, I discuss theoretical and practical considerations around the design of interactive artifacts, whether they are real, virtual, or augmented objects. Here, I will focus on the way information concerning the opportunities for action afforded by such objects is specified for an individual based on the environment, their bodily capabilities, and their current task goals.

If we can understand the sources of information that specify reachability of objects in unmediated everyday interactions, then such information can inform the design of digitally mediated environments as well.

5.4.3.1 Body-scaled, action-scaled, and nested affordances for reaching As with estimating the passibility of gaps or the interceptability of moving objects, estimating reachability requires an individual to perceive their own action capabilities and their action boundaries in relation to the layout of their immediate environment. As Choi and Mark point out, every reaching action is a scaling problem. To pick up or interact with any object in their environment, an individual must be able to perceive the arrangement of surfaces and materials relative to their own reaching capabilities and their posture. They must determine whether they can pick up an object from their current position or whether they will have to locomote to bring the object within arm's reach. In other words, an individual must estimate the distance between themself and the object with respect to their reaching ability. This ability, in turn, is constrained by both body-scaled measures, such as arm length, and action-scaled, biodynamic measures, such as strength and joint flexibility.[81]

Prior research by Carello et al. found that people are sensitive to arm length as an intrinsic, body-scaled unit when assessing reachability.[82] People's absolute critical boundary, the point beyond which they cannot reach a target without bending forward or standing up, was found to be a nearly invariant proportion of their arm length. However, more recent work examining reaching choices made by skilled climbers suggests that action-scaled, rather than body-scaled, information may be more relevant when studying the control of reaching movements that are goal directed. This is because action-scaled information can account for dynamic properties such as joint mobility and level of fitness, in addition to the invariant information provided by body-scaled measures such as arm or leg length.[83] When an interactive system incorporates motion capture, especially whole-body capture, it is tempting to estimate participants' reaching ability based on bodily proportions such as height and arm length. However, the work of Seifert et al. suggests that, in calibrating systems and designing interactions, estimates of reachability need to take into account not just static measures of bodily dimensions but also factors such as level of fitness and skill—measures that more accurately account for an individual's action capabilities.

Most of the time, when we perform reaching movements, we do not push our bodies to the limit of what is reachable for us within a given posture. Instead, we plan our actions to bring the things we need within a comfortable reaching range, such as when we decide to stand up from our chair to take a book from a shelf or when we get down on the floor to retrieve something that has fallen. The absolute critical boundary for a particular reaching movement, such as using only one's arm, does not determine whether a target is reachable because there are often other ways of moving to accomplish the same goal. Choi and Mark demonstrated that when people are asked to reach for targets using any combination of postures, they will often transition to more efficient strategies well before they reach their measured absolute critical boundary for a given action.[84] This "preferred action boundary" is likely to be more comfortable and energy efficient, even if it recruits more bodily resources.

In addition to favoring more energy-efficient modes of action, an individual's choice of reaching movements also depends on the degree to which they are attuned to the opportunities the environment affords, given their capabilities and the layout of surfaces and objects with which they must interact. In part, this includes the ability to perceive their absolute critical reach distance in a given context to determine when a transition to another action mode is preferable or required. To estimate transition points, an individual needs to be attuned not only to a single affordance, such as whether an object is reachable, but also to the opportunities afforded by the environment that will bring more distant objects within arm's reach. When I perceive that I can stand on a box to reach a high shelf, I am perceiving the fact that the box affords being climbed upon and that its height extends my reach sufficiently to retrieve the object from the shelf.

My ability to perceive information that specifies an affordance that, in turn, opens up a further opportunity for action represents attunement to a "nested affordance." In this case, I perceive that climbability of the box affords reachability of the shelf. As Wagman and Stoffregen have demonstrated, the role that a lower-order nested affordance plays in specifying information related to a superordinate affordance is distinguishable from the case where the same lower-order affordance is not nested.[85] In other words, my box's affordance of climbability when I only want to stand on it is not the same as its affordance of climbability when I want to reach a high shelf.

Wagman and Stoffregen proposed that when reaching with a tool, people do not perceive a superordinate reachability affordance as a sum of two subordinate affordances—that is, they do not perceive their critical reaching boundary without the tool *plus* the subordinate affordance of extendability equal to the length of the tool. Wagman and Stoffregen propose that people instead perceive a third superordinate affordance: the capability of the whole person plus tool system. To investigate this hypothesis, participants were shown an object positioned on a wall and were asked to tell an experimenter how to adjust its height until they estimated they could reach it either with their hand alone or with a grabber tool. They found that judgments of maximum reaching distance were not predicted by a model in which individual critical distances for arm-only reaching were added to the length of the tool (i.e., from a model in which both lower-order affordances were combined). They proposed that this superordinate arm/tool system's reaching affordance is consistent with the general tenet of the ecological approach—namely, that affordances within an individual's environment are perceived as arising from a current body–environment relationship rather than being inferred from constituent properties such as tool length, even if those properties themselves specify the availability of lower-order affordances.

Nested affordances are not unique to reaching. In fact, most of the actions we perform require us to perform one action that sets us up for a second action. In running to intercept a fly ball, a player plans their own trajectory with respect to the point at which they will be at the right place at the right time to catch the ball. In the same way, reaching actions that contain subordinate affordances, such as standing on a box to reach something, must be planned with respect to a current future in which one accomplishes their goal, assuming all their intermediate actions succeed.

Given that superordinate reaching affordances are not simply summations of subordinate affordances such as arm or tool length, Wagman and Morgan have proposed that nested affordances are perceived as action-scaled sources of specifying information relevant for accomplishing a future reaching goal.[86] To test this hypothesis, they asked participants to estimate their maximum reaching height when they planned to walk across a room and (1) reach for an object while standing on the floor, (2) use a step to reach the object, and (3) use a handheld tool to reach the object. They found that people's judgments of nested affordances were based on their understanding of their bodily capabilities (i.e., were action scaled), whether they planned to use the step or the tool or no additional object. The researchers also suggest that in planning actions incorporating nested behaviors that alter reaching capabilities in functionally equivalent ways (in this case, behaviors that all lead to acquiring the same object), people may perceive functionally equivalent changes in their maximum reaching heights.[87] This is particularly relevant for the second point I wish to raise here: how skilled individuals who have become attuned to nested affordances can draw on multiple sources of information and choose from multiple ways of accomplishing the same goal.

In a study that examined reaching behaviors of expert climbers, Seifert et al. examined the scaling of nested affordances with respect to arm span (body scaled) and maximum reaching capability (action scaled) estimates of reachability.[88] The researchers also compared nested affordances of varying complexity to determine whether estimated maximal reaching distances to handholds were more likely to be scaled by action capabilities or by body-scaled measures, such as arm span. Their hypothesis was that functionally equivalent behaviors are more likely to emerge for reaching movements involving more nested affordances in situations where climbers can perceive a greater range of possible actions. They found that action scaling of the relationship between climbers' bodies and available handhold and foothold configurations better captured reaching choices than did body-scaled reaching estimates. In climbing a rock face, climbers often reach for targets close to their absolute action boundaries. They do not think in terms of individual reaching actions because each choice of hand- or foothold must set them up for the next. For climbers, perceiving nested affordances as action scaled, as a function of action boundaries, is critical to planning and executing a successful sequence of actions that will bring them to the top of a rock face.

In short, the more complex the superordinate goal, the more likely it is that people will perceive subordinate nested affordances as action-scaled and not body-scaled relationships within the task environment. This is important when analyzing captured movement data, as it suggests that novices and experts may approach action boundaries and transitions between reaching movements in very different ways. When mapping captured motion data onto the response of an interactive system, it is often tempting to treat each movement of a user's body or limb as an independent action that should generate a specific system response. But if participants are organizing their movements in response to affordances that are nested on the basis of action capabilities relevant for specific task goals, then this suggests that interaction designers need to develop more sophisticated, context-aware models of what is and is not reachable for users with a range of experience and skill. Although novice users may break down tasks into smaller units, in time, they may discover opportunities for combining individual actions to accomplish larger task goals more effectively.

As discussed in the next section, the ability to discover nested affordances is often shaped by a very specific intent, such as the intent to use an object that is currently on a high shelf. Indeed, many immersive games take advantage of the fact that participants will discover nested affordances when presented with complex problems to solve. Often, success rests on their ability to evaluate both the action capabilities of their avatar and the objects in their environment to determine whether they can be of use in accomplishing their current goal. For example, can the rock over there be stepped upon to reach the opening of that tunnel? With experience, participants who are more familiar with a game will become more skilled at making choices that will lead to successful outcomes because they will become better at identifying nested affordances in the simulated environment. It could be argued, in fact, that interaction designs that leave open the possibility for discovering nested affordances are more supportive of skill acquisition and therefore are more engaging for their users.

5.4.3.2 The influence of intent on perceived action capabilities In discussing Gibson's theory of affordances in chapter 2, I noted how an individual's intentions determine their perception of opportunities for action. If they intend to knock an apple from a high tree branch, a long stick is perceived as extending their reach. If they intend to protect themselves from an attacker, the same long stick can become a weapon to extend the defendable space around them. When the stick is to be used as a weapon, it is not simultaneously perceived as a tool for harvesting apples. Perception of the affordance of the stick is shaped by *tone*, the individual's intent to act in a particular way with respect to a particular task goal.

In a study that examined the effect of intent to act on reaching performance, Witt et al. compared participant's estimates of reachability in conditions where they reached, with or without a tool, with conditions where they estimated reachability without moving and either held a tool or had empty hands.[89] Their targets were dots projected onto the surface of a table at which they sat. The tool was a 39 cm orchestra conductor's baton.

The distance between the participants' bodies and the projected targets was constant, but their reachability varied. With the tool, all targets were reachable by everyone; without it, only the closest targets were within arm's reach. The researchers found that out-of-reach targets appeared closer when participants actively used the tool to reach, but not when they passively held it in their hands while making oral reachability judgments. When passively holding the tool, participants compressed their estimates of distances to targets within reach compared with their estimates for unreachable targets when they were not holding the tool. This suggests that actively reaching for a target modulates an individual's perception of distance in egocentric space. The authors encourage us to perceive the environment in terms of how we intend to act on our ability to carry out our intentions.

In a related study, Osiurak et al. asked people to make reachability judgments while passively holding either a short or a long baton[90] (see also section 5.3.1.1). They found that participants underestimated distances when holding a long baton compared with judgments they made with empty hands, but that holding a short baton had no effect. The authors concluded that people's estimates of reachability are altered only if the tool they pick up substantially extends their action capabilities—that is, if it is likely to help them accomplish their intended action goal. However, neither of these studies takes into account how an individual's perception of both body-scaled and action-scaled affordances can change over time as they become more attuned to specifying information that can guide their intended actions and as their actions become better calibrated to the task-relevant properties of the environment. The impact of calibration and attunement on people's perception of their task environments becomes even more important when part or all of that environment is simulated because the quality of simulation can greatly influence people's perceptions of their action capabilities.

5.4.3.3 Reaching in virtual environments The reaching tasks we have considered so far have been situated in fully physical environments, but the designer of interactive systems must understand what happens when part or all of an environment is simulated. Does a virtual tool constitute a bodily extension in the same way a physical tool does? If so, are its affordances perceived as body-scaled relations or as action-scaled relations with the simulated environment? Does a virtual tool extend a user's action capabilities as effectively as a physical tool?

Research on the effect of virtual tool use on perception of body-scaled or action-scaled affordances in virtual environments is in its infancy. But preliminary studies suggest that users are sensitive to the affordances that arise from the relationship between their body, their environment, and their action capabilities when a virtual environment contains sufficient specifying information to support their action goals. Lin et al., for example, studied people's estimates of their action boundaries for reaching in a virtual environment following changes in their action capabilities, in this case implemented as changes in the length of their virtual avatar's arm.[91] Participants estimated their action boundary for horizontal reaching following a short period of calibration to one of three virtual arm lengths: long (extended reach condition, 50 percent longer than the participant's actual reach), short (constricted reach condition, 50 percent shorter), or variable arm length. Participants quickly recalibrated their reaching estimates to account for changes in their action capabilities, selecting action boundaries consistent with baseline calibration at the start of the study. They guessed that they could reach farther in the extended reach condition than in the constricted reach condition. When arm length varied randomly, participants' perceived boundaries for reaching more closely reflected the judgments they made for longer reaches than for shorter ones.

The researchers concluded that, in the presence of variability, participants did not make judgments based on average arm length. They made more liberal judgments because when arm length is unpredictable, overestimation of one's action boundary results in the greatest number of successful reaching attempts. This finding suggests that participants are sensitive to the presence of and changes in action capabilities, even when they estimate action boundaries for a virtual limb.

If people can flexibly recalibrate their action boundaries for a virtual limb, must they not be able to do so when they pick up a virtual tool? Evidence is somewhat sparse but points to the need for some form of physical continuity between the observer's body and the tool handle. Studies of extensions of PPS related to tool use demonstrate that when sensory continuity between the hand and the tool tip is not present, as when one uses a laser pointer rather than a stick to bisect a line, people's perception of reachable space does not appear to extend to incorporate the distal tool tip.[92] More recently, Ferroni and colleagues demonstrated that the remapping of reachable space that has been repeatedly observed with physical tool use does not transfer to the use of virtual tools in a simulated environment.[93] In their study, participants performed a pick-and-place task using a grabber tool to move small colored objects from one side of a table to the other in either a real or a fully simulated environment. In the simulated environment, the tool, the objects, and the table were all virtual. The table was always situated beyond arm's reach. Ferroni and colleagues found that participants who manipulated real objects with the physical grabber tool in a real environment had larger PPS boundaries after task completion than they had had before, whereas those who performed the task in the virtual environment did not. The researchers concluded that the lack of physical continuity for participants in the virtual condition resulted in a fundamentally different relationship between their bodies and the task environment, one that did not require an extension of their action boundary.

The study of tool use and its impact on shaping reachable space is of particular interest to designers of interactive systems because tool use appears to highlight a functional distinction between interactions with the digital world that incorporate haptic feedback (through tangible objects or simulated forces) and interactions that break the link between physical and digital worlds. In the line bisection study discussed above, the presence of a physical stick as a pointer functioned as a haptic proxy, a tangible handle that could provide reaction forces when gripped or wielded, but had no programmable features. And yet the physical continuity it provided with the virtual environment resulted in a recalibration of participants' action boundaries for the line bisection task when compared with a laser pointer that projected an intangible dot to mark the extent of its reach. It is possible that the laser pointer is not perceived as extending one's body, and hence one's reach, in the way a physical or even a simulated virtual tool might be.

In a related study, Sengül et al. compared people's perception of their PPS in the presence of both real and haptically/visually simulated golf clubs, whose shafts were either crossed or uncrossed, to determine whether crossing virtual tools reshaped action space.[94] Their work built on a seminal study by Maravita et al. that demonstrated how crossing tools in PPS creates a sensory conflict between tactile and visual stimuli when the tactilely stimulated hand holds a tool whose tip is visually located on the opposite side of the body.[95] Sengül and colleagues compared real golf clubs with robotically simulated clubs that had physical handles but visually simulated shafts and club heads, as well as vibrotactile feedback simulating contact with virtual objects.

They also compared conditions in which participants actively crossed and uncrossed golf clubs to conditions where these actions were performed by an experimenter.

They found that actively crossing clubs resulted in remapping participants' PPS so that they came to associate tactile stimuli in the hand with the tip of the tool, even when tool tips were crossed. They also found that, unlike remapping for physical tools, remapping for virtual tools occurred regardless of whether participants actively manipulated the tools or passively observed changes in their configurations. The researchers concluded that, during active use, the virtual robotic tools functioned to extend their user's arms into the virtual environment and produced a strong association between vibrotactile feedback on their hands and the visual position of tool tips, regardless of whether tools were crossed or uncrossed.

In a recent review of manipulation techniques for avatars in 3D virtual environments, Dewez and colleagues suggest that preserving continuity between the body of the VR participant and the hand of a virtual avatar is essential for promoting the participant's sense of ownership of the avatar's limb.[96] When objects are out of reach, it is better to elongate the avatar's arm or simulate a robotic arm extending from the user's position rather than to break the sense of continuity. If the choice is between extending the limb of one's avatar and selecting a tool such as a virtual grabber to do the same task, then the attentional switch required to find the virtual tool and assess its suitability may not be worth the cost of distracting the user from their current task. Where it makes sense to require a user to switch their focal awareness, as when starting the next step in a virtual assembly process, then asking participants to select a new tool, may be helpful in defining the boundary between tasks.

Although the ability of tools to extend PPS in VR does not require that those tools be perceived as part of the user's body in the same way that a limb might, this extension of PPS still requires that the tools be perceived as continuous with the body—that they be held in the avatar's hand as opposed to being detached.

Continuity also requires the simulation of congruent visual and tactile feedback in response to reaching movements. Given the evidence from both the line bisection and club-crossing tasks reviewed here, it seems that this is true whether haptic feedback is provided by a tangible proxy (the pointer) or using partially simulated haptic responses, as in the case of the club-crossing task. But do the presence and form of an avatar change people's perception of their action boundaries for reaching in a virtual environment? In general, the presence of self-avatars has been shown to improve a user's sense of presence in VR—that is, their sense of ownership of their virtual visual representation and the degree to which the avatar embodies their action capabilities. Ebrahimi et al. hypothesized that a higher-fidelity anthropomorphic visual avatar would result in more accurate estimates of action boundaries for reaching.[97] They asked study participants to estimate whether they could reach visual targets in one of four viewing conditions to which they were randomly assigned: high-fidelity self-avatar, low-fidelity self-avatar (where only joint position and motion were shown), VR with no avatar, and a real-world control condition. The researchers found that both verbal and action-based reachability judgments were least accurate for the no-avatar condition and most accurate for the real-world condition. In all cases, a five-minute calibration phase for VR conditions improved both action-based and verbal reachability judgments, achieving the highest accuracy in the high-fidelity self-avatar condition and the lowest in the no-avatar condition.

Ebrahimi et al. thus suggest that for VR applications that require only accuracy in reaching, the no-avatar condition might be sufficient provided that a calibration phase is incorporated into the task environment. In tasks where accuracy in both reach boundary estimation and reaching performance is required from the outset and where no calibration phase is available, such as in simulations of different body morphologies for surgical training, then high-fidelity self-avatars are likely to support better performance. As a user's representation in a virtual environment, an avatar becomes the medium through which users move through virtual spaces and manipulate virtual objects. The avatar's body becomes the user's primary spatial reference, shaping how they perceive themselves and how they understand their action capabilities in the virtual world. High-fidelity self-avatars can therefore bridge the gap between a participant's sense of embodiment, their perception of their action capabilities, and their accuracy in task performance, all promoting a tight sensorimotor coupling that supports perception of both body-scaled and action-scaled affordances for reaching within virtual environments.

The choice of whether to use an accurately calibrated self-avatar for a particular interaction design will therefore depend on the level to which a designer wishes a user to draw on their understanding of their own action capabilities. It may not always be necessary to incorporate costly calibration hardware and algorithms into a system if the reaching tasks to be performed are relatively straightforward. However, a well-calibrated self-avatar may be invaluable in reducing time and algorithmic complexity for environments where the aim is to train users to draw on their own action capabilities to learn skills that they can transfer to real-world contexts (such as surgery).

5.4.4 Nonvisual Sources of Information Specifying Action-Scaled Affordances

So far, this discussion has focused on potential sources of information in optic flow on which an individual can draw to guide current and future actions. But, as noted in section 5.2, the environment also contains other sources of energy to which we are sensitive. Like optic flow, acoustic and mechanical sources of energy can be structured by bodily movement and the movement of external objects. Once we become attuned to how our movements alter these sources of energy, they can provide information that specifies action-relevant properties of the spaces we inhabit. The following discussion will illustrate particular cases of action-relevant sources of nonvisual information identified for the acoustic guidance of locomotion and nonvisual guidance of reaching.

When we move through the world, our movements are guided by many sources of energy in the global energy array that specify our relationship to the objects around us in the environment. As designers, we tend to focus on what we can pick up visually, ignoring the ways in which auditory and/or tactile stimuli also guide movement. Think of how often you have walked out of your house and locked your door while looking down at your phone, of how you were peripherally aware of the sound of the closing door and the feel and sound of the lock as it slid into place. While we are less likely to recall the sounds our feet made as we walked through a familiar neighborhood than the visual style of the houses we passed, these sounds too form part of our bodily memory of a place. And yet, many immersive environments fail to simulate changes in the sound of footsteps as characters move past objects with different material properties (trees vs. walls, etc.). Taking account of such sources of information in the environment and understanding how they work together to guide our actions when we design interactive systems is likely to contribute to a more immersive and engaging experience, even for tasks such as manipulating digital artifacts embedded in augmented spaces.

Many examples cited here focus on studies where visual information was excluded, either because participants were blindfolded or because they have no useful vision. These observations depend on the degree to which people are attuned to sources of information other than optic flow. Although investigations of perceptual invariants used to guide action have often been studied independently, in real life, people do not consciously select the sense-specific source of information to which they attend. As Stoffregan and Pittenger point out, one action may generate structure in the global energy array that can be picked up by more than one sense modality.[98] Rather than being redundant, these different sources of information can make apparent higher-order relationships. The information can add up to more than the sum of its parts. In highlighting potential sources of specifying information identified for the acoustic and haptic guidance of action below, I intend not to suggest that these sources alone are optimal for guiding action. Rather, these cases demonstrate that in instances where, for example, visual information is not available, nonvisual information can be sufficient for guiding action, particularly when individuals are attuned to its potential to do so.

5.4.4.1 Acoustic information for moving and acting in space In chapter 3, I noted that information about the movement of sound sources in space is specified by a combination of variables, including the differences in timing of the arrival of sounds at both ears (the ITD), differences in the intensity of sounds at both ears (interaural intensity [IID]), and differences in the spectra of sounds, at each ear, that might arise because of the acoustic shadow created by the ear pinna, the head, and reflections from the shoulders and upper torso (typically referred to as the head-related transfer function [HRTF]). Together, these cues, when combined with head movements, allow us to pinpoint the location of a sound in 3D space, although our accuracy is much greater in the horizontal plane around the head than when identifying the height or elevation of sound sources.[99] In the same way that patterns of light and shadow in optical information tell us about the relative position and motion of objects in our surroundings, correlations between IID, ITD, and HRTF cues inform us about the locations of objects that emit sounds.

Most of the time, people do not pay much attention to sound reflections as they move about because vision is far more effective at conveying the information required to negotiate space. But there are times when people do notice how a space responds to sounds, such as when surfaces in a restaurant are *too* acoustically reflective. This raises the overall sound level in the space and blends information from many sound sources, making it hard to hold a conversation. However, as noted in section 5.2, sound reflections can be useful on occasion.

In humans, echolocation has long been known to be used by blind people as a means of detecting and avoiding obstacles. In 1749, the French philosopher Denis Diderot wrote of a blind acquaintance who could locate "silent" objects and judge his distance from them.[100] Although studies of human echolocation date back almost a hundred years, the challenge researchers face has been to identify the action-specific sources of information within the evolving spectra of reflected sound that are relevant in any given context. With the advent of high-fidelity hardware and software for rendering complex 3D audio environments, combined with the means for accurately tracking the movements of people's bodies and heads, it has become easier to systematically manipulate the way in which an environment responds acoustically to people's actions and to observe the action-relevant properties specified by acoustic information.

In their survey of this work, Kolarik and colleagues identified seven sources of information that have been consistently shown to contribute to the acoustic guidance of locomotion and action for both blind and sighted-blindfolded participants.[101] Most of these sources rely on people tracking the ways in which self-produced sounds, such as those from their footsteps, are transformed when reflected back from the environment. They also mention in passing a seventh cue, acoustic shadowing, on which I elaborate further in point 6, below:

1. Energy: When sounds you make as you move through your environment are reflected from nearby surfaces, the overall acoustic energy arriving at your ears is increased. When the surface is so close that the reflection happens in a very short time (e.g., tens of microseconds), people report that they hear a change in the quality of the sound, while larger distances creating longer delays (up to 10 ms) subjectively increase the overall loudness of the sound.
2. Time delay: When surfaces are farther away and the time delay between the emitted sound and its echo is in the range of 10–30 ms, the echo is often perceived as a discrete event, sometimes called a "repetition pitch."[102]

3. Changes in spectrum: In certain cases, the spectrum of the emitted sound and that of its reflection overlap, such that some returning wavelengths, add energy at particular frequencies, and others cancel each other out, taking energy away. The combined spectrum at the location of the ears will therefore have some peaks and troughs that relate to the distance from the surface. As you move toward or away from the reflecting surface, the length of the combined path from your noisy feet to the surface and back again will also change, changing the frequency of the resulting spectral peaks. This may be perceived as a change in timbre and/or pitch as the observer moves toward or away from the reflecting surface. It is often not clear in these cases whether people track changes in pitch or timbre.
4. Binaural cues: As noted above, differences in the relative intensity, timing, and spectra of sounds arriving at both ears are used as cues for localizing sound sources, particularly when sounds contain higher-frequency components. For reflected sounds, too, binaural cues can be useful in identifying, for example, the orientation of objects and surfaces relative to one's path of travel. When approaching a flat surface perpendicular to one's path, the timing, level, and spectra of sounds reaching both ears are similar. If the surface is at an oblique angle, as when one approaches an angled wall, for example, the sounds reaching each ear will differ, particularly at high frequency (as illustrated in figure 5.5).
5. Room reverberation cues: When sound propagates in rooms, it reflects off each surface with which it comes in contact. The relative timing of reflections, with respect to the original sound, relates to the path that sound has taken from its source to the listener's ears. If you can hear your footsteps as you walk through a room, the timing of reflections will provide information about the distance between you and each surface in the room. Moreover, the material from which surfaces are made—wood, carpet, glass, and so on—will shape the spectrum of the sounds' reflection as you hear them. This spectral variation therefore contains information about the material properties of surfaces within a space. Glass, for example, reflects high frequencies, whereas curtains do not.
6. Acoustic shadows: If a room contains an obstacle such as a pillar or large piece of furniture, the sound of your footsteps as you pass this obstacle will be altered in several ways. First, the material from which the obstacle is made will reflect or absorb the sound of your footsteps in ways that

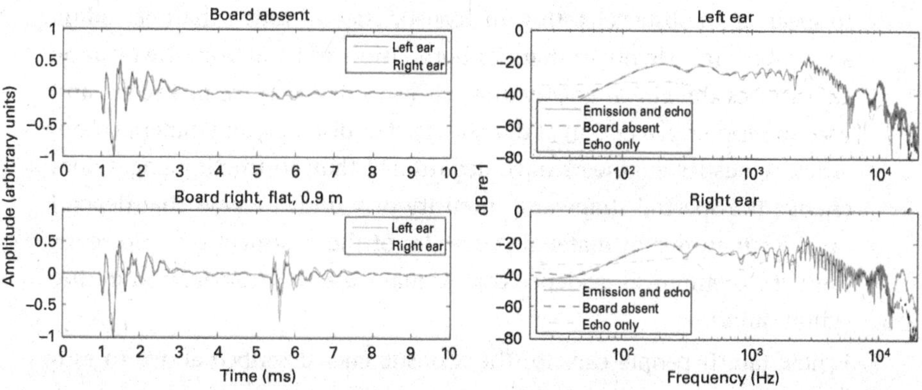

Figure 5.5
Waveforms and spectra of impulse responses at both ears during echolocation. Top left: Waveform response associated with the emission, which appears at 1:00 p.m. and decays in about 4 ms. Differences in ears represent asymmetrical head and ear geometry. Bottom left: A second echo response at 5 ms due to the presence of a board. The echo arrives at the right ear slightly before the left, with higher intensity. The graphs on the right represent the spectra of the impulse responses in each ear, representing the emission only, emission and echo, and echo only. *Source:* Rowan et al., 2013, figure 2. Used with permission from Elsevier Science & Technology Books. Permission conveyed through Copyright Clearance Center, Inc.

may be different from the room as a whole. Second, being closer to you, the obstacle will return reflections much more quickly than the walls, which are farther from you. Such changes will happen whether you move toward the obstacle or the reflecting surface moves toward you. Gordon and Rosenblum point out that the ways in which an obstacle positioned between a sound source and a listener absorbs, reflects, mediates, or refracts components of its spectrum provide information useful for someone navigating a space without vision.[103] Similarly, when you walk down a hallway, an open doorway creates an acoustic "hole"—a sudden change in the pattern of reflected sound where energy appears to be sucked into an adjoining acoustic space. As James Gibson noted, obstacles that cast an acoustic shadow against an otherwise consistent backdrop are not the same as obstacles that visually occlude other obstacles because some acoustic energy from the distant surface may be diffracted around the obstacle to reach the listener, usually by an indirect path (by a reflection from another surface).[104] Gibson suggested that it is therefore better to speak of an obstacle casting an acoustic shadow rather than occluding a sound source. He noted that, for both vision and audition, the obstacle restructures the global energy array in ways that may be picked up and used to support perception of the object's size, distance, and material properties. Acoustics engineers have determined that obstructing edges does change the spectral shape and intensity of sounds in ways that depend on the structure and material properties of the obstruction.[105] However, there is no empirical evidence that humans use this information during echolocation.

7. Echoic tau: If people can use the acoustic cues described above to estimate the distance between themselves and structures in the environment, then they may be able to estimate tau, the time it will take them to encounter the obstacle, based on their current movement velocity. In particular, the remaining distance to the source (here, the surface from which sound is being reflected) is given by multiplying tau (time-to-contact) by the moving person's current velocity. Acoustic tau therefore relies on tracking the rate of change of acoustic cues, not on estimations of their static values. David Lee et al. determined that bats are able to use acoustic tau to intercept moving prey and avoid obstacles, and he suggested that humans might have access to similar information.[106] Stoffregan and Pittenger have proposed that people may estimate their movement velocity from motor and kinesthetic feedback about their speed of walking, whereas distance to the obstruction may rely on the acoustic cues described in points 1–5.[107] If a person approaching an obstacle can move such that they can keep acoustic tau constant, then they will come to a halt just as they reach the obstacle's surface.

As noted at the outset of this section, the acoustic information specifying the position and motion of objects available in the global energy array, such as the information described here, is available to us whether we choose to draw on it or not. Often, acoustic cues direct visual attention to something happening nearby, such as the use of directional acoustic cues when someone behind us calls our name or the use of acoustic tau to judge the distance of a moving cyclist's bell when they approach us from behind. Other cues, such as the changing sounds of footsteps as one approaches a wall, can reinforce one's sense of place (e.g., in a virtual environment) because acoustic information can encapsulate information about the size and material properties of parts of the environment that are not immediately visible. In passing between spaces in a virtual environment, for example, well-designed acoustic cues can give a participant the sense that they have moved into a small, cramped place long before it is rendered visually by reducing the time reflections take to return from surfaces and changing the spectrum of reflected sounds to encode changes in their material properties. As we shall see in chapter 6, even in interactions with augmented objects, acoustic cues—such as sounds produced by tapping—can be manipulated to simulate changes in their material properties. In short, action-produced sounds are a rich source of information that can support designers in guiding the actions of participants in real and virtual task environments.

5.4.4.2 Acoustic specification of action-scaled affordances If movement and action can structure energy in the acoustic array such that it is sufficient for estimating distances and detecting obstacles, then the open question is whether such information is scaled by either the size and shape of the observer's body or by an individual's action capabilities. If so, then evidence for body-scaled and action-scaled affordances should be apparent in acoustically guided tasks that are similar to those used to uncover specifying information for visually guided action. Below, I will discuss studies that have uncovered evidence for specifying information in the acoustic array that is used to estimate gap passibility and to intercept and avoid moving objects.

As I have noted, people are remarkably good at identifying structures in their surroundings based on changes in acoustic information, conveyed by reflections of self-generated sounds such as footsteps. Reflections of other environmental sounds, too, can sonically illuminate an environment, revealing the sizes of rooms or the presence of obstacles. The growing body of work on human echolocation has identified many properties of reflected sound that provide information about objects' size, distance, and so on. Less well understood is how these sources specify what is possible for the individual navigating an environment using echolocation instead of vision. In the same way eye height or stride length serve as body-scaled affordances to scale information in optic flow, there must be equivalent scaling of information in acoustic flow, specifying possibilities for individual action based on bodily dimensions and action capabilities. Research in this area is sparse, but evidence suggests that action-scaled and body-scaled affordances for tasks such as passing through gaps and intercepting and avoiding obstacles can be specified by information in the acoustic energy array.

In an acoustic analog of Warren and Whang's gap-passibility study, Riehm and colleagues asked participants to estimate the passibility of static gaps rendered as virtual sound sources in a 3D acoustic virtual environment.[108] Their goal was to determine how blindfolded, sighted participants would use acoustic feedback—and which acoustic sources they would draw on—to judge gap passibility compared with how they had used visual feedback. Participants had to judge whether they would be able to walk between sound sources without needing to turn their shoulders. Researchers hypothesized that participants would base judgments on an informational variable both body-scaled and action-scaled, one that would make apparent the relationship between head rotation, forward locomotion, and gap width. By tracking body and head movement and systematically varying the gain between each participant's head rotation and the position of virtual sound sources, the experimenters determined that people could accurately judge the sizes of gaps relative to their shoulder width and that self-generated changes in acoustic flow, such as exploratory head movements, provided salient information specifying gap passibility.

However, manipulating gain of head rotation did not improve performance. Instead, participants who used lateral head movements, where they moved their heads side to side relative to the acoustic image, were much more accurate in judging passibility than those who only rotated their heads to estimate gap size. In rotating their heads, participants are, in effect, decoupling acoustic cues about head motion from the position of their shoulders and hence their bodies in space. It would be interesting to investigate whether this is the source of reduced performance in the head rotation condition. If so, this has implications for acoustic rendering in virtual environments because it suggests that tracking the position of both the head and body can lead to better performance in estimating the passibility of virtual gaps based on acoustic information alone.

In a second experiment, Riehm et al. asked participants to adjust the size of the gap between their two sound sources from a fixed distance.[109] Participants could freely move their heads but could not walk toward the acoustically specified gap. Gaps were presented in interleaved descending and ascending series to determine the critical boundary at which participants indicated they could pass through without needing to rotate their shoulders. The study revealed a critical boundary of 1.56 times shoulder width—larger than the value found by Warren and Whang for visual gap-passibility judgments. Riehm's team note that this might not necessarily reflect the accuracy with which experienced echolocators can judge passibility because participants in their study, who normally make such judgments visually, may be insufficiently attuned to relevant information in the acoustic energy array. Citing work by Jacobs and Michaels,[110] the researchers note that people placed in novel situations often attend to information that is non-specifying (i.e., they attune to sources of information that are not optimal for guiding their current movement). We will return to the question of attunement to task-relevant information in section 5.4.5.

Blind people often have to intercept moving objects based on acoustic information, such as when they cross a road that lacks a traffic control signal. As noted in chapter 4, David Lee and colleagues proposed that people crossing a road in traffic must couple the taus of two perceptual gaps: one specifying the time it will take to cross the road at their current rate of movement and one specifying the time it will take an approaching vehicle to reach them at its current speed or at a changed speed if it slows down.[111] In most situations, people make accurate road-crossing judgments based on visual information. However, as Hassan has demonstrated, people are less accurate when estimating crossability based on auditory information alone.[112] Hassan compared responses for sighted, low-vision, and blind participants and determined that all three groups underestimated their ability to safely cross roads when they could only hear traffic on a virtually simulated street, and that blind participants—who performed worst—were the most likely to be hit.

In the real world, acoustic judgments of crossability can be influenced by many factors, making these judgments even more difficult. In a comprehensive study involving participants who were experienced blind travelers, Emerson and Sauerburger asked them to observe a number of crossing configurations where no traffic control signals were present and to indicate when they first heard the sound of oncoming vehicles.[113] For these uncontrolled crossings, blind pedestrians are taught to wait until the street is "quiet" before starting to cross the road. Here, Emerson and Sauerburger wanted to determine the level of background noise considered "quiet" for this task and also to determine how judgments of crossability were impacted by structures in the environment that could alter sound level (the presence of hills, bends in roads, parked vehicles acting as baffles, or trees that absorb sound). Crossing locations and observation times were chosen such that there were distinct gaps between vehicles in a steady traffic flow.

The first measure of crossability was detection time, defined as the time between initial detection of the sound of a vehicle and the vehicle passing before the participant. This ranged from 2 to 35 s, with a mean value of 9.6 s. The second factor in determining crossability was a measure of the safety margin, defined as the difference between detection time and the estimated time it would take the participant to cross the road. All roads had two lanes, each 10 ft wide, and two safety shoulders of 4 ft, for a total width of 28 ft. The time required to cross the road was estimated as being 7 s based on a walking speed of 4 ft/s, the average walking speed of pedestrians crossing streets reported in the literature.[114] The average safety margin, across all environmental conditions, was found to be 2.63 s, although values ranged from 5 to 28 s. A value of –0.5 s would mean that the approaching vehicle arrived at the crossing point 5 s before the pedestrian completed their crossing. Higher safety margins were associated with straight, quiet roads where people could detect oncoming traffic from much farther away. The negative values for safety margins reported by Emerson and Sauerburger support their hypothesis that there are road configurations where crossability judgments can result in negative values for the safety margin (i.e., where participants do not hear vehicles in time to make safe road crossings, even in quiet conditions).

Of interest here is the finding that the greatest impact on safety margins was the level of background noise, not the speed of approaching vehicles or their engine noise. This suggests that acoustic cues used by people to navigate their environments are greatly impacted by the presence of other cues and by the degree to which the spectra of irrelevant sounds mask task-relevant acoustic cues. Although the goal of Emerson and Sauerburger's study was not to determine the invariant patterns in the acoustic array that people draw upon in making judgments concerning safety to cross roads, this work illustrates some of the environmental factors that influence people's estimation of the size of gaps in moving traffic deemed to be passable.

The interaction between sounds that structure the acoustic environment in ways that provide task-relevant information and those that do not informs the design of both real and virtual acoustic environments because it highlights the role that context plays in shaping how people attune to information in their surroundings. In their study of acoustic cues used by blind people in wayfinding in urban environments, Koutsoklenis and Papadopoulos list ten cues that were most commonly reported by blind participants as providing important acoustic information.[115] More significantly, the researchers report the kinds of information to which the blind travelers were attuned for each sound source. For example, a constant flow of traffic parallel to the line of travel when crossing a road serves as a guide—a continuous wall of sound along which one can walk to maintain a straight line. In a different context, such as when one needs to cross the street, this same flow of traffic indicates that the traffic light is green and the pedestrian light red, and that it is not safe to cross. The authors also point out ways in which acoustic cues in urban environments serve to structure spaces. For example, the sound of children playing is a permanent cue to the position of a playground, and the sound of people entering and leaving stores fixes the positions of doorways along a street.

Acoustic space can provide novel opportunities for action in VR gaming environments. If your goal in a game is to escape a pursuer and you are on foot, you can take advantage of loud places in the virtual world to hide the sounds of your feet from your pursuer in the same way you can use virtual shadows to conceal yourself from their view.

As previously noted, players running to intercept balls control their running trajectories on the basis of information specifying two sources of movement: their own motion and that of the approaching ball. We also noted that attunement to such specifying information in optic flow can lead to successful ball interception strategies—for example, the strategy of altering one's speed to cancel out acceleration/deceleration in the ball's trajectory using optical acceleration cancellation. Many researchers have sought to determine whether acoustic flow contains specifying information, similar to OAC, that can guide a person in intercepting a moving object such as a ball.

One challenge in comparing specifying variables between optic and acoustic arrays is that visual information specifying spatial dimensions of an illuminated environment is persistent, whereas acoustic information specifying spatial dimensions comes into being only when a sound source is present. This means that judging the position of an approaching object based on acoustic information assumes the object is emitting a sound. If it is, the question becomes one of determining what information is conveyed by changes in the structure of the acoustic array that specify the position and trajectory of a moving sound source, such that it is possible to estimate tau, the time-to-contact with the object. Gordon and Rosenblum have demonstrated that people can alternate between visual and acoustic sources of information present in the global energy array when estimating the time-to-contact of an approaching object, indicating that tau may be specified by higher-order invariants accessible through more than one form of energy.[116] However, even if acoustic information specifying tau is available, Vernat and Gordon suggest that it may not be sufficient to support the kind of online modification of action required to compensate for changes in the trajectory of moving targets that is possible when action is visually guided.[117] To test this hypothesis, they asked blind, sighted, and blindfolded-and-sighted participants to intercept moving targets by rolling balls toward them.

Because neither experienced blind listeners nor sighted but blindfolded participants achieved performance close to that of sighted participants, the researchers concluded that their task did not capture a source of information that could specify acoustic tau. Yet, in contexts where acoustic information is more salient with respect to a task goal, blind people have demonstrated that when they are attuned to acoustically specifying information, they can intercept moving targets as efficiently as sighted people using optic flow can. Mieda and Kokubu compared performance of experienced blind players who participate in the competitive sport of blind football with the performance of sighted but blindfolded participants in a task where all participants had to use their right foot to trap a ball that rolled toward them along the ground.[118] The researchers used an International Blind Sports Association–approved football designed to provide continuous auditory feedback in the form of a series of beeps as it moved.[119]

Mieda and Kokubu found that experienced blind footballers were more accurate in intercepting balls than their sighted but blindfolded counterparts. Blind footballers also consistently rotated their heads downward toward the ball shortly after it launched and kept it in that position until they intercepted the ball with their foot. By rotating their heads downward while monitoring the balls' trajectory, blind footballers could use interaural timing and IID cues to keep the ball centered in an egocentric frame of reference relative to their heads throughout its journey to their foot. They had learned to coordinate the movements of their heads with the movements of the ball using invariant properties of the acoustic array, just as baseball players attune to OAC as an invariant property of optic flow that can guide them to catch fly balls.

Mieda and Kokubu's research, like the other work cited in this section, has focused on using acoustic tau for guiding the movements of blind participants intercepting moving objects that have been augmented to produce sound output designed to make them trackable. Under normal circumstances, flying and rolling balls do not produce sufficient sound to be heard. But the fact that acoustic tau is a sufficiently robust source of information for intercepting moving objects may have application well beyond accessible sports. It is not hard to imagine how making the trajectories of invisible flying objects audible in immersive games, for example, could take advantage of the fact that participants could use acoustic tau to intercept them. Blind people are not exceptional in their ability to perform such tasks. Rather, they have just learned to draw on different sources of information present in the same global energy array that is available to sighted people. What these studies have uncovered is the capacity of participants to discover alternative salient cues to guide their actions—cues that could be used to extend the palate of available sources of information in the design of any augmented or virtual environment.

5.4.4.3 Haptic specification of information for moving and acting in space In chapter 3, I noted that touch is a complex sense modality whose receptors are not encapsulated in organs such as eyes or ears. Although the skin is the organ that contains the most touch receptors, the hands are used to pick up most tactile information. Gibson proposed that the hands are the organs of touch because they can both manipulate objects to sense their properties and act through objects to manipulate and sense the world.[120] As Berthoz's example of the man trying to stay upright in his boat (figure 5.3) demonstrates, our actions toward the world are situated in a body-centered reference frame that is normally oriented with respect to gravity. For this, we rely on proprioception, our sense of our body's position and motion, as picked up through vision and audition but primarily through receptors in the skin, muscles, and joints—our haptic proprioceptive system.

Much work on haptic spatial perception has been concerned with our ability to sense the relative position and motion of stimuli on the surface of the skin.[121] What concerns me here is not such tactile spatial perception per se but rather how the senses of touch contribute to our sense of space. In particular, what kinds of specifying information are present in the global energy array for the haptic guidance of action? In what ways might such information support us in getting about in the world?

Although proprioceptive feedback can serve as an additional source of information for moving through the environment, our relationship to it is flexible, allowing us to adapt rapidly when task constraints or environmental conditions change. This ability to recalibrate the relationship between our sense of self-motion and the feedback we receive about our actions from the environment, from vision or audition or touch, is central to our ability to adaptively cope with changes in our surroundings and to continue to acquire new bodily skills. In comparison to vision and audition, haptic information picked up from the environment is usually transmitted to the body by a mechanical coupling. In many cases, the body is coupled to the environment directly via contact with hands or feet, but often haptic feedback is picked up through mediating objects such as handheld tools. Unlike sources of auditory or visual information, which can be far from the body, sources of haptic information tend to be located in peripersonal space—even in teleoperated environments where the source of the haptic stimulus is remote, the force-reflecting apparatus that the operator uses is located in PPS.

Current research suggests that the haptic system makes flexible use of mechanical invariants present in the stimulus flow. This poses a challenge for current haptic technologies, which typically render a single form of stimulus: tactile dots and textures that stimulate mechanoreceptors in the skin, vibrotactile patterns that stimulate receptors in the skin and joints, or grounded force-feedback devices that deliver reaction forces sensed by receptors in muscles and joints in response to hand or body motion. To provide a sufficiently rich haptic energy array to support users in exploring many available sources of information so that they can attune to those most salient, it is currently necessary to combine different technologies and ensure that they deliver colocated stimuli in response to the tracked movements of the hand or body. In the study of a virtual cane guiding blind users through a virtual environment discussed in section 5.5, this colocation is accomplished by mounting a vibrotactile display within a proxy object, the handle of a physical cane, to render sliding contact with the surface of a virtual path directly to the hand of the user. In addition, a body-mounted force-feedback device renders contact forces when the virtual cane penetrates virtual walls or the boundaries of virtual objects. Finally, the sounds of the contact between the virtual cane tip and the environment are rendered using binaural audio to orient the egocentric space of the haptically encountered environment within the allocentric reference frame of the acoustically rendered virtual world. It is only when both modes of rendering are combined and both egocentric and allocentric reference frames are aligned that the blind virtual cane user can selectively attune to different sources of relevant information as they move about in the virtual environment. As in the real world, the sounds of contact with objects in PPS will echo from objects beyond arm's reach, acoustically illuminating the world and allowing the person to anticipate what they are about to encounter.

Like far and near vision, acoustic and haptic information combine to support the virtual traveler in detecting objects in their local surroundings while still planning their path through the global virtual world. Stoffregan and Pittenger go so far as to suggest that echolocation, like active touch, is fundamentally dependent on an individual's ability to correlate their actions with responses from the environment.[122] In echolocation, a self-generated sound results in reflections from the world that convey information about its layout. In active touch, one explores the world by moving one's body to stimulate the haptic senses, picking up information about the texture, form, spatial layout, and motion of things in one's immediate environment. It is no surprise, then, that the blind cane user relies on synergy between acoustic and haptic information arising from self-motion, and that, combined, these two sources can support an integrated, nonvisual view of personal, peripersonal, and extrapersonal space.

The success of the virtual cane project is not in developing new technology per se but rather in realizing that grounding the project in the action capabilities of blind users would result in a solution that prioritized the task-relevant sources of information for blind people navigating through the world. Furthermore, the abundance of information-specifying properties like the size and material composition of the VE, such as coherent spectral and tactile renderings of hard surfaces, supports a user in selectively attuning to the properties of the environment that are most salient for their current task goal.

5.4.5 Attunement and Calibration: An Addendum

Designing spaces where users can actively explore sources of available stimuli to discover invariant structures in a simulated global energy array is as important as developing the hardware and software to render the stimuli in the first place because it is through such active exploration that people will become skillful users of the digitally augmented tools and environments we design. As I noted at the outset of this discussion, the spatial properties of the environment are specified for an individual on the basis of discoverable opportunities for action. Further, as the examples of gap passibility, object interception, and reachability have demonstrated, the opportunities for action the environment affords are relational, arising from coordination between the structure of the environment, the individual's bodily capabilities, and the goals and constraints of the current task. For instance, in reaching for an object, the task goal is to acquire the object, and the task constraints arise from the action boundaries of the reaching person, given their current posture, level of fitness, and so on.

In most of the studies cited so far, the perception of specifying information for an affordance, such as passibility, has been posed as a task goal, with passibility judgments based on short-term encounters with a relatively constrained task environment. Such studies do not provide opportunities to observe longer-term changes in performance that might arise as people learn to distinguish between relevant specifying information for guiding action and information that is correlated with a given movement but is nonspecifying in relation to one's action goal. Further, these studies do not typically account for people's abilities to actively rescale their actions in response to a dynamic task environment. As we noted in discussing Bernstein's work with blacksmiths, people readily adapt their actions in response to changes in task constraints, reconfiguring their bodily resources as they attune to the changing requirements of their goals.[123]

Withagen and Michaels suggest that such perceptual learning is supported by two processes: attuning to specifying sources of information and recalibrating one's actions so that they are appropriately scaled to map to the information that is perceived.[124] People tell you that once you learn to ride a bike, you will never forget. In other words, you have learned to attune to the information that specifies how to keep the bike upright. Similarly, when you learn to estimate how quickly to cross a road, you have learned to calibrate your actions to close the spatiotemporal gap between you and the far side of the road while avoiding traffic. If you purchase an e-bike, you have to recalibrate your actions to account for an altered ratio between your rate of movement and the speed of the bike. But you most likely do not need to attune to different specifying information to guide your actions. Such recalibration is necessary simply to ensure a consistent relationship between the units in which relevant specifying information is perceived and the units in which your actions are executed.

Withagen and Michaels, among others, point out that the process of acquiring a new skill is a process of discovering and attuning to appropriate specifying information for a given action and then learning how to recalibrate one's actions to account for perturbations in the task environment. When learning a new skill, a novice may initially attend to a non-specifying informational variable that has an ambiguous relationship to the property relevant for their current task. For example, children who are learning to ride bicycles are more likely than adults are to misjudge the relationship between the width of a road, a gap in traffic, and their action capabilities when accelerating to cross roads.[125]

A large part of improving one's accuracy in performing a task, Withagen and Michaels argue, is in converging on specifying variables that have one-to-one relationships with the property of interest. This convergence toward specifying informational variables is supported by feedback gained through experience, but it can be accelerated through instruction, as in the case of supervised training in sport. Wagman and colleagues discovered that when a participant was asked to perceive the width of wielded objects through touch alone, they initially attuned to mass as the variable they thought specified the object's width.[126] When guided by the experimenter, they gradually learned to attend to the third moment of inertia, the resistance to motion of the distributed mass as the object is wielded in the hand. Once they were attuned to the appropriate specifying information, the next step was for participants to calibrate the perceived length to measurable units so that they could more accurately describe the differences in width in terms of, for example, inches. Withagen and Michaels argue that to make accurate judgments, individuals had to be attuned to appropriate specifying information and had to calibrate their judgments to accurately reflect the relationship between bodily and environmental units of measurement.

In the foregoing discussion, I have noted how attunement to specifying information impacts perception of space, particularly as the perception of information relates to the guidance of action. The work of Withagen and Michaels would further suggest that, with training, people's accuracy in judging task-relevant properties of the environment can improve. This also appears to be true for the perception of spatial properties of the environment. As already noted, the space of reachable objects is extended when we are holding a long tool. As Witt notes, "Tools do not just expand our capabilities.

They change what we can do, and in doing so, they change who we are. Serena is Serena because of what she can do with a tennis racket. Tiger is Tiger because of what he can do with a golf club. In changing what we can do, tools also change the very way we perceive the spatial layout of the world."[127]

For Serena Williams and Tiger Woods, the space of possible actions is defined by their attunement to specifying information and shaped by their understanding of their bodily capabilities. They can outperform their competitors because they can both perceive and act upon their environments in ways that others cannot. Their perceptual motor experience and their highly developed bodily capabilities allow them to discover task-relevant affordances specific to their domains of expertise. As I noted in chapter 2, discussing the concept of enactive landscapes, this is true for any skilled performer: the chef whose enactive landscape is shaped by food-preparation tasks and the tools she has at hand, or the potter whose enactive landscape is shaped by the clay he is working and the form of the pot he intends to create (see section 2.4 for further discussion).

Both attunement and calibration to information specifying the affordances of a task environment are dynamic. People's sensitivity to a particular source of specifying information can emerge and disappear as the constraints of a given task unfold. Chow and colleagues, for example, demonstrated that novices attempting to kick balls toward targets changed their strategies over the course of twelve practice sessions.[128] They first focused on the height of their kick, then switched to focusing on distance to target, before returning to a focus on kick height. Learning to attune to specifying information and to calibrate and recalibrate one's actions as the constraints of a task evolve appears to result in improved accuracy in judging the affordances of the task environment. This allows for exploration of different sources of specifying information and supports convergence toward those sources that will give rise to the best action outcomes.

The topics of attunement and calibration of perception and action to properties of a task environment are of increasing interest to designers of augmented and virtual environments because as technologies for capturing movement and rendering immersive spaces improve, it has become apparent that failing to account for the role of action in shaping perception leads to errors in people's judgments of affordances, including distance, reachability, and passibility, and of properties such as object size. For example, when passively viewing virtual environments, people underestimate egocentric distances by up to 50 percent. When people are permitted to walk through the same virtual environment with continuous visual feedback, the accuracy of their distance judgments quickly improves, indicating that walking through virtual environments serves to rescale perceived space.[129] Interestingly, Researchers conducting this study did not find that space was rescaled when participants reached for targets in the same virtual environment.

If action scaling of affordances in virtual environments by walking through them improves egocentric distance judgments, is it possible that accurate calibration to body-scaled measures such as eye height or shoulder width might also improve performance in virtual environments? As Ponto et al. point out, many viewing models for virtual environments place virtual cameras by the viewer's eyes to reproduce a viewing experience that is as natural as possible.[130] Such models include a measure of the displacement between the position of the head tracker and the center of each virtual eye (the center of projection offset [PO]), and a measure of binocular disparity that accounts for the distance between the center of the virtual cameras. When combined, PO and BD are intended to allow for the visual perception of attributes such as shape, depth, and object motion. In addition to calibrating the positions of virtual cameras, it is necessary to calibrate accurately the motion sensor on a head-mounted display to ensure that the pitch of the virtual world is aligned with the direction of gravity and the optical axis of the display to maintain a stable representation of virtual space.

Although PO and BD are certainly body-scaled measures that give rise to the perception of, for example, depth, they represent a subset of available sources of body-scaled information that are used to guide action in real-world environments. As we have noted, parameters such as eye height and shoulder width, which are not available from PO or BD, are also used in judging affordances. For example, the height of objects in an observer's visual field is always relative to some ratio of the visible horizon measured with reference to their eye height, a ratio that does not change with their distance from the object. Therefore, if an object occludes half the height to the horizon, it is half the height of the individual, regardless of the intervening distance. And as we have already noted, eye height–scaled information also serves in estimating the distance to be covered to pass through a gap. Having access to eye height–scaled information for individuals using VR headsets would therefore seem to open up a wide range of information-specifying task-relevant affordances.

To test this hypothesis, Leyrer and colleagues asked participants to estimate egocentric distances to objects in a virtual room while manipulating their virtual eye height.[131] The study compared conditions where participants viewed the virtual environment with and without a self-avatar. The researchers found that manipulating eye height had a significant impact on people's estimates of both the dimensions of the virtual room and egocentric distances. They also found that the influence of eye height on distance was greater when eye height was raised than when it was lowered. Other researchers have found that using cues from auditory or haptic modalities can help calibrate perception of space in virtual reality. Huang's team, for example, asked participants to make relative depth judgments between two scenes that included a virtual character positioned at different distances from the observer.[132] The experimenters manipulated the reverberation time (RT) of the audio in each scene. They found that RT had a significant impact on depth judgments, with participants associating shorter RTs with shorter distances and vice versa. Haptic feedback, too, has been found to improve performance when combined with visual feedback for reaching tasks in virtual environments. As noted in section 5.4.3.3, the presence of passive haptic feedback resulted in improved accuracy for reaching in VR when compared with visual feedback alone.[133] More significantly, Altenhoff et al. demonstrated that the presence of continuous haptic feedback during calibration to a virtual environment resulted in improved accuracy in judging distances for reaching tasks.[134]

In many virtual or augmented environments, it may not be possible to simulate the specifying information that a participant might use to guide a comparable task in a real-world environment, even if the designer is aware of the nature of such information for a given user of their system. If we ask a soccer goalie to track ball spin but we are unable to simulate the texture on the surface of the ball, the goalie may have to rely on information that is non-specifying but still correlated with a curved trajectory of a ball. This might be the changing shape of the ball's image on the goalie's retina. In the absence of specifying information, goalies in virtual environments may become attuned to alternate non-specifying variables. Although this may result in reasonable performance, a novice VR user, even if they have become attuned to non-specifying information, will lack the real-world experience to make a good goalie in virtual soccer.

Even in cases where specifying information is simulated, the nature of display technology may itself introduce ambiguity, providing feedback that inadvertently results in users picking up on non-specifying information. Instead of simply hoping that participants in augmented and virtual environments discover correlated non-specifying sources of information on which they can base their judgments, researchers including Singapogu and colleagues (whose work on simulating haptic object properties is discussed in section 6.5.1)[135] suggest that it might be possible to train users to attune to specifying variables even where information is sparse or ambiguous. By introducing carefully designed training, they argue, designers can help users discover properties of simulated environments that may specify higher-order, task-relevant information. In other words, appropriate training can turn participants into skillful perceivers of the information our augmented and virtual environments can support. Thus, the study of attunement and calibration to specifying information in simulated environments is one of the fastest-growing areas of research in augmented and virtual reality.

5.5 Getting About in Virtual Environments Using Sound and Touch

Throughout this chapter, I have argued that spatial perception is not synonymous with visual perception and that much of the information we glean about the spaces we inhabit is provided by sensory modalities other than sight. The global energy array contains patterns in sound and tactile stimuli to which we are also attuned. As we interact with the spaces we inhabit and the objects in them, we are continually modifying our actions in response to information we are picking up without necessarily being aware of the sensory sources of this information. This is true whether we are climbing mountains, navigating a grocery store, or playing a fully immersive online game. The only difference in the latter two cases is that the information available to be picked up has been designed with our attentional mechanisms and bodily capabilities in mind. Although the designer of a grocery store can take advantage of physical infrastructure, such as shelves and aisles, to place items and signage so it falls within the visual fields of shoppers, designers of immersive environments have to track the movements of participants and then construct relationships between their actions and the visual, auditory, and haptic imagery displayed in response to what they are doing. Then, system developers have to ensure that this information is simulated at the appropriate points in time and space so that it is available to participants to guide their actions. To do this, designers need to understand the perceptual motor capabilities of users. They must discover the information that is relevant within the *Umwelten* of their participants.

Starting from the perspective of an individual's perceptual motor capabilities and working outward to a design that supports their everyday modes of encountering the world is a process that can only be rooted in an individual's experience and bodily skills. This is another way of framing Slater's claim that a virtual environment can be said to be immersive if it reproduces for an individual the same sensorimotor contingencies they encounter in the real world.[136]

When the *Umwelt* of a designer and an end user are vastly different, it is the user's world that must take precedence, not the designer's. Consider the challenge of designing a fully immersive environment to support blind participants in whole-body navigation of a complex virtual world. Most readers of this book are probably not used to navigating the world without vision. The starting point for a fully immersive design for blind people must therefore be determining the information they use as they navigate in real life. In an essay titled "Navigating Without Vision," Nicholas Giudice points out that most sighted people do not think about their process of navigation. They depend primarily on a visually guided sensorimotor coupling to move through their surroundings. For people who can see, "vision provides simultaneous and rapid access to highly precise distance and direction information about the 3D position and interposition of both nearby and distant visible landmarks, allows apprehension of geometric information about spatial structure, affords easy recognition of objects over a large field of view, and provides access to precise motion cues about changing self-to-object and object-to-object relations that occur during navigation."[137]

Although much of this information may also be accessible through other sense modalities, Giudice notes that, in nonvisual forms, such information can be ambiguous and unreliable. He points out that "hearing and touch convey far less information than vision about self-motion, self-to-object distances and directions, inter-object relations, and global spatial structure."[138] Tactile and haptic stimuli convey proximal low-resolution information within a field of view, limited to the world within arm's reach or to what can be reached using a long cane (1.5–2.2 m, depending on a person's height and walking speed). Points of contact with the environment are also limited to the feet and the fingers of the hand holding the cane. Most objects proximal to the traveled path are not detectable through touch because they are too far from the path, too large, or too dangerous to physically encounter.[139] Although auditory perception provides information about more distal objects and events, sounds are transient and the spatial resolution of hearing is less precise than vision when judging direction and distance to localize objects.[140]

The advantage of hearing, however, is that we can very accurately judge the timing of events and we can attend to sounds coming from any direction—important attributes for alerting us to potential danger. In his essay, Giudice uses the following thought experiment to illustrate how an experienced blind traveler draws on nonvisual cues to negotiate their environment—in this case, using a long cane. Imagine that you decide to take a walk from your hotel to a bakery you noticed on the next block. From the hotel's door, you can simultaneously see a nearby trash can, several outside tables at an adjacent café, the traffic light at the corner about 100 ft down the sidewalk, and the colorful awning of the bakery in the distance. In this scenario, blind and visually impaired travelers might access the same information, but it would be derived from very different sources and perceived in a sequential and less gestalt manner. For instance, they might use echolocation from their cane or footsteps to determine the position and distance of the trash can as they pass, the sound of people eating to identify the location and direction of the tables at the café, the feel of texture changes from a tactile warning strip to indicate the presence of the intersection and the correct orientation to adopt for a safe street crossing, the auditory flow of passing cars to assess the state of the traffic signal, the smell of the bakery as a cue that they are nearing the destination, and, finally, sensing the shade of the awning to indicate that they have reached their desired location.[141]

As Giudice notes, nonvisual sources of information are present in the environment for anyone to pick up. For a blind traveler, they are the primary sources of information in the *Umwelt* to which they are attuned in the same way as sighted people are attuned to changing patterns in light falling on the retina. So, what does Giudice's thought experiment imply for the design of our proposed collaborative virtual environment in which our blind and sighted participants are equipped to participate on equal terms? Certainly, the subset of the global energy array of interest to a blind VR participant has a different shape than the subset of information salient for a sighted participant. Of course, not all people have equal access to visual information. So, even assuming uniform visual capabilities for the population at large oversimplifies the process of designing virtual worlds. The purpose of focusing on a totally blind participant is to think afresh about the kinds of information that can be simulated to guide the actions of an individual in a virtual environment and demonstrate how people can accomplish similar tasks, even when they draw on invariant structures present in different forms of energy within the global energy array.

There are many examples of audio and haptically rendered virtual environments specifically designed for blind participants, including applications to convey the layout of complex traffic intersections,[142] to introduce students to the 3D structure of molecules,[143] and to render the 3D structure of virtual game worlds.[144] The Canetroller stands apart because it exemplifies Mel Slater's observation that promoting presence in VR involves closing the loop between a participant's actions and the response of the system in a manner that reflects their real-world experience so that they can move and act as they would in real life.[145]

The Canetroller is a body-worn device developed by a team at Microsoft research and designed to support experienced blind cane users in navigating virtual environments. Canetroller, shown in figure 5.6, attempts to reproduce the haptic and echolocation cues that blind people use to navigate. Proximal haptic and distal auditory information are combined to specify salient information about the virtual environment, such as the topography of a path to be traveled or the location and orientation of objects to be located and those to be avoided.

In the Canetroller system, proximal information, such as reaction forces generated in response to contact with objects in the environment and vibrations arising from moving over textured surfaces, are simulated using a wearable VR controller that renders vibration and reaction forces at a body scale. Distal information, such as the sound of the participant's virtual cane taps, their footsteps, reflections from nearby virtual surfaces, and sounds of nearby moving objects are simulated using spatialized audio rendering methods.[146] Audio rendering algorithms also take into account the material and geometric properties of objects and surfaces in the environment because, as Giudice points out in the thought experiment cited above, these environmental properties shape the spectral cues that provide valuable information for a blind user.

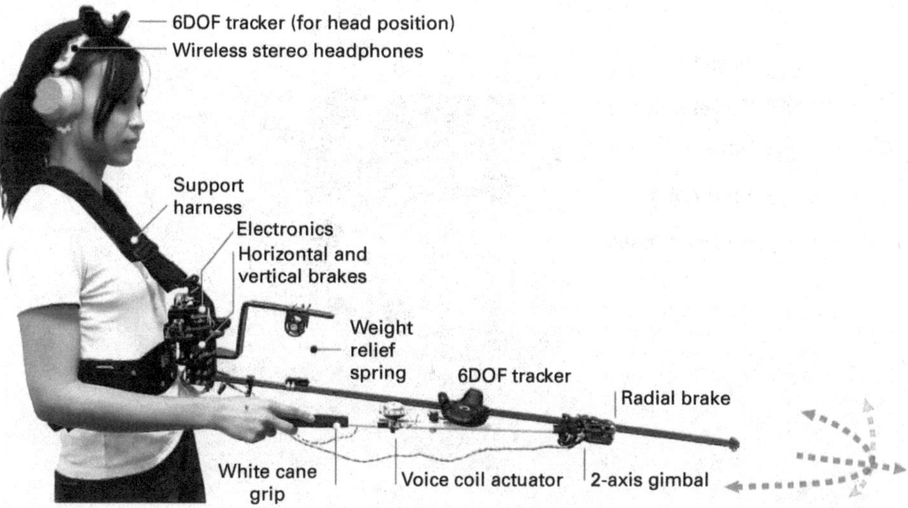

Figure 5.6
The Canetroller, designed to study how blind cane users might transfer their navigation skills into virtual environments. The controller utilizes force feedback in three axes, tactile feedback through a voice coil actuator, and spatialized audio through headphones. *Source:* Siu et al., 2020, figure 2. Reprinted with the permission of Alexa F. Siu. Copyright © 2020 by ACM.

The goal of the Canetroller project was to determine whether experienced blind long cane users could transfer their real-world navigation skills into a virtual environment if appropriate cues were present. In determining which cues to render, Siu et al. note that, in the hand of a moving person, the white cane performs three functions: to detect obstacles in the traveled path, to determine the surface topography, and to preview the path so that the user might plan foot placement. Distal information from surfaces in the environment—the sounds of footsteps, of the cane, and of other sounding objects—helps the user estimate distance and direction to objects and surfaces beyond reach. The Canetroller integrates a wave-based acoustic simulation that renders geometry-aware sound effects such as acoustic shadows, enabling the detection of obstacles that partially occlude audio sources.

To evaluate the potential of the Canetroller system to support navigation for blind participants, Siu et al. designed a scavenger hunt in a virtual environment combining indoor and outdoor spaces connected by a doorway

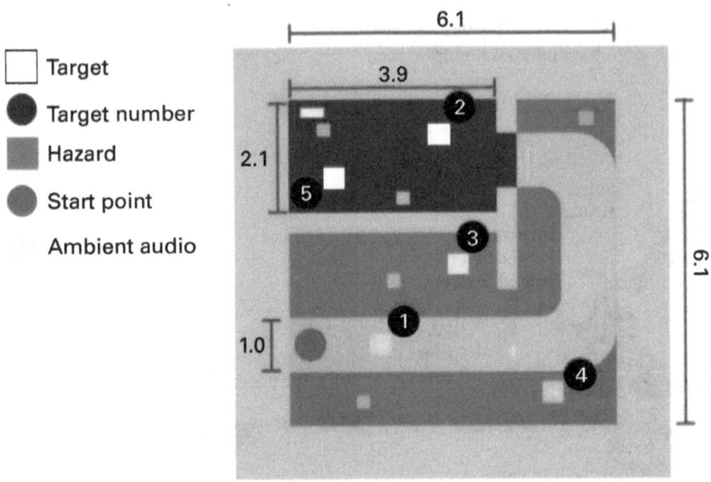

Figure 5.7
Floor plan of virtual room for the Canetroller study. Dimensions are shown in meters. *Source:* Siu et al., 2020, figure 10. Reprinted with the permission of Alexa F. Siu. Copyright © 2020 by ACM.

(shown in figure 5.7). The indoor space was a carpeted room measuring 3.9 m×2.1 m and containing a localizable sound source (a virtual radio playing a podcast). The outdoor space represented an acoustically open environment that incorporated a winding concrete path leading to the doorway to the room. All walls were rendered as drywall, and the total walking area of the VE measured 6.1 m^2. Participants were asked to navigate the environment and acquire five targets. They were assured that the winding pathway contained no obstacles but that the targets, in the form of sounding beacons, could appear anywhere. Sometimes they were situated out of the direct line of hearing, around corners, or partially occluded by obstacles. They were rendered as cubes that played techno loops and that sounded like drums when tapped. The user "collected" each one by walking over it. The environment also contained hazards, rendered as metal cubes that did not emit sound until they were tapped, whereupon they emitted the sound of an explosion. While gathering targets, participants had to avoid hazards and avoid walking into walls. Targets, which were presented in sequence as the participant walked, provided a guided path through the environment, whereas walls and hazards encouraged participants to pay attention to their surroundings and move carefully.

The researchers reported that seven out of eight participants successfully navigated the environment and collected all five targets and that navigation performance improved with repetition of the game. Siu et al. concluded that the participants in their study integrated audio and haptic/tactile cues to successfully explore and navigate a virtual environment by drawing on their skills and experience of navigating with a white cane in real life.[147] In other words, blind participants with well-developed cane skills could successfully navigate this virtual environment because they were already attuned to these sensorimotor couplings. Therefore, audio and haptic effects rendered in response to their movements contained information related to sensorimotor contingencies to which they were already attuned from years of navigating the physical world with a cane.

Now, imagine our blind participant is playing this scavenger game with others, most of whom use vision as their primary mode of guiding action. What aspects of the rendered virtual environment would need to be altered? If everyone is to have access to the information they use to navigate in real life, the virtual building would need to be rendered visually, and sighted participants would need to have an alternative controller incorporating visual feedback—for example, through a head-mounted display. Would anything else need to change?

Let us consider the spatial scale of our virtual world. Recalling Giudice's observation that audition and touch are less accurate than sight at conveying information about self-motion, self-to-object direction and distance, and global spatial structure, we might expect that the sighted player in the scavenger game would have more immediate access to a global view of the building's structure to plan routes and avoid obstacles. Drawing on Montello's functional classification of spatial scale into figural, vista, environmental, and geographical levels,[148] Giudice notes that people relying solely on nonvisual information find the environmental scale most challenging because it requires more spatiotemporal integration when compared with visually guided movement to apprehend the world beyond the traveled path.[149] At the figural or desktop scale, he says, acquisition of nonvisual information to guide action for learning and skill transfer is like that for visually guided action because maps, models, and object arrays are situated in a reference frame that is within arm's reach. Little extra work is therefore required to apprehend the geometric properties, relative positions, or orientations of those models and arrays. At the vista scale, where vision provides a stationary vantage point from which to view a larger scene, it is more difficult to perceive non-visually the relative position and orientation of scene elements. At the geographical scale, which cannot be apprehended through motion over time, the information to be conveyed is less dependent on the modality of presentation and can take the form of visual or tactile maps or of audio or visual turn-by-turn guides.[150] Thus, if our game world were to expand beyond its one virtual building, if it were to require apprehension of large scenes or navigation to unseen parts of virtual worlds, then the nonvisual information that would best support our blind gamer might not be haptic information conveyed via a virtual cane but rather a tactile map or spoken turn-by-turn guide.

A second consideration for our scavenger hunt players is the perspective their different modes of access provide for them within the virtual environment. Blind and sighted players are both using systems that track whole-body movement, so everyone's motion through the game is accomplished by walking. The sighted player wearing a head-mounted display typically has a first-person view of the world. The virtual camera is mapped to their eye height and calibrated to track the movements of their head. The visual scene is rendered to align with their perspective as they move, with images scaled as objects advancing and receding within their field of view. The Canetroller uniquely provides a blind gamer with a first-person perspective "view" of a virtual world. Like the visually rendered world for the sighted player, the haptically rendered environment changes in response to movement, bringing the blind player into the rendered fabric of the game.

It is not clear from Siu et al.'s work whether the cane's length is calibrated to account for the height and walking speed of each player, which is necessary for a timely preview of a path's topography to plan foot placement.[151] Incorrect cane calibration can become a source of breakdown of presence for the blind gamer because, as noted in discussing the work of Merleau-Ponty in chapter 2, a cane in the hand of a blind traveler extends their body schema by extending their point of contact with the environment beyond arm's reach. If the length of the virtual cane is not appropriately calibrated or, worse, if its projected point of contact with the virtual environment is inconsistent, there is potential for the cues it provides to be misinterpreted as an unstable topography that is unsafe to walk across.

So, to return to the question previously posed—how we might render the virtual scavenger game to provide equal access for blind and sighted players—our primary consideration would be to ensure the game world contains information specifying opportunities for action that can be picked up through more than just the visual modality. Further, all players should have access to devices that convey any combination of modality-specific cues available within the rendered world. They can then use their individual bodily skills to pick up whatever task-relevant information they need to guide their actions and accomplish their goals.

Note that this approach is not the same as sensory substitution. In sensory substitution, the patterns in the global energy array that convey information about the changing state of the world through visual cues (such as expanding images on the retina) are not *converted* into stimuli presented to the blind user as sound or touch. Instead, the sensory milieu of the virtual world is enriched by *adding* information normally picked up through touch or sound, the senses on which a blind person draws to guide their actions. In Slater's terms, as long as the information available to a blind VR participant closes the loop between their actions and the response of the system by conveying patterns of energy that guide their movements in real life, the blind participant will be fine. After all, they have a lifetime of experience of their *Umwelt* and of relying on nonvisual information to get about in the world.

5.6 Conclusion

The spatiality of our world is a lived space, brought forth through our actions. Its structure is determined for us by what we need or want to do. We perceive things as being near or far, present or absent, not on the basis of their physical proximity to our bodies or whether we can experience them through one or more sense modalities but rather on the basis of what those things can do for us and whether they are reachable.[152] The relevant measures of spatiality in our lifeworld are therefore not exact units such as feet and inches but rather sources of information scaled by bodily dimensions and action capabilities. What matters is not how many feet away from me an object is but whether I can plan and execute a sequence of movements that will bring it within reach should I need it. An individual's understanding of space is therefore based on what their current environment affords them with respect to their bodily and action capabilities and within the context of their current task goal. Can I fit through this gap? Can I run fast enough to catch that ball?

People's ability to successfully move through their environments is determined by how well they can attune to sources of information that specify task-relevant opportunities for action. The baseball player who knows how to use the optical information generated by a fly ball to predict where to intercept that ball will be more successful than her teammate who is not attuned to this information. This is no less true for people navigating virtual environments. A user of the Canetroller can avoid virtual obstacles because they have learned to couple their movements to the sounds of their cane and to their footsteps, reflected back to them as simulated auditory and haptic events, colocated in space and time.

In this chapter, I have argued that the relevant spatial information to be rendered is not the geometry of a virtual world but rather a simulated energy array sufficiently complex for users to discover sources of specifying information, invariant patterns in multiple forms of energy that can guide their actions in partly or wholly simulated worlds. If the technology at hand is not able to render sufficiently specifying information for a task or if a user initially latches onto non-specifying but correlated information, then, as designers, we have the opportunity to guide users toward the best available task-relevant invariant patterns of energy in our simulated energy array.

to guide their actions. Space, real or simulated, arises not from geometry but instead from the relationship between our bodies, our action goals, and the information we pick up from the objects and surfaces around us and the role it plays in supporting us in going about our daily lives.

6 Objects

6.1 Introduction

In the world of interaction design, objects are challenging because they force us to directly confront the boundary between the virtual and the physical, between the world of bits and the world of atoms.[1] In this world, it is no longer even clear what objects are. The Collins dictionary defines an object as "anything that has a fixed shape or form, that you can touch or see, and that is not alive."[2] However, the kinds of objects we can now create can change their form (as with shape-changing interfaces [SCIs]), be seen but not touched (as with holograms), or touched but not seen (as with ultrasonic haptic renderings of object forms). In this world, we are forced to reconsider what objects are and how we even know what to do with them. I have saved this discussion until the end of this book because, in addressing how we interact with objects, I will draw upon much of the material that has been presented in prior chapters, including how we capture and use the movements that guide our interactions with objects, how the spaces we inhabit provide context for those actions, and how the temporal unfolding of our actions is guided by the information that is present for us as we seek to accomplish our goals.

This chapter takes as its point of departure three assumptions grounded in our discussion of embodied cognition in chapter 2:

1. Objects take on meaning when embedded in their use contexts.
2. Digital artifacts, like any other artifacts, can be both objects of inspection (viewed from a distance) and objects of action (like tools that are grasped and brought into the body's schema).
3. If the digitally augmented objects we design are intended to be used as tools that couple the user to the augmented or virtual environment, they

should "withdraw" from perceptual awareness (i.e., become transparent in use) to bring forth the task-dependent perceptual invariants that are relevant to the work at hand.

Accomplishing these design goals is no small feat, given that the objects of interest are so mutable.

In this chapter, I offer a perspective on our perception of objects that focuses less on their physical properties (e.g., their color, size, and weight) than on how perceptual experiences of objects are shaped by the contexts in which they are encountered. This shift in perspective, I argue, is necessary to understand why we can design objects whose form and material properties are so malleable—objects that can be partly or wholly simulated using visual, auditory, and/or haptic displays and rendering techniques. As interaction designers, our aim is always to ensure that the users of our systems can figure out what the objects or elements of an interactive environment are to be used for. It is our job to understand what, for them, is likely to be perceptually salient in the context of the task they are performing (i.e., which sources of information they will likely need to be attuned to at any point in time in order to succeed). In an interactive game, this might mean tracking the shape and speed of a falling object. In the case of Pong, for example, color becomes a less salient cue. In fact, the early computers for which Pong was developed could not render color, so designers focused on dynamic cues, such as speed and size, which they could render. Even though today's interactive environments can capture and use a vast range of bodily movements (Pong could only rely on fingers pressing letters on a keyboard), the need for interaction designers to understand the relation between a user's actions and a system's response, in revealing to users the salient information for their current task, has not changed.

Throughout this book, I have argued that the computationally enhanced interactive systems we build are not separate from the world we inhabit but rather exist alongside all of the other objects we use every day. In this hybrid world, actions may have consequences in the physical world, in the digital world, or sometimes in both. The argument I present here is that to design for users who inhabit this blended world, we must move beyond our current understanding of digital artifacts as physical representations of digital information. In order to do so, we need to develop a deep understanding of our perception of and interaction with the objects we encounter in everyday life and of how these objects facilitate our ability to accomplish our goals. Then, instead of thinking of digital artifacts as portals to digital worlds, we need to think of the things we design as augmented structures that support us in skillfully negotiating our new hybrid lives.

The threads I draw together to support interaction designers in accomplishing this shift in thinking come from fields including phenomenology, embodied cognition, object perception, and interaction design for AR and VR. These threads reveal the richness of the hybrid world that can be brought forth through interaction design if we take time to understand how a person's actions bring forth their world of experience, mediated by the objects in their environment.

6.2 Objects of Experience

In July 1688, the Irish scientist and politician William Molyneux, prompted by his wife's loss of sight, wrote to the philosopher John Locke with a question that still exercises the minds of philosophers and scientists today:

> A Man, being born blind, and having a Globe and a Cube, nigh of the same bignes, Committed into his Hands, and being taught or Told, which is Called the Globe, and which the Cube, so as easily to distinguish them by his Touch or Feeling; Then both being taken from Him, and Laid on a Table, Let us Suppose his Sight Restored to Him; Whether he Could, by his Sight, and before he touch them, know which is the Globe and which the Cube? Or Whether he Could know by his Sight, before he stretch'd out his Hand, whether he Could not Reach them, tho they were Removed 20 or 1000 feet from Him?[3]

At first, Molyneux's question seems trivial—surely globes are so different from cubes that even if the blind man had only ever felt these items, he would have sufficient cues to make the judgment. However, when one remembers that he is supposed to make his judgment on the basis of vision alone, Molyneux's question becomes more challenging.

Molyneux is asking if a lifetime of feeling objects (such as those in figure 6.1) can inform a newly gained ability to see them but without touching them. The question then becomes, What exactly is Molyneux questioning? Most obviously, the transfer of percepts from one modality to another. Also, perhaps, perception of distance? Perception of space? Of form? Or perhaps a prior experience relating to spatial memory of objects?

Those who cite Molyneux's question often stop here. But the meat of the problem is in the second part of the question, where Molyneux asks

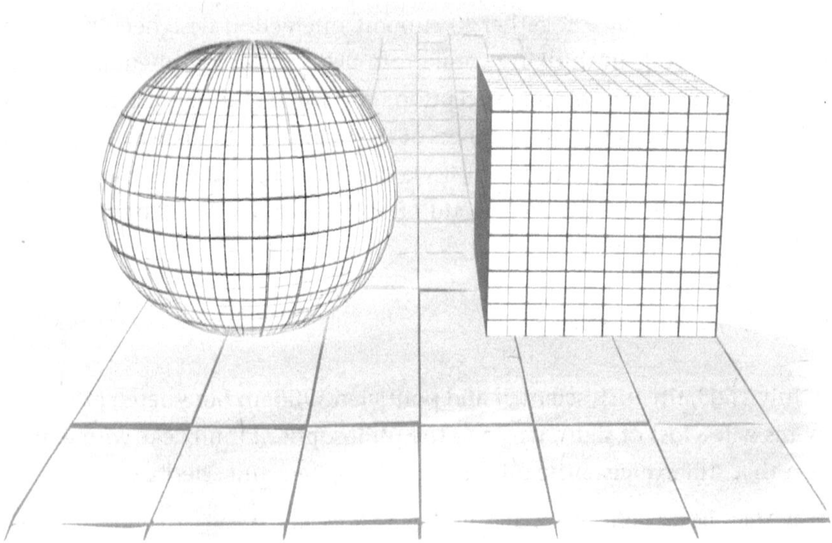

Figure 6.1
Perspective drawing of a globe and a cube. When viewed, a globe and a cube cast two-dimensional images on the retina, which include perspective cues that inform the viewer of their three-dimensional shape. When felt with the hand, these extra parts of the projected image are not present. For this reason, blind people feeling tactile versions of perspective drawings find them very confusing. *Source:* Original drawing by Hans Anderson.

whether the blind man, on having his vision restored, would be able to gauge how far to reach to grasp the objects: "Whether he Could know by his Sight, before he stretch'd out his Hand, whether he Could not Reach them, tho they were Removed 20 or 1000 feet from Him?"[4] In perceptual terms, Molyneux is asking whether the newly sighted man would understand visual cues pertaining to distance—the relationship between the size of the image projected on his retina and the position, relative to him, of the globe and the sphere. Are they small and only 20 ft away or large and 1,000 ft away? This part of the question relates to cues that are specific to the visual sense. There is no equivalent of size constancy when we consider the sense of touch, and thus there is no way in which prior haptic knowledge can inform vision in this regard.

Molyneux highlights a distinction between information that might be shared across modalities and information that is sense specific. "Molyneux's Problem," as it has come to be known, continues to challenge philosophers, psychologists, and, more recently, creators of virtual environments because it questions our fundamental understanding of how we relate to objects around us. In this chapter, we review discussions of Molyneux and draw upon the literature on interacting with objects—real, hybrid, and virtual—to develop an understanding of what we mean by the word "object" and how objects are represented in real, virtual, and augmented environments.

You might ask, What is the answer to Molyneux's question? In *How the Body Shapes the Mind*, Shaun Gallagher provides a reasoned discussion based on centuries of empirical case studies and on the related issue of whether the experience of a person with newly acquired vision as an adult is equivalent to the first visual experience of a newborn baby.[5] Gallagher's conclusion is that it depends on how the question is posed. If the question is posed empirically (i.e., if the assumption is that experience in one sense modality does, in fact, inform experience in another), then empirical findings have supported the hypothesis that a newly sighted person cannot recognize a cube and a sphere using vision alone, at least not at first.[6] Gallagher points out that if the answer to the Molyneux question is no, this is further confirmation that the newly sighted adult can never acquire the visual experience of one who has always been able to see.[7]

In this respect, the Molyneux question is insufficiently constrained because it assumes that the "first perception" of the newly sighted man is equivalent to the "first perception" of a newborn baby. This, in turn, assumes that there are no significant differences in the visual cortex and consequently in the visual performance of:

1. a newborn baby, whose visual cortex is primed to acquire sensory experience and that continues to exhibit an increase in cortical connectivity for the first few years of life;
2. a sighted adult, whose visual cortex has been processing visual stimulation since (and perhaps even before) birth;
3. the hypothetical "subject" (taken to be the sighted person lacking visual stimulation, as if wearing a blindfold); or

4. the more likely participant, Molyneux's newly sighted man, whose visual cortex has lacked stimulation since birth; on regaining vision, therefore, the blind man is likely to lack perceptual capabilities that develop in the first few years of life.[8]

However, if the Molyneux question is posed not from the perspective of a detached observer, but from the perspective of an engaged actor, then Gallagher suggests the answer is likely to be yes. He then poses a further question: Even if the Molyneux patient is unable visually to recognize the difference between the cube and the sphere, is it possible that their grasp, informed by vision, can differentiate between them? One would, he supposes, be able to tell from the shape of their hand as they reached to grasp the object.[9] Although this study has never been performed, the existence of separate but interdependent visual pathways related to object recognition and goal-directed action could provide a reason to test this hypothesis.[10] In particular, as O'Regan and Noë point out, seeing for the purpose of guiding action does not depend on having a complete and unified representation of the objects seen.[11]

Gallagher, citing a growing body of evidence pointing to the existence of a well-developed prenatal proprioceptive system and body schema in which different sense modalities are already in communication, suggests that the newborn baby is not a sensorimotor tabula rasa. Rather, the newborn already has a capacity to acquire experience on the basis of action. In other words, the newborn has begun the process of acquiring knowledge through interaction with the prenatal world.[12] Perception, Gallagher argues, "is less the result of an internal processing of sense information, and more the result of an interaction between the body and its environment."[13] If the doctors who reported that they had restored vision to patients and then subjected them to Molyneux's test had only observed them reaching out for the objects before requiring them to name them, then, even if the patient never touched the objects, one could determine whether the visual cues guiding action would—as Gallagher believes—provide evidence for a positive answer to the Molyneux question.

What can the issues raised by Molyneux's question and Gallagher's response tell us about how we interact with objects in mixed and virtual environments? I propose that interactions with augmented and virtual objects are, like interactions with real objects, framed by the contexts in which they are encountered. It follows that the associated perceptual experiences of such encounters, in turn, are shaped by the action goals such contexts define.

6.3 Encountering Objects in Context

Just as every scale needs a reference point to give its values meaning, objects in our environment are given a purpose by the contexts in which we encounter them. As I argued in chapter 2, the active nature of perception means that such encounters are not static but rather dynamically shaped by the actions and intentions of the individual who moves in their environment. Context plays a role in shaping our encounters with objects, and such encounters, in turn, reshape the fabric of our world. Indeed, in augmented and in virtual reality, the very notion of an object and its context of use are malleable. What does it mean to encounter an object and, in the case of an object that is a tool, to understand what it is and what it is for?

In chapter 2, I cited Heidegger's observation that, for the most part, we use the things in our environment to accomplish particular, situated tasks, and that our places of work are equipped for the purpose of accomplishing these tasks.[14] The desk of a writer is organized around the task of writing and so may contain a laptop, pens, paper, and the like. The workspace of a chef will contain knives, chopping boards, pots, pans, and so on. In each case, the tools present gain meaning through what they are to be used for—that is, what they are is understood in terms of what they are for.[15] In this reading, an object and the context in which it is encountered are not separable but instead are bound together by the presence of the individual for whom their coexistence takes on meaning. "We never find ourselves in 'the' world but always in 'our' world. Things are what they are by virtue of our relations to them, just as we are what we are in terms of our relations to things. Humans and world are inseparably bound to each other and constitute each other in this bondage."[16]

Don Ihde, in describing what he terms the "gestalt features of technology," points out that, for Heidegger, the relationship between a tool and its context is always defined by the intentionality of the person who will use it and for what purpose.[17] In other words, equipment is encountered "in order to." In this model, there is a built-in assignment of something with reference to something else. The carpenter picks up the hammer and nail with a clear intent "in order to" hammer the nail into a wall. Here, the roles of hammer, nail, and wall are all clear. But, Ihde points out, this raises a problem for Heidegger. The tool or equipment in use becomes the means, not the object, of the experience of hammering. Having withdrawn, a tool cannot simultaneously be the object of reflection. The carpenter's understanding of what the tool is for cannot be copresent with her understanding of what it is—she cannot be aware of the object properties of the hammer, such as its overall size and shape, at the precise moment when she is focusing on the task of hammering and on the feel of the contact between the hammer and the nail. For the carpenter, the withdrawal of the technology from her "direct experience" is what Ihde terms "an embodiment relation," in which, in Merleau-Ponty's language, the carpenter extends her body schema to incorporate the hammer, so that her boundary of sensation with the world is no longer between the hand and the hammer but between the hammer and the nail.

Ihde highlights two further aspects of the relationship between the body and the technological artifact. First, the ability to incorporate technologies in use into the body schema is polymorphic—that is, the body-plus-tool schema takes the shape of the coupling. Second, the schema is malleable—that is, the body–tool schema is assembled and disassembled on the fly. The reconfigurable nature of this embodiment relationship arises because the body–tool schema is given its form by the current task. In other words, the act of picking up a tool "in order to" determines the unfolding spatiotemporal configuration of the carpenter's perceived body schema. We have already encountered this concept in chapter 2, where we discussed the idea of enactive landscapes—landscapes of possibilities for action that are constructed on the fly by individuals who pick up artifacts with the intent of accomplishing specific goals.[18]

There are two ways the context of an encounter with objects in the environment can shape embodiment relations. First, as noted in chapter 3, the more skilled a person is in using a tool, the more possibilities for action are available and the more resilient that person's actions can be in the face of unpredictable perturbations in the task environment. In other words, the range of possibilities available to the carpenter when she picks up a hammer in order to accomplish her goal is also shaped by how skillfully she can wield the hammer. The range of tasks she understands the hammer to be for is shaped by both the current task context and her past experience in picking up hammers "in order to."

My second point concerns the unfolding nature of the physical encounter with things that lie at the center of the embodiment relationship. As I noted in discussing the third embodied paradigm in HCI in chapter 2, the tools (artifacts or interactions) that we create, in turn, shape the environments with which we interact; or, as Culkin says, "We shape our tools and thereafter, our tools shape us."[19] Lambros Malafouris, whose work on the anthropology of tool use led him to develop Material Engagement Theory, goes even further. In describing the entanglement between the way in which people and their artifacts have coevolved from their earliest adoption of objects as tools, he posits that the very engagement of our bodies and brains with the materials around us gave rise to our ability to think and imagine new possibilities[20]—that is, it was from the synergistic engagement of bodies, brains, and things that our minds emerged.[21] In the language of phenomenology, this connection between humans and their environment is referred to as "intentionality." But intentionality does not imply that any being actively chooses to be connected to their world. They are involved with the world because they cannot choose to be otherwise.[22]

For Malafouris, such connectedness with the world always involves intention in action, where the intent does not emerge inside the head of the acting person but instead is co-constituted by the individual and the things in their environment. He takes the example of a potter manipulating clay. To make a pot, the potter moves his hands to shape unformed clay, but as he does so, he also feels the resistance of the clay to being shaped. To successfully make a pot, he must both act upon and react to the clay. Malafouris refers to this unfolding conversation between the potter and his clay as the "dance of agency," noting that both the clay and the potter are afforded active roles in shaping the final pot.[23] For Malafouris, engagement with the material world comes from the ability to act on things, whereas the ability to act on things, in turn, has already been shaped by the environment that contains both potter and clay. "Without the clay and its physical affordances as the basic element of the potter's background, the potter would not be able to form the intention to form the pot."[24] In other words, there is a context for the work environment that goes beyond the immediate availability of clay and hands and perhaps tools to contain a broader cultural understanding of what pots are and what they are for.

The question raised by Ihde and Malafouris, and many others who study the development of technology from a philosophical perspective, is what it means to be co-constituents of the world we inhabit. Unlike the makers of clay pots, whose material environment could be shaped by their hands into the things they required for holding water and cooking, many of the things that are the object of our work are completely intangible, existing only as digital algorithms made apparent to us through representations constructed and deconstructed on the fly. Lacking a material context, our engagement with such artifacts no longer can be said to be an embodiment relation but rather becomes, in the language of Don Ihde, a "hermeneutic relation."[25] Here, artifacts do not withdraw from our experience, like the hammer, but provide a representation of the world, which must be "read" or interpreted. We cannot readily perceive the current temperature of our immediate environment, but by means of a thermometer or other such technology, we can access this value by "reading" the symbols from its display. The world is not perceived through the artifact but instead is perceived "by means of" the artifact. The involvement here is not with the thermometer itself but rather with an aspect of the world revealed by the thermometer.[26]

For such relationships to have meaning, they must exist within a broader cultural context, where there is a common understanding of what thermometers are for. The implication for individuals who inhabit this world of virtual or partly rendered objects is that the understanding of what such objects are for—the context that frames their use—is no longer given by the world in which they are situated. For example, an interaction might involve picking a drawing tool (such as a paintbrush) from a menu, but what does this interaction signify if one's surroundings do not contain cans of paint or a wall to be painted? The contextual framing that defines what the tool is for relies on the user knowing something about painting in the physical world. In other words, lacking materiality, the objects of use and their use contexts are co-constructed by interaction designers but without the engagement of the individual who will ultimately have to grapple with them in use. As we shall see later, such a decoupling of real and simulated environments, even when such environments contain tangible objects to support physical interaction, has consequences for an individual trying to figure out both what things are for and how to interact with them in order to accomplish anything.

If designers want digitally augmented artifacts to be used as tools, coupling the user to an augmented or virtual environment, then these tools should, as with Heidegger's hammer, "withdraw" from perceptual awareness (i.e., become transparent in use) to allow their user to discover the task-dependent perceptual invariants that are relevant to the work at hand. This is what happens in the world of the carpenter—the forces and torques she perceives as she works guide her in adjusting to the changing constraints of the task environment. In taking on the role of designing both the tool and its context of use, designers therefore need to understand not just how to simulate the static physical attributes of the world of a task but also how to identify and render the perceptually relevant and dynamic information that we rely on to guide our actions in performing tasks in the physical environment.

In interacting with and through objects, whether real, virtual, or augmented, the information we seek is provided by the relationship between the perceivable properties of the environment that change in response to our actions and those properties that are, in Gibson's terms, invariant. In designing interactive systems, we typically focus on rendering properties of objects for modes of inspection, where they are present at hand, but we overlook the properties of the environment users seek out once they pick up the same objects to grapple with them in use—that is, when they must be ready to hand. If we are to take into account the active information-seeking behaviors users engage in as they interact with and try to make sense of things we design, then it seems we must first understand what kind of information we are attuned to when we seek to act with and through tangible objects such as tools. Beyond this, given that we can now create objects whose properties are mutable, what are the structural and transformational invariant properties within our designed environments that provide the kinds of information users seek in order to accomplish their task goals?

6.4 Objects of Action and Perception

That we can accomplish anything in an environment that is a simulation of the physical world is remarkable. And it arises only because we are adaptive, goal-directed beings who seek sensory support for the actions in which we are engaged and contextual framing to accomplish these goals. As I pointed out in chapter 2, even in "real" experiences, our sensation is contingent upon the capabilities of our sense organs, and our perception is shaped by factors such as attention and goal-directed action. This does not render the laws governing physical objects mutable, but it does point to a relationship with such objects that may be far more malleable than we think. This is what allows us to co-construct the highly mediated virtual and augmented worlds we increasingly inhabit.

6.4.1 Objects and Their Properties

In chapter 2, we noted that in any description of our environment based on physical properties—such as specific electromagnetic wavelengths that we sense as light, or a specific range of acoustic wavelengths that we sense as sound—these sensations, in and of themselves, do not constitute perception. What is perceived, Gibson asserts, are not light or sound or forces applied to the body directly but rather higher-order variables of information specified by the structure within the global energy array, in light, sound, and other environmental forces.

For an animal successfully negotiating their environment, Gibson claims that the invariant properties of that environment further specify for that individual the opportunities for action available to them. Concerning environments that contain objects, he states, "When the constant properties of constant objects are perceived (the shape, size, color, texture, composition, motion, animation, and position relative to other objects), the observer can go on to detect their affordances."[27] In other words, once an individual can discern the ways in which the objects in an environment structure the global energy array to which that individual is attuned, that individual can determine what the environment affords them. For a human, it might be that the flat, low, horizontal surface in front of them affords walking upon and the slightly higher, cushioned surface affords sitting on.

The designer of an augmented or virtual world, then, has to take care of several things:

1. The constant properties of objects (specified as visual, auditory, and/or haptic properties)
2. The ways in which the presence of these constant properties structure the partly or wholly simulated global energy array for the user (the structural and transformational invariants to which they can become attuned)
3. The way the system behaves in response to the actions of the user (the lawful relationship between their actions and the systems response)

As I noted in chapter 2, perceptual modalities are structured not just by invariant properties of the energy array but also by the stability that arises from the actions of the skilled individual seeking information to guide them toward a goal. Without taking into account our ability to adaptively cope with these new environments, to become skilled users of the new tools and equipment we can now create, we would continually be encountering these worlds as unfamiliar places, and it is unlikely we could accomplish anything in or through them. When we talk about perception of object properties, therefore, it may be useful to talk about perception not in terms of specific attributes such as color or texture or weight, as these things are mutable in augmented and virtual environments. It is perhaps more useful to talk about how object properties such as color, texture, or weight serve to structure our interactions with such environments and how the information we pick up through these interactions guides us toward accomplishing our goals.

6.4.2 Modes of Encounter

There is a long tradition in research of objectifying the object—studying perceivable attributes of objects that are abstract in form or removed from their familiar contexts of encounter. We are all familiar with moments when we encounter objects in modes of what Heidegger would term "gawking," times when objects are "present-at-hand," i.e., when they become artifacts to be inspected rather than things to be used. Of course, we sometimes purchase artifacts purely for their aesthetic properties, in which case their function is ornamental. But we also choose functional objects such as tools or furniture based on how well they fit our bodies and how well they support us in accomplishing tasks. At the moment of choosing such an object, we might consider what it feels or looks like. Often, such inspections are carried out at a distance—through a store window or as online images. Such indirect forms of inspection inadvertently privilege vision. Indeed, most objects we encounter in the world are not things we can directly touch. It is therefore not surprising that much work on object perception has focused on visual object recognition—on indirect encounters with objects of inspection, not direct encounters with objects of action.

It is not that the active role of vision in guiding action was entirely neglected. I have emphasized the work of many researchers in uncovering the role of the body, including its visual resources, in shaping our interactions with the world. It is more that work on active perception considers objects within their broader contexts so that the perception of object properties arises within, not apart from, the perception of a broader landscape of opportunities for action. With the advent of 3D interactive virtual environments, though, this is changing. Although it is far beyond the scope of this book to provide a comprehensive summary of research on object recognition, I will draw attention to some waypoints that are significant in understanding the active nature of the perception of objects and their properties. Through our current lens of rendering interactive augmented and virtual environments, we begin to recognize the role of movement and action in structuring our experience of handling objects, and we recognize ways in which movement provides information to guide our actions.

In his paper titled "Observations on Active Touch," James Gibson proposes that the kind of tactile exploration we engage in when we are exploring properties of objects, such as their weight, shape, texture, and so on, which he termed "active touch," is a mode of scanning the tactual environment that is equivalent to visually scanning our surroundings. He points out that the kinds of hand movements people use in this mode of touch are distinct from mechanical interactions with the environment that cause haptic sensations, such as stimulation from reaction forces arising from lifting or moving objects. "When a person touches something with his fingers he produces the stimulation as it were. More exactly, variations in skin stimulation are caused by variation in his motor activity. What happens at his fingers depends on the movements that he makes—and also, of course, on the object that he touches."[28]

In active touch, the movements a person makes do not modify the environment. Rather, they only modify the stimuli coming from the environment. This is why, in Gibson's view, the hand movements used in active touch, like the eye or head movements used in visual scanning, uncover the lawful relationship between a person's actions and the corresponding changes in the global energy array. Gibson further argues that the particular ways in which the fingers move in exploring objects serve to enhance some features of the available stimuli and reduce others.[29] His stance is in sharp contrast to more traditional views of the sense of touch that, he points out, conflate two sources of sensation: proprio-specific stimulation, arising from the feeling of moving one's own body, and extero-specific sensation, arising from contact with the environment. In active touch, Gibson argues, exploratory movements isolate and enhance the extero-specific components of stimulation, as these specify the shape and other properties of objects being touched.

Building on this hypothesis, Lederman, Summers, and Klatzky demonstrated that when asked to describe haptic properties of a series of everyday objects, such as their shape, texture, and weight, the hand movements used were very consistent among their participants, as can be seen in figure 6.2.[30]

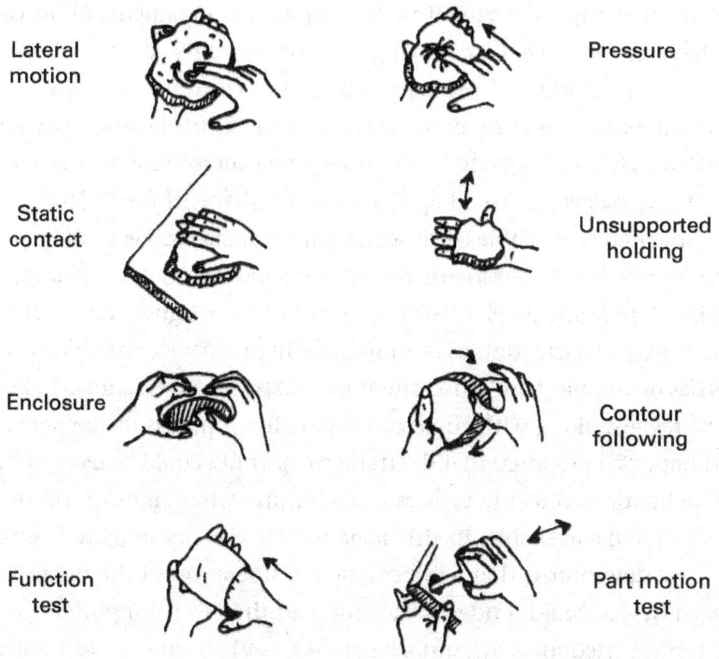

Figure 6.2
Haptic exploratory procedures and their typical movement patterns. *Source:* Lederman and Klatzky, 1987, figure 1. Reprinted with permission from Roberta Klatzky. Copyright © 1987 by Academic Press, Inc.

In making judgments about surface texture, for example, people rubbed their fingers back and forth; when asked to make judgments about shape, participants either enclosed the object with their hands or followed along its edges with their fingers. Lederman, Summers, and Klatzky called these hand movements "exploratory procedures" (EPs) and suggested, like Gibson, that their purpose was to selectively enhance particular sensations that could reveal information, pertaining to attributes of objects, salient in the context of discerning a particular property.[31]

In light of the fact that simultaneously rendering multiple haptic properties of objects remains a significant technical challenge, the concept of EPs has gained significance because it provides a way to determine, for a given haptic device, what object properties can and cannot be rendered. For example, what if the only device available is a point-based force-feedback device such as a Phantom?[32] Where interaction with the rendered world happens by means of the tip of a virtual probe, any haptic judgments of an object's shape have to be derived by moving a point along its contours to infer its form from the motion of the hand (using the contour-following EP). But judging the object's texture or softness is easier, so these properties tend to be noticed first (i.e., they are likely to be given more weight or salience by users). If one has access to a haptic glove, the shape of the object becomes more immediately available because the same rendered object can be grasped with the whole hand (employing the EP of enclosure). In this scenario, shape might be more immediately perceivable and therefore given more salience.

In this way, the capabilities of individual haptic devices determine which properties of an object are most salient for a user in a given context of use. In a recent review, along with three other visually impaired researchers in the field of haptics, I proposed that the framework of EPs could be used to classify available haptic and tactile devices to determine what kinds of information they could make available, in this instance for visually impaired computer users.[33] We determined that a systematic consideration of the haptic effects that each device could render with respect to the EPs it supported—normal force, surface friction, 2.5D surface features, and so on—could be usefully employed to match device capabilities with the kind of information required in a given application. We further suggest that the technical devices we use to render properties of virtual environments have a significant bearing on the kinds of information that users of our systems can reach. If a graph is rendered using only a single-point device such as the Phantom, then its contour has to be experienced by sweeping the point of the device along any of its lines and constructing the shape of a line in one's mind from the temporal arc of the movement of one's hand. If the same graph is presented on a tactile array, the whole curve is persistent in time and available to be explored by one or both hands using an increased range of EPs—lateral motion, whole-hand exploration of 2D shape, and so on. It also makes comparison of values and line trajectories much faster. With a Phantom, for example, one would have to go back and forth between line paths and try to remember the feel of trajectories that are immediately available on the tactile surface—a task that is slower and also carries a much greater cognitive load. In Lederman and Klatzky's terms, while the contour-following EP alone would be considered sufficient for extracting information about lines on a graph, being able to explore the spatially rendered image with both hands to discover its global shape draws on more optimal EPs for extracting this information.[34]

As we noted in chapter 2, perception is an active process in which our movements reveal structure in the flow of sensation being picked up from the environment. This structure, in turn, is information that we use to guide our further action. What Gibson and, later, Lederman and Klatzky have demonstrated in the context of touch is that we can observe people actively seeking out structural invariants in response to a particular task goal. In describing the texture of a cushion, for example, we may pick up on the regularity of patterns of stimulation of the fingertips as we move laterally across its surface and determine, from this regularity, that the material we are feeling is corduroy.

In short, in designing interactive systems that include augmented and virtual objects—objects that may be situated within real, augmented, or virtual environments—our goal is to maintain lawfulness in the relationships between how people are moving and what they are perceiving as they actively explore and work with our systems. And we must do so while at the same time understanding the ways in which these systems shape the kinds of information that can be provided in response to a user's actions. If we can achieve this, it will be possible for the users of our systems to discover the perceptual invariants within our constructed environments, persistent experiences on which they can depend as they come to skillfully bring forth, in perceptual terms, the new kinds of worlds we have designed.

6.4.3 Perceiving Invariant Properties of Objects

In *Action in Perception,* Alva Noë, like Gibson before him, argues that far from being passive and detached from immediate contact with the environment, vision is an active sense, engaged in the same kinds of intentional and exploratory activities that we usually associate with touch. For Noë, "seeing is much more like touching than it is like depicting."[35] But what are the invariant properties of the environment that can be picked up by vision when considered as an active sense? In answering this question, Noë uses the example of viewing a humble penny. He points out that the stimulus falling on your retina will vary in shape in accordance with how you move your head or eyes. So, the penny's outline might be circular when viewed straight on or elliptical when viewed from an angle. And yet, Noë points out, it is still perceived as being a penny. Noë argues that this is because the person perceiving the penny knows how to move in order to bring it into one of these perspectives—that is, they understand the sensorimotor contingencies relevant to viewing pennies from multiple perspectives so that the concept of the penny persists regardless of its position relative to the eyes. In the language of Gibson, one might say that the penny persists because its shape continues to structure the optic array such that it is invariant, even under the transformation associated with viewing it from a different angle.

Although Noë and Gibson might appear to be saying the same thing—that perception arises from action and is not composed of raw sensory stimuli but of invariants—there is a subtle distinction when we think about perception of objects that are partially or wholly rendered in digital form. In his discussion of the penny, Noë focuses on how the sensory stimuli arriving at the retina are transformed by movement and how we understand these transformations because we have established sensorimotor contingencies that allow us to make sense of such changes. In the sensorimotor approach to perception, perceptual invariants are said to be motor specific. That is to say, perception of a given invariant is associated with a very specific pattern of muscular activity, and the assumption is that perceptual systems can extract information in the form of the invariant relationships between co-occurring sensory and motor patterns. "Whenever a sensory transformation co-occurs with a movement, those properties that remain unchanged in this coupling constitute a sensorimotor invariant."[36] In other words, movement plays a constitutive role in defining such invariants.

Gibson, too, notes that perception arises because of action and is not of sensory stimuli per se but rather of invariants: "The active observer gets invariant perceptions despite varying sensations."[37] For Gibson, perception of the environment guides action, and, in turn, action upon the environment generates the information needed to perceive opportunities for future action. Information, for Gibson, takes the form of invariants that are specified by our intention toward the objects in the environment, not simply by the invariant aspects of the relationship between sensory and motor patterns, as with SMCs. What is important here is how the information available for the individual in the ambient energy array is specified by what they want to accomplish at this moment and in this context. Perception here is not of simple physical constructs (such as brightness or weight) but rather of what Gibson terms "higher-order invariants," information that is useful in guiding action. A frequently cited example of a higher-order perceptual invariant is the moment of inertia, where the amount of torque required to rotate a handheld object remains constant, despite the involvement of many muscles applying different forces across their individual ranges of motion.[38] In this interpretation, perceptual invariants are said to be transformation specific and motor equivalent. Many patterns of muscular activity will give rise to the same perceived invariant property of the object or the environment. Movement is therefore instrumental in determining such ecological invariants. For Gibson, such invariant structure in the ambient array represents the way the world "shows up" for an individual and affords opportunities for them to accomplish the actions required to achieve their goal.

This distinction points to the underlying bias toward vision that has dominated research on object perception for so long. In emphasizing the systematic way in which visual images on the retina are transformed by movement, Noë invokes the language of visual perspective inherited from the painters of the Renaissance. They were striving to understand how to "fool" the eye into believing that the judicious distribution of lines, shapes, colors, and shadows on a canvas could suggest a three-dimensional scene. They sought to determine which of the invariant properties of light cast upon such a scene were most likely to specify the presence of and relationship between three-dimensional objects—shadows, partly or wholly occluded edges, and so on. Their strategies for representing perspective were successful because they could assume their viewers had sufficient experience of encountering objects and scenes in the real world such that they could recognize the structural invariants shaping artificially constructed light in the painted world without having to act in that world. Painters had to understand the distinction between the distal stimulus in the world and the proximal image falling on the retina. They invented machines to assist them in drawing in perspective. Leon Battista Alberti, for example, invented a frame with a grid of wires to be placed at a short distance in front of the subject to be drawn. The artist then transferred the image before him, square by square, onto a sheet containing a corresponding grid.[39]

The issue of whether visual experience is defined by distal or proximal stimuli falling on the retina is one we still deal with in the study of visual perception today, and it pervades even Noë's discussion of vision as an active sense. This is seen, for example, in his explication of how SMCs impose meaning on the relationship between the circular and elliptical views of the penny. In short, Noë's view is a perspective more akin to Heidegger's concept of objects that are *vorhanden*, to be encountered as objects of inspection, rather than *zuhanden*, being grappled with as objects in use.

Designers of virtual and augmented realities must take users on a journey from the world of familiar experience to a world where new modes of interaction require new approaches to designing and rendering information sources on which such interactions are founded. Technologies for rendering such environments are coevolving with the capacities of participants to interact with what is being rendered. Just as Renaissance artists discovered how they could design their paintings to create the illusion of three-dimensional objects and scenes on two-dimensional surfaces, so designers of VR have discovered how to render the illusion of three-dimensional virtual objects that appear to be untethered from any physical substrate. Where images rendered in early virtual worlds were flat and appear as distal objects to be inspected (*vorhanden*), objects rendered in 3D are proximal (*zuhanden*), intended to be grasped and grappled with as means of dynamically interacting with the environment. In VR, one reaches out for a tool rendered in 3D, grasping it to bring it into proximity with the body in order to use it, much as one would for a physical tool in a workshop. When one is doing so, the tool is brought into a body-based coordinate system, and all actions relate to its position and motion with respect to the body of the user. In contrast, 2D renderings of tools, even in VR, often require one to approach a distal surface, on which they are rendered, and to interact with them through surface-based gestures such as pinching, dragging, and dropping. These interactions are performed within the 2D coordinate system of the interactive surface, e.g., the screen, on which the tool is displayed.

Three-dimensional, visually rendered virtual reality relies on the constitutive role that movement plays in making apparent the SMCs that bring forth objects and environments, on the assumption that participants can co-constitute the relationships between sensory and motor patterns to bring these worlds into being. But the motor specificity of this assumption is limiting. If someone does something different, they may perceive something different. For example, if a participant moves at a different speed through a VE they have visited before, the movements and positions of objects will be rendered at a different rate in response to their speed, and the SMCs that couple their movement to changes in the environment will be updated to account for this altered relationship with their surroundings.

One way to circumvent this challenge might be to start with the premise on which the ecological approach to perception is based—that movement does not constitute invariants but rather is instrumental in discovering task-relevant information, specified by properties of the environment, that are structurally or transformationally invariant in the presence of otherwise fluctuating energy in the global array. Arguably, such an approach would be more robust, supporting a broader range of strategies for exploration and leaving open the possibility that more opportunities for action would "show up" for users inhabiting augmented and virtual worlds, if designers focused less on large-scale realistic rendering of entire worlds and more on the timely delivery of haptic, auditory, and visual information to support the manipulation of the rendered objects that are the focus of the current task. For example, in rendering a virtual soccer game from a goalie's perspective, as discussed in chapter 5, a virtual environment designer might use gaze-tracking to determine when the goalie is following the movement of the ball and ensure that salient information, such as the changing size and shape of the ball's image in the goalie's field of view, are accurately rendered to reflect its incoming angle of approach. When this cue is present, the goalie can use this information to make more accurate judgments about how to move to intercept the ball, in comparison to cases where the ball is just rendered as an expanding circle or sphere that does not change shape.

6.4.4 Detecting Object Affordances

As we noted in chapter 3, the process of acquiring skill in manipulating objects depends on one's ability to physically interrogate one's environment—on the ability to discover, from all possible ways of accomplishing a task, which strategies are going to be the most successful. Furthermore, acquiring expertise in a task environment depends on developing strategies for accomplishing a task that are adaptable—that is to say, strategies that are robust to unpredictable changes in that environment. If interactions with objects in virtual and augmented environments are to be successful in the long term, designers of simulated environments must anticipate the kinds of exploratory strategies their users will employ. Addressing this question inevitably returns us to the question of affordances and, in particular, to the question of how Gibson's concept of affordance has been interpreted within the field of interaction design.

Don Norman first introduced the term "affordance" (later restated as "perceived affordance") to the field of interaction design.[40] He suggested that designers could design objects (physical or virtual) whose form would indicate to a user how they should be used—that the shape of a door handle, for example, should indicate to an approaching person whether the door should be pushed or pulled. Notably, Norman's understanding only partially captured Gibson's theory of affordance.[41] Norman's interpretation seems to focus on emphasizing the designer's role in representing, through the form of designed artifacts, how individuals should interact with the habitat of their designed environment. But, it neglects to account for the intentionality of the user, for the way in which the affordances of their *Umwelt* "show up" as opportunities for action within the unfolding context of accomplishing a given task goal. In other words, as van Dijk points out, affordances are not cues encoded in a product's shape that determine how it should be used. Rather, they "develop from moment to moment as part of the skilled ways of perceiving (how the world 'shows up') in situations."[42] Van Dijk notes that the challenge for designers of technologies is whether the augmented environments they create will support users seeking to develop new perceptual skills, which they will need in order to interact with the new worlds these technologies bring forth.

Let us return to our earlier discussion of embodiment relations as defined by Don Ihde.[43] In the physical world, the ability to incorporate technologies into the body schema is polymorphic. When we pick up an object, such as a tool, in order to do something, the tool becomes incorporated into our body schema. When we pick up a screwdriver, we have both extended and reshaped our body schema, providing it with the ability to reach into tiny spaces and turn screws. Ihde notes that the body–tool coupling is given both its configuration and its meaning by the task at hand. Furthermore, this coupling is malleable. If we take up another tool to do another task, our body schema is adaptively reconstituted in response to the requirements of the new task. Van Dijk and others ask how malleable this configuration can be while still allowing users to make sense of what such tools are and what they are for (see section 6.6.5).

One approach proposed by Kaptelinin and Nardi is to reframe affordances as "possibilities for human action mediated by cultural means."[44] Gibson's theory of affordances, they suggest, is limited in its usefulness when we think about our interactions with technology because it does not account for the kinds of mediated actions with the environment that characterize today's augmented realities. Kaptelinin and Nardi propose an alternative—instrumental technology affordances: relational properties of a three-way interaction among a user, a mediating technology, and the resulting environment. This proposal integrates two related spaces of action possibilities: handling affordances (opportunities for action afforded by the technology) and effector affordances (the possibilities for having some effect, through the technology, on an object within the environment). For traditional tools, such as screwdrivers, the link between handling and effector affordances is clear: You grasp the handle and, with the other end, engage the screw. In the case of a digitally mediated action, such as a scroll bar in a GUI, the handling affordance involves operating on the scroll widget by dragging a pointer, with a mouse or finger, whereas the effector affordance operates on the contents of the underlying window. In terms of the screwdriver, it is as though you cut the shaft of the tool in half and introduced some form of digital mediation where the relation between the handle and the tool tip is now malleable because it is programmable. As with non-mediated tools, Kaptelinin and Nardi note that the relationship between handling and effecter affordances must be learned or mastered and that the time taken to do this will depend on the complexity of both the mediation and the associated task environment. For Kaptelinin and Nardi, the concept of an instrumental technology affordance is grounded in a sociocultural perspective, which acknowledges that the activities of humans are always mediated by sociocultural tools, including their technology. This is in line with Heidegger's concept of "equipment," with the idea that the carpenter's activities are given their context by their workshop, which physically and culturally situates them within their broader community.[45]

By extending Gibson's concept of affordance to incorporate Heidegger's concept of technology as equipment, Kaptelinin and Nardi provide a sociocultural framework that accounts for how we can become attuned to new kinds of information present in our technology-mediated environment. Kaptelinin and Nardi do not ask how, in flexibly coupling handling and end-effector affordances, users can come to understand the relationship between the two. For a user to learn this relationship and become skillful in working with such tools, there must be some regularity or isomorphism between handle and effector. The idea of technology affordances already exists in telerobotics (e.g., between a local control interface, or leader; and a remote manipulator, or follower). Much attention has been given to how the local operator can model the linkage between their input gestures, as mediated by the leader, and the actions of the remote follower, which might be operating in very different conditions—for example, on the seabed. Findings from this work indicate that as long as there is a predictable and stable mapping, users can become skillful operators, even when the mapping between their hands and the remote robot's effectors are very different.[46] As a result, users of such systems can bring to these encounters their skill in discovering the invariants present in these mappings. In the next section, I outline ways in which it is possible to support users in attuning to invariant properties of such interactions, even when the available technologies cannot render the most perceptually salient higher-order invariants that would support interactions with physical tools.

Given sufficient feedback and the presence of reliable perceptually salient information, it is reasonable to assume that the concept of the instrumental technology affordance could be a means of supporting a flexible coupling to a virtual or augmented environment. Rather than gawking at 3D renderings of visual objects from a distance, we can then truly grapple with objects that are ready to hand, rendered by fully developed algorithms for structuring optic, haptic, and auditory stimulus energy arrays. As we enter this world, though, we have to address challenges that prior research on object perception could not have anticipated—namely, that virtual objects and their properties are now completely malleable, such that they can be reassigned at will and reconstituted on the fly. Fully understanding the implications of such dynamic interactions with objects of perception and action will require us to completely rethink our notion of objecthood—of what it means to act with and through artifacts that might come into being for only a moment.

6.5 Coming to Grips with Digital Equipment

At the heart of our ability to do anything in augmented and virtual environments is the capacity of those worlds to render their places and objects and to capture our movements as we engage with the worlds they present. The greatest challenge facing designers lies in the fragmented world of hardware and software platforms with which they must contend. Although there is a growing infrastructure to support cross-platform development, a glance at the unresolved differences in visual frame rates, audio and motion-capture sampling rates, the number and disposition of input sensors, haptic formats, and audio output channels—as illustrated in table 6.1—demonstrates that there is a long way to go before content creators can work with a platform that is even as stable as HDTV and 5.1 surround sound for today's home theaters.

For this reason, the following discussion of rendering object properties and grappling with them in AR and VR is not concerned with specific hardware and software solutions. It will instead focus on higher-level opportunities

Table 6.1
Overview of current major head-mounted display devices. Source: Martin et al., 2022, table 1.

Manufacturer and Model	Resolution per Eye	Positional Tracking	Max. Refresh Rate (Hz)	Field of View (Degree)	Display Type	Integrated Audio	Price ($)
FOVE	1,280×1,440	Yes	70	100	OLED	No—Jack 3.5 mm	600
HP Reverb—Pro Edition	2,160×2,160	Inside-out	90	114	LCD	Built-in headphones	649
HTC VIVE Pro	1,440×1,600	Yes	90	110	AMOLED	Built-in headphones	799
HTC VIVE Pro 2	2,448×2,448	Yes	120	120	LCD	Built-in headphones	800
Oculus Rift S	1,280×1,440	Inside-out	80	110	LCD	In-line speakers	399
Oculus Quest 2	1,832×1,920	Inside-out	120	104	LCD	Stereo speakers	399
Samsung Odyssey	1,440×1,600	Inside-out	90	110	AMOLED	Built-in headphones	500
PlayStation VR	960×1,080	Outside-in	120	100	OLED	No—Jack 3.5 mm	299
Valve Index	1,440×1,600	Yes	144	130	LED	No—Jack 3.5 mm	999
Varjo VR-3	2,880×2,720	Yes	90	115	uOLED	Off-ear speakers	3,195

and challenges facing designers of interactive objects, independent of the particular device or platform through which they are encountered.

6.5.1 Rendering Virtual and Augmented Objects

When we talk about rendering objects in AR and VR, what do we mean? Most literature on rendering properties of objects has focused on rendering their visual properties. It is easier to render visual images than it is to render haptic properties of objects or to colocate sounding objects precisely in space with their visual or haptic counterparts. But with the advent of high-fidelity VR systems integrating holographic displays, movement tracking, sound spatialization, and haptic feedback, the opportunity to render properties of objects that are perceivable through multiple sense modalities, once limited to specialized systems such as medical simulators, is now becoming a reality for consumer products as well.

I have argued that the properties of objects that we attune to in accomplishing tasks in our environment are those that provide information specific to what we are doing at a particular moment—that is, those properties that specify the opportunities for action afforded by a given object. The question then becomes, Which of all possible properties of an object must be rendered such that they structure the global energy array to provide the information a user needs? Recall that, for Gibson, properties of objects are specified for an individual by invariants—higher-order variables to which they are attuned. For people catching balls, this might be the rate of expansion of an image on the retina, the invariant specifying the time-to-contact of an approaching ball. The question for designers of augmented and virtual objects is whether simulated object properties can serve as either structural or transformational invariants within the global array, such that they, too, provide the information about affordances of objects to which we can attune. As Dessing and Craig have demonstrated in their study of rugby players intercepting goals, discussed in chapter 5, when such information is made available to participants in VR who are already skilled and attuned to the opportunities it specifies, they can successfully intercept virtual balls too.[47]

One challenging aspect of simulating physical objects and of understanding how such simulations might structure the global energy array is in determining the possible sources of information to which users might become attuned. For material objects, physical properties are those properties that can be measured without changing the object's chemical or atomic structure. Such properties can be intrinsic, remaining the same independent of the conditions under which they are measured, or extrinsic, changing in accordance with some external condition. For example, the mass of a bowling ball is intrinsic, being dependent on its material composition, but its weight is extrinsic because it depends on the gravitational force where it is weighed. The color of an object can be intrinsic to that object (e.g., a blue gemstone) or extrinsic if it depends on, say, the angle from which it is viewed. In simulating objects, we can decouple them from their intrinsic and extrinsic properties. We can use projection mapping to give our bowling ball a new color, and combined kinesthetic/tactile rendering to alter its apparent weight.[48]

Not all the simulated properties of objects are virtual. The behavior of light as emitted by an HMD or screen is still constituted by photons, as with any light source, and the acoustic waves emitted by loudspeakers or even headphones still reach the ear as changes in air pressure, just as with any other sound source. These rendering methods still rely on the physics of the electromagnetic, acoustic, and force energy arrays, but they also function as stand-ins, by rendering a subset of the properties of real objects. It is "as if" the bowling ball were visible or audible or wieldable by virtue of the information about it that could be picked up by our sense organs. If the bowling ball were thrown at us in a virtual environment, we could anticipate intercepting it with our hand using the same invariant properties of the visible energy array (the rate of expansion of the retinal image) as we would in real life. If we failed to catch it and it fell to the ground in our virtual environment, the simulated sound of its contact with the ground would reach us as a disturbance in the acoustic energy array, just as in the real world.

But we need not render all the physical properties of our bowling ball, such as its capacitance or the amount of electromagnetic radiation it emits in the infrared range. We *cannot* render others, such as the combined tactile properties specifying the material it is made from—its surface texture, its compliance, and the amount of heat it absorbs from our skin when we hold it in our hands. In virtual environments, then, some of the possible sources of information available to us are real sources of energy (although their behavior might be under algorithmic control as for sound synthesis or animation). But other object properties are simulated, as in the case of the interaction forces we experience through our hand as we lift the virtual bowling ball.

In short, some of our interactions with virtual objects might relate to aspects of their material composition, whereas others might relate to properties of the object in a given context. The object's physicality could be considered as the sum of possible interactions between the object and different energy media. A subset of this sum may be perceivable, whereas a subset of these perceivable properties (ways that the object interacts with energy/force) may have relevance for behavior of different sorts. This subset of information specifies relevant behaviors for an individual within their environment and gives the object meaning or objecthood (doorstop, sports equipment, heavy-looking thing, and so on). When rendered as a virtual stand-in, the only possibilities for action the object can make available to the participant are those that can be acted upon. Thus, the goalie in a virtual rugby game can intercept a virtual ball, but they cannot tuck it under their arm and walk away with it because we still lack the means of rendering an untethered 3D haptic hologram with a virtual mass.

As I shall discuss in section 6.6, one of the constraints on thinking about virtual and even augmented objects as entities in their own right is that we cannot think of the media we encounter in AR and VR independently of the technologies we use to render and manipulate such objects. Although the use of physics engines makes it possible to derive visible, tangible, and audible behaviors for an object from the same physical model, the technologies used to render those behaviors vary so radically in their form, their fidelity, and the number of degrees of freedom they support (on input and output) that a user's experience is often dictated by the technology, not by the experience of interacting with the media. For example, a haptic device that only supports single-point interaction with a virtual object, such as a Phantom, allows you to feel its surface texture or push on it to judge its compliance, somewhat like feeling the world through the end of a pencil. In comparison, a haptic glove supports whole-hand interactions such as grasping virtual handles or enclosing small virtual objects in one's hand. Both display haptic properties of objects, but they impose very different constraints on how a user can interact with a virtual object through touch.

Just how disparate this experience has become is apparent from a recent review by Marshall et al., who draw attention to some of the many sources of sensory (mis)alignment that arise for users of today's immersive media environments.[49] Some of this challenge is apparent in table 6.2, which maps possible means of digitally stimulating sense modalities to available display technologies. Looking at this table, it is apparent that each time you add a means of rendering a new sensory cue, the process of mapping object properties to sense-specific display devices increases in complexity in a nonlinear manner. This is VR's version of the curse of dimensionality.[50]

Table 6.2
An overview of ways to stimulate different sense modalities under digital control for use in virtual reality. Source: Adapted from Marshall et al., 2019, table 1.

	Remotely	Mechanically	Chemically	Electrically
Vision	Screens, head-mounted displays, holographic displays			Implants
Auditory	Speakers/headphones	Bone conduction		Implants
Touch	Ultrasonics, blown air	Physical actuators		Electrode stimulation
Kinesthetic	Visual Balance	Motion platforms		Vestibular stimulation Muscle stimulation
Taste			Delivery of chemical mixes or foodstuffs	Tongue stimulation
Smell			Delivery of fragrances in air or in nasal tubes	

The question that remains is what properties of objects we can and should render in AR/VR. If the information relevant for users is that which specifies relevant properties of the environment in the context of a particular situation, then the choice of what and how object properties should be rendered is not as simple as rendering form, color, or surface texture. Of course, there will be situations in which these attributes specify relevant information, but this is likely to be the exception rather than the rule. Similarly, there will be situations where the most salient object attributes *cannot* be rendered because of current technical limitations, so others might have to "stand in" to convey the relevant information.

As a case in point, let us take the task of estimating the weight of an object in VR. In a 2021 review, Lim et al. summarize the many cues that we use to estimate the weight of objects in our physical environment.[51] They then present a systematic review of methods that have been used to render the weights of virtual objects. Citing the work of Lederman and Klatzky on EPs, studying the hand movements we perform to extract properties of objects through touch, Lim and Lee note that we typically estimate the weight of an object using "unsupported holding," where the object is picked up and held in one or both hands.[52] As Lim and Lee point out, lifting objects requires muscular exertion to counteract the object's mass and the effect of gravity. So, efforts to render object weight have typically focused on haptically rendering reaction forces using robotic force-feedback devices. As their name suggests, these devices apply force to the hand or body in response to a user's movements, typically using motors. In the case of the Phantom, mentioned above, a user holds a pen-like handle or manipulandum and moves it around in a virtual space. When the virtual tip of the manipulandum penetrates the surface of a virtual object, a motor turns on to push against the user's hand, giving them the impression that they have encountered a physical surface. If the tip of the manipulandum is a hook, and the virtual object can be picked up, its weight can be simulated by opposing the direction of motion of the hand, applying appropriate forces to mimic the effect of gravity.

But real objects have a specific center of mass, which can impact the direction and relative magnitude of forces that are perceived when it is lifted, further complicating the task of conveying a realistic sense of their weight. Moreover, sensing force involves more than stimulating receptors in muscles and joints. The heavier the object, the more it deforms the skin of the hand, stimulating cutaneous and subcutaneous mechanoreceptors as well. Therefore, stimulating both cutaneous and kinesthetic receptors by simulating combined force and tactile feedback results in a more accurate perception of the weight of virtual objects, particularly for heavier weights.[53]

In interacting with everyday objects, the perception of weight is influenced by many factors other than the forces that resist the motion of the hands when holding or lifting. In the well-documented size–weight illusion, people presented with objects of different volumes but the same weights report the larger object as being lighter because its mass-to-volume ratio is smaller.[54] Material, too, can impact perception of weight, at least for objects with small masses, where small grip forces are required for lifting. In the material–weight illusion (MWI), metal objects are perceived to be lighter than those with identical size and weight but made from a less dense material.[55] In later work, Ellis and Lederman suggested that the MWI is primarily because of an interaction between haptic perception of weight/mass and haptic judgments of material properties because visual estimates of the MWI produced markedly weaker results.[56] Surface texture, too, can influence perceived weight. Objects with smoother surfaces are perceived to be heavier than those with textured surfaces because of the greater grip force required to prevent them from slipping from the hand.[57]

In short, the fact that multiple, distributed touch receptors are involved in the perception of weight—and that its judgment can be impacted by information-related properties other than actual weight (such as surface texture, material, and size)—means that accurately conveying the weight of objects to participants interacting with them in VR represents a considerable challenge. In their review, Lim and colleagues identify several approaches to rendering object weight in VR,[58] including methods for haptically rendering normal and sheer forces experienced by the muscles, joints, and skin during static holding,[59] and dynamic forces because of the inertia experienced when wielding virtual rods of different lengths and diameters.[60] Lim and Lee also identify several "pseudo-haptic" methods for conveying the apparent weights of objects visually, including delaying the motion of a visually rendered object being lifted to increase its apparent weight[61] or using visual cues to direct users to increase or decrease their movement velocity, which, in turn, can be used to calculate the magnitude of the forces they are applying to lift visually rendered dumbbells.[62] Audio feedback in the form of an increased heartbeat can also be added to provide a sense of increased effort when lifting virtual weights.[63] Users can even come to associate sound and color as cues with the weights of virtual objects—lighter objects being associated with both higher pitch and greater color intensity.[64]

But what if we thought of rendering information relating to the weights of objects not in terms of basic forces, torques, and masses but rather in terms of higher-order task-relevant invariant properties available in the global energy array? In haptic interaction with objects, we may sense many mechanical properties that change over time, some of which are correlated with a given property (specifying variables) and some of which are uncorrelated (non-specifying variables). In the presence of feedback from the environment, we become attuned to specifying variables because they more accurately predict the outcome of our actions, and we pay less attention to non-specifying variables. With respect to haptic perception of the weight and shape of occluded objects, for example, one specifying property that has been extensively studied is the second moment of mass distribution, known as the moment of inertia (or "I"), which remains invariant under rotation when wielding objects with the hand (see discussion in section 5.4.5). In determining whether such a higher-order invariant could be simulated and, more importantly, whether it could be "picked up" by participants wielding virtual objects, Singapogu et al. asked participants to estimate the lengths of haptically rendered virtual rods.[65]

In prior work, Withagen and Michaels had demonstrated that, given sufficient feedback during training, participants could become attuned to (I) as a specifying variable that they could use to judge changes in lengths of occluded rods wielded by the hand.[66] Moreover, with practice, they could also judge the magnitude of (I) to make more accurate predictions of the lengths of rods—a process referred to as "calibration." Crucially, invariant properties of objects become apparent only under conditions where other aspects of the environment are changing. In wielding objects during dynamic touch, as Singapogu et al. note, muscle forces and torques are produced to set an object in motion, producing reaction forces and torques that are correlated with the object's mass distribution. Singapogu's team hypothesize that in simulated environments where the variables that optimally specify perceptually salient object properties may not be available—for example, (I) in the case above—users may become attuned to non-specifying variables instead. Even in cases where such properties are available, the nature of display technology itself may introduce ambiguity, which may provide feedback that inadvertently results in users picking up on non-specifying information. In our weight estimation example above, for example, people's estimation of the weight of objects based on parameters such as brightness and pitch would represent an attunement to non-specifying information that was artificially correlated with changes in object mass by the system's designer.

Instead of simply hoping that users discover such correlations, Singapogu et al. suggest it might be possible to train users to attune to specifying variables, even where information is sparse or ambiguous. Using Withagen and Michaels's method of haptic training, based on the perceptual framework of attunement and calibration, Singapogu's team demonstrated that their participants were able to pick up on simulated information about the distribution of virtual masses and could accurately predict the length of haptically simulated rods. They concluded that when kinesthetic properties of virtual objects are rendered and systematic feedback is provided, participants in VR can attune to properties that specify information relevant to their task goal, even if these would be non-specifying properties for equivalent interactions with tangible objects in the physical world. This finding is valuable when considering how people learn to interact with real and augmented objects because it suggests a route by which we can attune participants to those properties in simulated environments that may specify higher-order, task-relevant information. In other words, appropriate training can turn participants into skillful perceivers of the information our environments provide.

Even given the promise of attuning users to the kinds of specifying and non-specifying information we can simulate, the task of matching available rendering resources to perceptually relevant properties of objects, at least when it comes to their weight, demonstrates that we are far from the point where we can think about simultaneously rendering multiple intrinsic and extrinsic properties of objects without encountering the curse of dimensionality and the roadblocks of technical capability and platform specificity. And that is not even addressing how participants in VR might pick up the object in the first place. As we shall see in the next section, even the task of grasping and picking up objects in augmented and virtual reality is subject to its own world of hardware, software, and conceptual challenges.

6.5.2 Grappling with Objects in Simulated Environments

If, as Heidegger suggests, our environment is equipped with the tools we need to accomplish our goals, then what does such an environment look like when our goals are to perform tasks in augmented or virtual spaces? What are the tools we use and how is the environment equipped to support our work? There are so many "handles" (mice, game controllers, etc.) and so many ways to map the body's movements, mediated by these devices, onto behaviors of objects such that each encounter with the AR/VR is a process of relearning—a new tool picked up "in order to." Moreover, although the "handles" we use do not change, the way they are mapped to the task environment is entirely programmable. In such an environment, can the tools we pick up "in order to" truly withdraw from our awareness? In Heidegger's terms, there is one hammer with which to do everything. And, as the adage goes, "When all you have is a hammer, everything in the world looks like a nail." So, what do the hammers we are currently constrained to working with look like and what do they enable us to do?

Interacting with objects in AR/VR starts from movement. The actions of our fingers, hands, arms, and bodies have to be tracked, and the resulting data has to be mapped onto the behavior of rendered objects and environments. The challenge for designers of AR/VR systems, as depicted in figure 6.3, is that each kind of "handle" (or manipulandum, mouse, game controller, etc.)

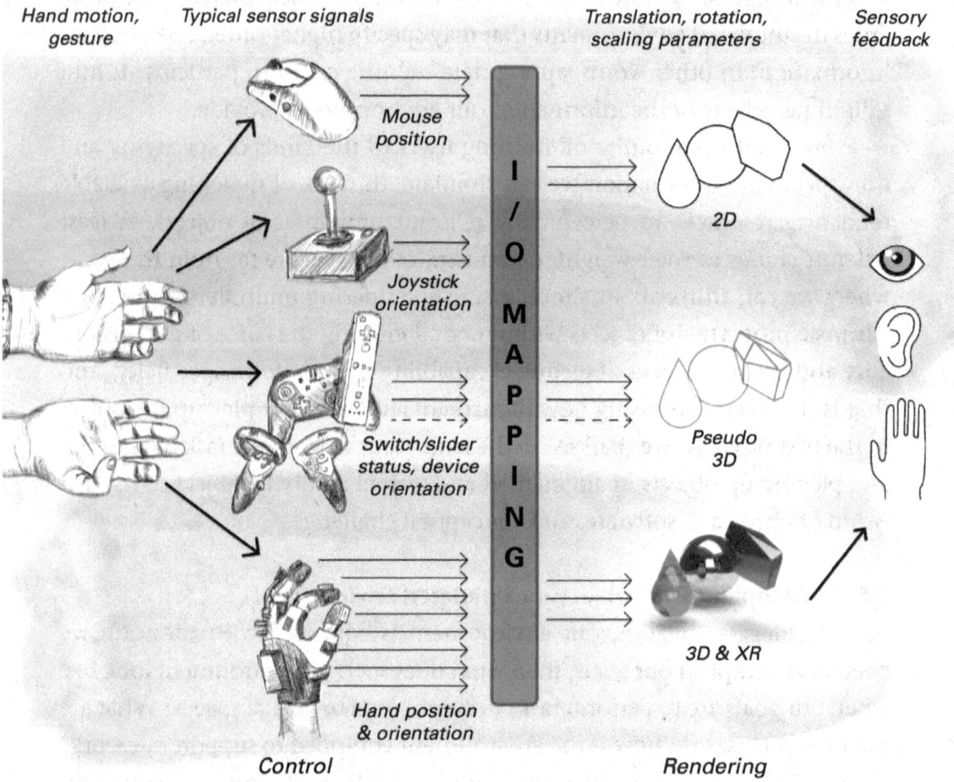

Figure 6.3
Input/output bottlenecks when converting hand motion to sensor signals used to manipulate virtual objects. *Source:* Original drawing by Hans Anderson.

contains a different number and arrangement of buttons and different sensing technology. A mouse, for example, might have two buttons and a trackball that allows its X/Y position on a surface to be sensed.

For the purposes of interaction design, such a manipulandum is said to have four degrees of freedom: four independent sensors (two switches and an X/Y or 2D trackball), whose state can be independently mapped to some behavior of the system. Most game controllers have many buttons, one or two joysticks, and often a 6-DOF IMU (measuring acceleration and rotation in all three axes), making it possible to track the motion of, for example, the hand in free space as well as the state of the joysticks and buttons.

The ways we can manipulate objects in AR/VR are defined by the constraints of the devices we are using. Even if we use cameras to track the movements of hands directly, the amount of information we can extract about, say, fine-grain finger movements is constrained by the spatial resolution of the camera-based tracking system and its temporal resolution (or sampling rate), not to mention issues arising from occlusion of one part of the hand by another. Although sensing gloves resolve some of these issues, their utility is also determined by which of the hand's twenty-three available DOF they can track and by the resolution of their sensing systems.

A second constraint is the number of available DOF of the system for which the objects are rendered. As we have noted, objects rendered on flat surfaces are typically 2D, with the potential for using perspective cues to create pseudo-3D renderings. Holographic renderings in more fully immersive AR/VR environments are truly 3D, allowing the viewer to move around them and view them from different angles. Objects in these environments can be translated, rotated, and scaled or resized in all three planes (X, Y, and Z) and are said to have nine available DOF. This continues to frustrate designers of AR and VR environments because they need to figure out how to map the movements they can capture from the range of available manipulanda onto the behavior of the objects and worlds they are rendering.

In a 2019 review, Mendes and colleagues addressed both sides of this issue.[67] In reviewing input technologies and manipulation techniques, they divided methods for 3D object manipulation into three categories: mouse/keyboard, 2D interactive surfaces, and midair interactions (including handheld controllers, wearable devices such as gloves, and hand-tracking systems such as cameras). Not only do these devices' modes and degrees of sensing constrain what movements can be used, they also vary considerably in the amount of control and precision they support. In the context of entertainment applications such as gaming, this is not such an issue. But in applications where manual dexterity is required, such as in surgical training or manufacturing, bespoke devices with high-precision position tracking need to be developed. This is exacerbated in systems relying on AR/VR headsets because their relatively low-fidelity visual rendering, combined with issues associated with occlusion of hands or objects during manipulation tasks, reduces their effectiveness for many potential task domains.

A third constraint, which Mendes et al. do not address, is how the form of the device influences what users will and can do with a given manipulandum. A gaming device with a gun grip, for example, maps well to pointing gestures and to grasping objects using a scissor or trigger grip. In contrast, a bimanual game controller, such as Microsoft's XBox Wireless Controller, can support a broader range of two-handed fine-grain motor actions through buttons, joysticks, and 2D trackpads.[68] The result of such disparities is that, for cross-platform games, multiple strategies for picking up and manipulating rendered objects must be supported.

In addition to the challenge of working with so many manipulanda, the second part of the challenge facing designers concerns how objects behave in response to what the user does. As Mendes et al. point out, it might not always make sense simply to mirror combined translations and rotations tracked at input into similar movements of a simulated object, particularly for renderings constrained to 2D surfaces. Therefore, alongside the evolving space of controllers and rendering systems and methods, there has been a parallel evolution in mapping strategies to support these systems. The solutions range from mappings that completely decouple translation, rotation, and scaling (TRS) of objects (performed as separate operations) to those that mirror the motion of the hand in 3D space.

Many design metaphors have been developed to create meaningful frameworks for users to make sense of manipulation strategies. For example, metaphors have been developed to overcome the challenge of manipulating out-of-reach objects, such as the Go-Go Immersive Interaction Technique.[69] Here, the user's virtual arm grows to extend their reach in a linear way until it reaches a predefined distance, after which the rate of extension is scaled by a predetermined factor. This allows for an easy transition between the manipulation of close and distant objects. A second approach to reducing the complexity of the mapping problem is that of smart mappings—ways to combine TRS functions that reduce the cognitive load for the user, for example when mapping the motion of a midair 7-DOF controller such as a 3D mouse to TRS movements for an object rendered on a 2D surface.[70] Although mirroring hand movements used to manipulate physical objects for midair interaction might appear to be the most desirable solution, its usefulness is limited by users' lack of precision and dexterity in manipulating objects in midair without haptic feedback.[71] In addition, direct mappings may result in users of AR/VR environments having to move more to reach objects or becoming frustrated when performing complex manipulations that the system cannot accurately track (e.g., controlling translation and orientation of objects in multiple dimensions). In such situations, providing independent control for each "DOF separation" has been shown to increase accuracy but at the cost of increased task completion time.[72]

In short, when considering how movements of the hands are translated into manipulations of virtual or augmented objects, four things must be considered:

1. How does the form of the manipulandum constrain the hand movements a user can or will use?
2. What kinds of movements can be captured by the sensors in the device (or the camera observing the hand)?
3. How does the form of the rendering system (flat screen, fully 3D VR, etc.) constrain how objects can be rendered?
4. What is the best way to map the movements that are tracked onto the behavior of the simulated objects to support the task goals of the user?

Each of these considerations represents a potential point of disruption or breakdown between the hand of the user and the response of the system. The manipulandum may not feel like the right handle for the task (e.g., picking up and putting down an object using a trigger grip). Even if a user can overcome the unwieldy tool handle, the tool might not seem to respond in the way they predict. (It might be difficult to align the virtual tool tip with the object to be grasped because of system latency and lack of spatial resolution.) Finally, on picking up the object, it might be difficult to reorient it in a way that makes sense, given the form of the physical tool handle and how well its movements map to the desired movements of the rendered object.

As I will argue in section 6.6, the challenge of "coming to grips" with augmented and virtual environments only arises because those environments are currently defined by the technologies through which we can access them, not by the media they render. In fact, we have no means of thinking about the new kinds of worlds they bring forth that are independent of technology. The medium of rendering is not yet independent of its content. We still have too many kinds of hammers, too many kinds of nails, and no shared understanding of what kinds of materials the walls we are trying to construct are made of.

6.6 Designing Objects for Entangled Worlds

The foregoing discussion has assumed that the properties of objects encountered by a person in their environment are associated with objects physically present in that environment. Moreover, it assumes that the perceivable attributes of encountered objects are exclusively allocated to those objects—the color associated with a coffee cup is a material property of that cup and arises from the pigment of the glaze applied to its surface.

But in augmented and virtual reality, the objects of our experience need no longer be constrained by materiality. Lighting, color filters, and projection mapping can "recolor" the surfaces of objects. By using haptic gloves, we can augment forms and apply new tactile textures to surfaces. If we combine holograms with 3D midair haptic rendering techniques, we can conjure rabbits, hats, and many other kinds of objects. On the other hand, using sophisticated methods for tracking the movements of physical objects, we can lend them behaviors that blur the boundary between the physical and the digital. We can entangle the world of atoms with the world of bits.[73] Sometimes, the boundary between the digital and the physical is drawn at the boundary of the body; sometimes, it lies on the other side of a physical object that represents, or stands in for, an object whose behavior is realized as a fully digital algorithm.

Although the locus of the physical/virtual boundary is somewhat movable, certain conventions have been adopted to describe its typical configurations. For the purpose of this discussion, I will adopt the following definitions:

1. Virtual reality: Users are immersed in a fully artificial digital environment. Objects and scenes in that world bear no relationship to anything in the immediate surroundings of the user.
2. Augmented reality (sometimes also called "extended reality" or "XR"): Users interact with virtual objects overlaid on the real environment, as in heads-up displays used in manufacturing, where a virtual ruler might be laid over the user's workpieces.[74]
3. Mixed reality (MR): Virtual objects are anchored to objects in the real world and can be interacted with. An example might be a MR-enhanced room with a virtual TV screen projected on a wall, complete with menus and buttons with which the user can interact. A more blended MR might allow you to paint a physical wall virtually using a brush widget.[75]

The increasing enmeshment of the world of bits and the world of atoms represents an inflection point for our understanding of the relationship between reality and physicality. This point is hard to articulate because we are living through this moment. The very fact that we must differentiate terms such as "virtual," "augmented," and "mixed reality" underscores how our understanding of this entanglement is still defined by its technical accomplishments rather than by our ability to comprehend what it means to exist in a hybrid world.

Baranyi et al. define reality as "a set of conceptions and perceptions that form an integrated unit of comprehension, and create an understanding of what is possible, desirable and actual."[76] In this definition, many components comprise reality, some originating in physicality and others from the realm of algorithms. At the higher level of reality, which they call "the integrated unit of comprehension," a new meaning emerges from the combined subcomponents, a meaning that could not exist at the level of the components themselves. A real-time map application on your phone can function only if your phone, your phone network, and the GPS all exist and are connected. Conversely, GPS data as a means of locating your phone in space is useful only if you have the correct and updated map application on your phone to situate it accurately. Your understanding of what is possible for you arises because your goal of getting to a particular location is now mapped onto real geo-positioning data by the cell network that tracks the position of your phone.

Baranyi et al. further argue that the axes of real–virtual and physical–digital that characterize their definition of digital reality do not overlap but instead are orthogonal: Just because something is said to be "real" does not necessarily mean it is physically present, nor does "virtual object" exclusively refer to things rendered digitally. If files exist virtually on the computers of two people who are unconnected except that they must both use the information, the files belong to the independent realities of their independent lives. If I have the same information on my computer and on paper in my filing cabinet, both exist in my reality but in independent forms. Baranyi et al. argue that we can only speak of multiple realities in cases like these, where a constellation of possibilities, desires, and manifestations of the world does not overlap with any other such constellation.

The reasons we grapple with these axes and why we talk about virtual, augmented, and mixed realities as different concepts are not because they are inherently different but rather because our current digital reality has multiple, fragmented technological approaches, each with its own medium of interaction. Lacking any common, pervasive modes of interaction, we are not able to manipulate and organize information independently of the digital medium we are using. When we can do this, then the axes of the real–virtual and of the physical–digital will increasingly overlap, and the verbal distinctions we now recognize will disappear.

For now, like Molyneux's patient at the outset of this chapter, we find ourselves in an underspecified world where we reach for objects we can see but not feel or where we use haptic displays to feel things we cannot see. We can separate, manipulate, and reconfigure object attributes in ways Molyneux could not dream of. These capabilities allow us to deconstruct notions of proximal, tangible objecthood and reframe what it means to encounter "things," in our environment, that we use every day. In the words of Stewart, Khatchatourov, and Lenay, these new technologies bring forth new worlds.[77] But what kinds of worlds? What kinds of bodily resources are required to interact with the objects in these worlds? What are these new kinds of objects?

We could be drawn into a further discussion of the phenomenology of objecthood, but although such a discussion would be instructive, it is beyond the scope of the current chapter. Here, I will build on Gallagher, who proposes that *how* the Molyneux patient reached for the globe or the cube would reveal what shape they anticipated grasping.[78] The way in which they prepared to interact with the object on the table would anticipate a coupling between their hand and the object. In an enactive sense, the visual perception of the object would guide the reaching movement, and once they made contact, the haptic properties of the object would generate sensory feedback that would help them refine the grasp. So far, so enactive. But let us examine the Molyneux question from a different perspective. Let us consider, for worlds containing virtual, augmented, and even shape-changing physical objects, where the ambiguity for today's participants in such interactive environments might lie.

6.6.1 Virtual Objects and Sensorimotor Contingencies

In his letter to Locke, Molyneux mentions that the cube and globe presented to his newly sighted observer exist within a context. They are sitting on a table, in a scene fully specified in all its spatial dimensions. The table sits within a room, and its position can be specified by any number of world-centered coordinate systems. For the fully sighted observer, the objects are immediately contextualized.

Now enter the blindfolded Molyneux participant, who has reached adulthood without having ever had sight. They are led to a table. If they have good echolocation skills, they may already have a sense of the size and shape of the room, as reflected sound is structured by the surfaces with which it comes into contact, the number of people present, and perhaps something about the material properties of the space. But compared with the experience of a fully sighted observer, this would represent a sparse specification of the scene. It might specify where things are but not necessarily what they are. On removing the blindfold, then, the Molyneux observer is bombarded by light structured by the surfaces and edges around them, and for them, this structure lacks meaning.

On being asked to describe this scene, the newly sighted participant is simply confused.[79] However, if asked to perform a familiar action, to reach for something on the table in front of them, they can begin to draw on more familiar sensory cues—the proprioceptive information arising from their own movements. It would be interesting to know whether participants in such studies have been observed closing their eyes, to return to the sensorimotor milieus in which they know they can succeed. Even if they continue to look at the progress of their hand within the scene, it is likely that vision will not be a salient cue in the control of action in this initial stage of seeing. I have lived my life with only limited perception of light, and I know that, in this type of situation, I would not trust my visual experience for some time. I would approach learning the relationship between these new perceptual cues and my own actions (these new sensorimotor contingencies) in the same way I would approach the acquisition of any new skill, assuming my early strategies would at best be inefficient but that, over time, my performance would improve. In short, for the Molyneux participant, what is missing are sensorimotor contingencies driven purely by visual cues.

Now let us imagine that the missing sensorimotor cues are not visual but instead haptic; that the scene containing the table, cube, and sphere is a hologram; and that the participant is a fully sighted user of a 3D VR system. Of course, there are some differences between the Molyneux participant and the user of our hologram. For one, the VR user has always been able to see. So, for her, the virtual image of 3D objects contains sufficient cues to structure light—cues relating to shading, form, depth—to elicit appropriate gestures for grasping and moving even intangible objects. Recent work by Pham et al., for example, elicited gestures from participants asked to imagine themselves completing tasks involving the manipulation of holographic objects embedded in three scenarios: midair, surface, and room.[80] Each of these incorporated a series of tasks, or "referents," relevant to a particular scale of interaction. In

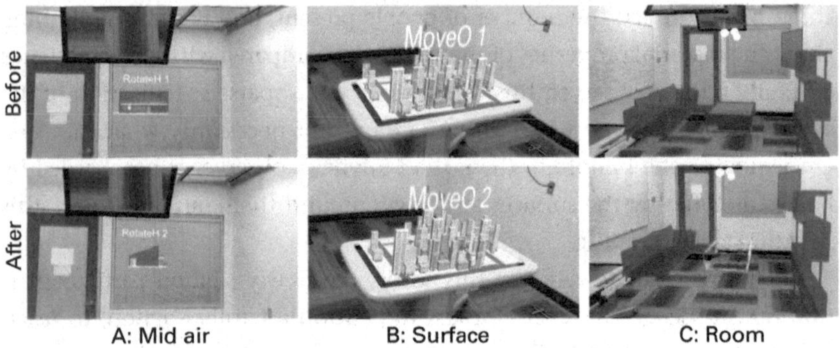

Figure 6.4
Three augmented reality scenarios used in Pham's study, depicting the state of the virtual environment before and after the user performed a series of tasks. Scenario A (before) represents a virtual house model in midair, with a "rotate scene" referent. Scenario B represents a city model on a tabletop, with a "move building" referent. Scenario C represents a physical room with virtual furniture, with a "select coffee table" referent. *Source:* Pham et al., 2018, figure 1. Used with permission from Anthony Tang. Copyright © 2018 by ACM.

each case, the apparent visual size of the virtual objects differed within each scene, and the volume of the virtual scenes, shown in figure 6.4, differed as follows:

- Midair: 14 cm × 31 cm × 25 cm model of a home
- Surface: 26.5 cm × 97 cm × 56 cm small city model with buildings and streets situated on a physical tabletop
- Room: 549 cm × 303 cm × 255 cm physical three-walled room (with an open side), containing a virtual couch, coffee table, and shelves

The same task referents were used to elicit gestures across all three interaction scales: select object, move object, rotate object, duplicate object, undo, delete object, and scale object.

The researchers found that the scale of the holographic object impacted how participants expected to interact with it and the gestures they performed in each case. They noted that the physical nature of many gestures was influenced by the opportunities for action that their size and form appeared to afford. For example, the rotate task in the midair scenario required participants to turn the house around, which three participants did by grasping the house like a handle and rotating their wrist as if they were removing a light bulb. The same task in the cityscape involved rotating a building, which two participants did by placing their hand on top of the model and turning it like a bottle. In the room scenario, the equivalent task involved rotating a bookcase, which two participants did by grasping it on either end and turning it using their whole body as if it were a heavy box.

The size and scale of each scene also impacted how people moved, including how they used perspective-based gestures for larger scenes. For example, when asked to select an object, some participants walked up to an object and tapped on it. Others relied on perspectival cues, changing their body position or viewing angle so that their hand appeared between their eye and the object. In short, rather than using the same gesture at different scales, participants generated new gestures for each scenario that mirrored the physical requirements of manipulating equivalently sized real-world objects within the simulated scenarios.

Even though the objects being manipulated here were holograms, participants appear to have used sensorimotor contingencies co-opted from paradigms for interacting with objects in real-world environments. They chose actions based not on what was possible in the virtual environment but rather on how much effort they expected would be required to perform the same actions in the real world, using two hands to turn the large hologram of the bookcase in the room-scale scenario and a single hand to rotate the small hologram of the building in the surface or tabletop model of the city. However, in light of our earlier discussion of users picking up on specifying versus non-specifying task-relevant information in rendered environments, it is possible that, as novice users, their initial approach is to attune to variables that would be perceptually available in accomplishing the task in the real world. They would do so even if the only information in the virtual world is non-specifying but coincident with the real-world situation—for example, information relating to visual size, which might indicate mass, weight, and so on. This, in turn, would indicate to the novice user the bodily resources required to accomplish the task. With time, though, users of VR may attune to other variables that more accurately specify appropriate manipulation strategies. They might learn that large objects have no mass in VR and can be pushed around with a single finger. In Chalmers's words, they might become sophisticated users of these new kinds of objects: "Given that the user knows they are using VR, they will not form the belief they are interacting with non-virtual objects in physical space. They will know full well that they are interacting with virtual objects in virtual space."[81]

Pham's study is enlightening with respect to our reframing of the Molyneux question because, in both cases, the participant is presented with a world where prior experience must be reframed to account for new SMCs.[82] The participants in Pham's study acted upon visible objects with no tangible form and no mass, producing movements that are appropriate for manipulating a given object but unconstrained by its material properties, much like the movements of a mime. The SMCs that are relevant for Pham's participants are not those of the holographic objects but rather those imposed by the constraints of the AR/VR system with which they interact. As Mel Slater points out, "The SCs (or SMCs) supported by a system define a set of valid actions that are meaningful in terms of perception within the virtual environment depicted."[83]

In other words, if the action of turning your head is to be valid in a given interactive VR system, the visual and auditory images rendered by the system should update, so that there is a lawful relationship between the velocity and angle of your moving head and the way in which the rendered auditory and visual scenes around you change. If no head tracking is enabled, no such updating can occur, and the structure in light and sound can bear no relationship to your movements and be of no use in guiding your action. Sensation cannot be contingent on movement, which is the very basis of SMCs. And the same is true for all other movements. If the movement of a tracked hand or arm or foot does not result in consistent changes to haptic or auditory or visual images, then the perceptual responses to such movements will not change in meaningful ways.

The Molyneux problem is similar to this VR scene because, in the VR scene, some of the cues that guide the hand in manipulating objects are also absent—in this case, the tactile cues—so the VR user, too, has to discover new sensorimotor contingencies, new ways in which perceived stimuli can guide their actions in a world devoid of the reaction forces and material properties that usually constrain movements during the manipulation of physical objects. For both the Molyneux patient and the user of a virtual environment, the task is not merely one of adapting existing SMCs to account for new sensory experiences. Rather, it is of discovering entirely new SMCs that are appropriate for performing familiar actions within a different sensory milieu. Like the user of the tactile visual substitution system, described in chapter 2, these SMCs can only arise when the user of a system can discover the structures in the sensory milieu that are invariant, that are reliable cues for guiding action, such as the angle of expansion of the retinal image cast by an approaching object or the pattern of skin deformation arising from the sliding contact of an object slipping from the hand. Simulating the sensory milieu that will support a virtual environment is as much about simulating the structured invariants arising from interaction with virtual objects in that environment as it is about rendering their static shape, color, and material properties.

In *Experience on Demand*, Jeremy Bailenson describes how a news anchor, visiting his lab, reacted to a simulated earthquake.[84] In the scenario, the participant was standing in a virtual factory, with piles of desk-sized crates precariously stacked just 10 ft in front of and behind them. There is a steel table nearby under which one can shelter. The news anchor put on a head-mounted display, and the floor of the lab began to shake. The boxes began to sway and slide, and it became obvious that they would fall on top of him. He was so present in the experience that he immediately dropped to his knees and crawled under the virtual table, putting his head on the floor and his hands over his head. But one of the boxes came tumbling toward him and collided with him under his table, whereupon he jumped to his feet, screaming, and ran for the factory's exit. Although his virtual path was clear, in the real world, he was heading for a wall. The lab director intervened and prevented him from being injured.

Why did this man act so instinctively, when he knew he was participating in a simulated scene? As Bailenson points out, VR is so compelling for some participants that they respond as though driven by physical reality. For such "high-presence" users, simulated perceptual information guides action in the same way its physically grounded counterpart would. Real boxes in a real factory would fall, regardless of anyone being there, but the simulated boxes only appeared to fall, an appearance grounded in the efficacy of the rendering of relevant perceptual information. The sensory environment, rendered on demand, contained sufficient relevant visual information to constitute a sensory milieu that enabled the participant to avoid the virtual boxes. This resulted in their protecting their physical body. But the perceptual invariants they needed to avoid running into the wall of the lab were absent from the VR milieu, so the man's actions could not be contingent on the information those invariants would have provided.

If a virtual tree falls in a virtual forest and there's no one there, not only will the tree not be heard, seen, or felt, its demise will not even be rendered. It will simply disappear from a database of objects in a coded world. In Baranyi's terms, encountering virtual objects in this virtual world cannot be considered a fully formed digital reality because the realm of the virtual and the realm of the real are not bound together in a common set of perceptions and conceptions to form an integrated unit of comprehension.[85] The simulated factory preserves this separation of worlds because it preserves the orthogonal axes of the real–virtual and the physical–digital, supporting two independent realities. Because it is fully rendered as a 3D digital world, it can, like a file on a computer, be independently accessed by many individuals who exist in fully independent realities.

At the same time, the man in the factory is inhabiting two forms of space that, like a file and its printout, both exist in his reality. If the information in the virtual factory was presented in a medium that could meaningfully coexist in the man's immediate physical surroundings, then he would not have been in danger of running into the physical wall of the lab. In developing the virtual rendering of the factory, no attempt was made to blend the real and the virtual, to merge the world of the digital with the world of the physical. The perceptually salient information available specified the approach of an object, picked up visually and perhaps aurally as well. The haptic information available specified the earthquake that caused the boxes to fall. The non-specifying information relating to the static lab environment was not available. So, without this, he reacted to what was immediately perceptually relevant and sought cover, albeit under a holographically rendered table. Had he been a more sophisticated user of VR, he would have accounted for the lack of blending between his rendered and real surroundings and would have allowed the boxes to pass right through his virtual avatar. He might have known how to negotiate the intertwining of specifying and non-specifying information that constitutes fully immersive 3D-rendered worlds.

The challenge of blending virtual and physical environments is further complicated when the designer of the virtual environment does not have access to the details of the physical world in which a participant will be situated when they are using their system. Sometimes this can be mitigated by using "see-through" renderings of virtual objects, so that the physical world is always present in their peripheral awareness. As long as there is sufficient specifying information available in such transparent renderings—about their form, size, and so on—then users of blended worlds can attend to the specifying information that is necessary for them to accomplish a given task goal while still having peripheral awareness of non-specifying information about their physical surroundings (such as the presence of immovable walls.

As I noted in discussing the concept of nested affordances in chapter 5 (see section 5.4.3.1), people are very good at discovering salient information to guide nested action goals when they need to do so. As long as we, as designers, pay attention to how the information we encode in simulated environments becomes incorporated within the global energy array for our users, then the opportunities for action that users discover can be more agnostic as to the source of specifying information. In this way, manipulating a virtual ruler to measure a real object on top of a real workbench is perceived as one coherent task where the ruler just happens to be a draggable image instead of a physical tool. True, the sensorimotor contingencies may not be exactly those required to place a physical ruler on the object, but they are guided by information specifying the same task goal: to bring a tool into the workspace that can reveal the object's size. I will return to the topic of designing for hybrid, blended worlds in section 6.6.7.

6.6.2 Augmented Objects, Overlaid and Blended Worlds

Virtual objects are rendered inside worlds that completely immerse their participants and decouple their perceptual and motor experiences from their physical surroundings. In AR, by contrast, virtual objects are digitally rendered as visual (or sometimes haptic) images inside of or laid over a user's physical surroundings. When rendered visually, augmentations may take the form of see-through overlays on real-world visual scenes, presented on the screen of a mobile device, as in the game world of Pokémon Go, or a head-mounted display, such as in an augmented assembly task,[86] or they may be images projected onto the surfaces of real-world objects such as projected clocks.[87] Haptic augmentations may take the form of virtual fixtures, haptic guides rendered using force-feedback devices that, like a physical jig, assist in accurately guiding the tip of a real or virtual tool.[88] In either case, the goal is to provide a user with colocated, context-relevant information that is not physically present in that environment. In addition to being copresent with objects in the real world, Azuma et al. point out that, to be considered augmentations, virtual overlays must be interactive and must be accurately registered or aligned with the real-world objects they augment.[89]

One of the challenges of augmented environments has been the degree to which rendered information must be processed in parallel with the information already present in the world. For example, in navigating a city using a see-through map presented on augmented glasses, the more proximal 2D map imagery is updated by a predetermined software algorithm and is superimposed on distal 3D visual information arising from the movement of the user through their physical environment. In the case of the displayed map, changes in the visual image are updated in response to the movement of the user, as tracked by some form of sensing (e.g., IMUs or GPS), and are structured according to a predetermined algorithm defined by the system's programmers. For information arising from the real world that is "seen through" the display, changes in the structure of the visual field arise only in response to how the user moves through their environment. Information presented on the map is symbolic. Lines and symbols on the map refer to roads, buildings, and other locations in the user's environment. To get where they are going, the user must read the map. At the same time, the information present in the structure of the real-world visual scene on which the map is projected is concerned with structures in light that the user relies on to walk in straight lines, avoid collisions, and so on. This information does not need to be read but is causally and lawfully related to how the user moves through the world.[90]

Raja and Calvo point out that although such maps fulfill AR's requirements that information should be colocated, context-relevant, and otherwise not present in the environment, the result is a visual milieu in which the user's attention is continually alternating between what is on the display and what is in the world.[91] The solution, Raja and Calvo suggest, is to design the overlaid information so that it, too, changes in ways that are congruent with the kinds of perceptual invariants that are relevant to a person moving through their environment. In the case of an overlaid map, they argue that the desired route could be augmented as an illuminated path, with all irrelevant turns being grayed out so that they are less perceptually salient features within the user's visual field as they move. This precludes the need for symbols to be read en route and yet accomplishes the goal of guiding the user to their destination. They refer to this approach as "ecological augmented reality" (E-AR). In E-AR, they suggest, the virtual layer of information superimposed on the world does not have meaning on its own but instead is given meaning by its juxtaposition with the structure of the physical environment.

The symbolic map, on the other hand, will still retain its meaning, even when the physical environment is not present.

The central tenet of Raja and Calvo's approach is that the world does not need to be augmented with information that is "missing" by superimposing layers of symbolic information. Rather, the information to be communicated can be embedded within the environment and structured to suggest opportunities for action that are relevant within the unfolding landscape of the user's goal-directed actions. In the example of the haptic virtual fixtures above, the haptic ruler is available when a user wants to draw a straight line. When they want to place the tip of a tool on a workpiece, a funnel-like force field is available to guide their hand. Opportunities for action are presented not as menu options or buttons on a display that overlays an otherwise underspecified task environment but rather as rich sources of environmental information that can be detected in the flow of the task.

The key to choosing what information to present within E-AR is determining the informational invariants that are relevant for a user undertaking a particular task. Recall that, according to Gibson's definition, perceptual invariants are sources of information in the ambient energy array that are constant, despite changes in sensory stimulation. As Raja and Calvo note, perceptual invariants are characterized by three main features:

1. They are higher-order variables such as ratios and proportions (i.e., they are not describable by basic attributes such as length, weight, roughness, etc.).
2. Detection of perceptual invariants requires action. The expanding angle on the retina from an approaching ball, the perceptual invariant that specifies the angle of approach, is perceivable only because the ball is moving toward the observer.
3. The way in which perceptual invariants structure sensory stimuli specifies, for an organism, what opportunities for action the environment affords them.

Raja and Calvo's argument is that traditional implementations of AR prioritize the alignment of the semantic content of the virtual overlay with the semantic framework of the real-world environment (e.g., substituting a paper map for a colocated overlaid map). Meaning arises from the superposition of these layers of semantic content. E-AR, on the other hand, emphasizes the role that action plays in the user's interaction with the ecologically augmented environment, where new opportunities for action arise because of the blending of real and virtual layers of information. Such affordances are not present in either environment alone.

Recalling the map example we just explored, the route highlighted in the E-AR layer is meaningful only when it is projected onto a particular real-world environment that the user is traversing. Similarly, the route to be traversed is not visible in the real environment; it is only indicated in the E-AR overlay. Raja and Calvo propose that by blending real and virtual layers, it is possible to develop E-AR environments where reality itself is augmented. They point out that no new hardware is required and that, in some ways, environments that integrate physical objects (such as tangible user interfaces) already accomplish this blending. This is because the physical objects on which they depend, like the highlighted routes on the map, shape the kinds of actions that are available in the interaction space. For example, an overlaid map route for a driver is very different from the same route for someone walking. The pacing of information provided by the overlay must be adjusted to the speed of travel in each case. Drivers need information far sooner than walkers. So, the algorithm that updates their map (e.g., to inform them of an upcoming turn) should be attuned to a different relationship between the driver's actions and the system's response than when someone is walking the same route. In other words, the pace at which information is delivered will shape the actions available to the map user. This is most noticeable when one picks the wrong set of directions and is surprised, for example, by the time a map thinks it will take to arrive at one's destination. Raja and Calvo further suggest that any system designed to enable sensory substitution, inasmuch as it requires users to discover the invariant structure of information within a different energy array (e.g., to feel the invariants in structured light), already augments one environment by overlaying ecologically grounded information from another source.

Raja and Calvo's E-AR emphasizes Gibson's approach, where many motor-equivalent actions could give rise to the same transformations in the global energy array. There are many ways in which one could organize one's body to intercept an incoming ball, but the invariant pattern of the expanding optical angle on the retina is an invariant form of information that specifies when the ball will reach you. Since AR aims to support participants who are active within real-world environments, designers cannot fully predict how a user's sensory experiences will relate to their actions. Unlike the action-specific invariants that are more desirable for VR, an ecological approach to augmenting the real environment, with its ability to account for many action paths for extracting one perceptual invariant, seems to lend itself much more readily to guiding action in such systems.

Although Raja and Calvo highlight the role of perceptual invariants in guiding a user's actions in E-AR, they do not consider how these actions reveal the structural invariants within the blended information layers in the first place. They do not discuss the user's role in discovering how the actions they perform relate to the invariants they observe. As Slater notes, it is not the content of the visual (or haptic or auditory) imagery rendered within an augmented or virtual environment that defines its affordances but rather the sensorimotor contingencies that are supported by the system.[92] If the whole visual world moves with you when you turn your head, you cannot use head movements to home in on the perceptual invariant that specifies the angle of approach of an incoming missile.

Although augmenting objects and environments holds much promise for blending real and virtual sources of context-relevant information, such augmentations should not simply overlay layers of semantic information but also present new information that reveals blended affordances—opportunities for action not perceivable from the real or augmented layers alone. In developing E-AR systems, designers should further acknowledge that this information is available to the users of their systems only to the degree that they can deploy their bodily resources within the augmented environment to reveal the structural invariants within what is essentially a blended energy array.

But does the emerging class of MR environments, with its emphasis on overlaying semantic layers of information on physical objects in the world, promote the kind of blended design that Raja and Calvo seek? Shavel notes that the degree of enmeshment or embedding in MR happens at multiple levels.[93] In an enhanced environment, the participant's physical surroundings still play a central role in MR. Recall the interactive TV projected onto the bedroom wall. This kind of MR environment is distinguished from AR by the degree to which you can interact with augmented objects, but at least these objects are now "attached" to objects in the real world (in this case, a wall). Interactions are therefore grounded by a physical context, which can constrain actions in ways that support the discovery of relevant sensorimotor couplings. You tap on the virtual screen or drag a menu item much as you might on an interactive surface, such as a tablet. The irony is that the relevant

SMCs here, learned from interacting with flat screens, are easily transferred to the wall, which is a simulacrum of an already familiar but haptically impoverished "real" interaction with representations of physical buttons.

Blended MR seems to promise the most in realizing Raja and Calvo's vision of E-AR, as it allows for the integration of physical and digital objects in the same view, and it may also integrate tactile feedback. If, for example, blended clay and holographic models can be manipulated in real time, a designer can engage in a "dance of agency" with this blended materiality, as described by Malafouris[94] and demonstrated by projects such as Illuminating Clay.[95] In the Illuminating Clay project, an image is projected onto a malleable surface, a tray of sand, which is also being captured by an overhead camera. If the projected image is a topographic map, then reshaping the surface with one's hands can cause the map to be updated. Using Illuminating Clay, geology students could interactively discover the impact that altering a watershed might have on shaping a landscape by physically remolding it with their hands. Not only are these participants in the blended MR of the map observers, but they are also actively shaping their blended environments. This degree of agency, when anchored to a tangible interaction with the physical world, provides a means to discover the kinds of higher-level perceptual invariants on which E-AR depends.[96]

Shavel's final category is immersive MR, which might reference attributes of the physical world but which integrates these into a fully immersive experience. For example, a system that can scan the physical environment might reshape the virtual world and constrain a participant to moving within a space free from obstacles. Physical characteristics of the real environment are "captured" and used to shape the virtual environment but without the user being aware of this integration. MR headsets that integrate "inside-out tracking" can monitor the space around the participant, and as these become more available, this kind of MR is likely to supplant VR for fully immersive games. However, as previously noted, since the user of immersive MR is unaware of their physical surroundings, this cannot be considered a blended environment. With no means of accessing the perceptual invariants available in their non-rendered environments, the most that users of such a system can hope for is that the invisible hand of their MR system designer is skilled enough to keep them from injuring themselves.

The foregoing discussion suggests that designing a successful MR environment depends on how sources of information that are simulated can be integrated within the global energy array such that they structure light, sound, and reaction forces in ways that are discoverable by users moving and acting within the blended spaces we design. In the Illuminating Clay example, pushing on and moving the sand changed the structure of both the haptic and visual information users could discover, which, in turn, continued to guide their actions as they reshaped the simulated map topography. Only when the malleable material and the programmable projection were coupled through the captured movements of the system's users could this interactive augmented environment come into being.

6.6.3 Tangible Objects and Their Affordances

Tangible objects and the associated field of tangible embedded and embodied interaction (TEI) can be considered a particular kind of augmented reality that prioritizes the form and function of physical objects and the ways in which they are coupled to their underlying digital substrates. Broadly speaking, interactive systems that use tangible objects, also called TUIs, have the following characteristics:

1. An input to the system is triggered by a participant using one or both hands to manipulate a physical object.
2. The movement of the object is sensed by a computer, and, in response, the computer changes its state.
3. Feedback is provided to the user through some change in the physical state of the system; for example, it might change color, provide haptic feedback, or make sound.[97]

In reviewing the field of tangible interaction, Hornecker and Buur note that three approaches have contributed to the design of TUIs, each of them shaping the design space in unique ways.[98] The first is the data-centered approach of HCI, which focuses on the physical manipulation of underlying representations of information—for example, physical handles to move elements around on a digital scheduling grid.[99] The second approach is rooted in product design and how people interact with tangible objects. This takes into account the user's own embodied skill and experience in a domain and focuses on supporting rich, expressive modes of interaction with augmented artifacts designed for a specific task environment. This approach considers the skillful and nuanced movements of users interacting with objects and environments as raw materials to be incorporated into a product's design process.[100] To this category belong interactive systems, such as digital piano keyboards, where players bring to bear years of skill acquired in playing a physical instrument.

Hornecker and Buur's final category is composed of tangible objects that mediate interactions with situated content in architectural or public spaces.[101] For example, Behrens et al. have developed the concept of media-augmented spaces, where people interact with digital content projected onto the facades of buildings by manipulating tangible objects at street level.[102] Such designs prioritize the spatial nature of bodily interactions with content, often involving multiple participants engaging in collaborative exploration or play. Hornecker and Buur note that all three of their approaches share some common characteristics: they are predicated on tangible interaction with material objects, they physically embody some form of data, they are designed to incorporate bodily movement as a fundamental part of their interaction, and they are embedded in a real space that provides the contextual framing for interaction.[103]

Designers need to understand how users make sense of what these systems are *and* what they are to be used for. A tabletop TUI designed to support manipulation of architectural models, such as Urp,[104] has a very different set of action goals from Reactable, for example, which was designed for interactive music performance.[105] Whereas Urp is a large tabletop system intended as a shared workspace where architects can collaborate to develop a hybrid model of a building and its surroundings, Reactable is a musical instrument with tangible objects that can be manipulated by one or more people and controls that determine the evolving sound of the instrument, in real time, in response to the creative and performative goals of its players. To the uninitiated observer, these both appear as a set of objects sitting atop a surface among projected visual imagery. To further complicate things, different kinds of TUIs build on different kinds of relationships between the actions users perform and how and where responses to those actions are situated.

Thus, for users to understand how TUIs work, Hornecker and Buur point out, the designed causality that couples tangible objects to their underlying digital representations and the causality that couples a user's actions to the system's response must both be apparent.[106] This will help users understand the relationship between what they do and how the system responds. Hornecker and Buur suggest that designers of TUIs should address the following:

- Representation: Are both the physical and digital representations of information equally important in supporting a user's interaction with the system?
- Externalization: Does the external representation support users in being able to think with objects to support problem solving? Do the objects of interaction support discussions and can they represent a record of decisions?
- Coupling: Is the coupling between the physical representation and the digital representation clear for the user? Is there a clear link between what a user does and how the system responds?

Ishii et al. also suggest that achieving a tight coupling between physical and digital representations can be more easily accomplished if a system's input (in the form of sensed movement of objects or hands) and output (in the form of transformations of associated visual representations) are both colocated in space and tightly coordinated in time.[107]

Designers of TUIs must predict how users will perceive the relationship between a system's tangible and digital representations, between the locus of their actions and the locus of the system's response. Do users perceive that the object they are handling is augmented, or is it a token that controls something else? To aid designers in negotiating this space, Fishkin developed a taxonomy that organized existing TUI examples along two axes, which he termed "embodiment" and "metaphor."[108] The embodiment axis measures the user's understanding of where the response to their actions is situated. In a highly or fully embodied system, changes resulting from physical manipulation of an object are represented by a change in the object itself, in the same way that pushing on a piece of modeling clay will result in a change in its shape. An example of a fully embodied TUI might be a cube containing a motion sensor that makes different sounds when moved. At the opposite end of this scale is "distant embodiment," where the locus of the system's response is more distant, like projections on building facades.

Fishkin's metaphor axis addresses *how* a user's actions relate to what the system does—to what degree manipulations of augmented TUI objects build upon or preserve analogies to manipulations of objects in a real-world context. He groups metaphors as "metaphor of noun," or those concerned with the shape of objects, and "metaphor of verb," or those concerned with their motion. The more a design exhibits these metaphors, the higher it sits on this axis. Categories include "none," for interfaces that employ no obvious metaphor or analogy (e.g., squeezing a ball to turn on a light); "noun," where one object might serve many purposes, such as a cube with an RFID tag, which can be brought close to a sensing surface to trigger any of a number of different functions; and "noun and verb," where an action performed on an object within the interface is analogous to an action performed on an object in the real world. Fishkin cites Urp, the urban planning interface mentioned above, in which blocks represent buildings and moving them causes their digital shadows to be cast on a virtual landscape.[109] Moving a block in the interface "is like" moving a building in a real cityscape. The user must understand the metaphor to make sense of the analogy. The highest value on this axis, called "full metaphor," requires no analogy because the physical system *is an analog of* the virtual system, and manipulating physical objects has the direct, desired impact on the real world. Fishkin points to Illuminating Clay,[110] mentioned earlier, where shaping clay with one's hands directly alters parameters within a projected landscape; changing the contour of a hill, for example, will change the path of a stream running down its side.

In addition to accounting for how and where actions have their effects, Fishkin's taxonomy captures two more aspects of the relationship between the objects of interaction in TUIs and their constructed contexts. First, the higher the level of embodiment of an interface, the less distance there is between the actions performed on objects and the response to those actions. The relationship between the two is more transparent, promoting a sense in which they are, in Heidegger's terms, more "ready to hand." Second, interfaces that lie higher on the metaphor axis are those that need no analogy to ground their metaphor. We might say the objects of interaction are situated in a context that indicates both what they are and what they are for. Like a carpenter's tools, such objects are, in Heidegger's terms, the "equipment" that provides the physical, social, and cultural context for the work an individual needs to perform. Thus, the architectural planners who work with Urp have a shared understanding of why it is important to "come to grips" with the impact of the design of their buildings on the inhabitants of the environment in which they will be located. The manipulations afforded by the TUI provide a means of reorganizing the augmented cityscape in real time as they discuss the trade-offs of their designs.

Fishkin's taxonomy helps designers understand how users might interpret the relationship between their actions and a system's response, but it does not address how users understand what the tangible objects in a system are for. This is particularly true when the metaphor does not stem from an established body of practice, as with the urban planners, but rather is developed as part of the system itself. Are the physical objects they encounter tools that can process or change information in some way, are they simply physical representations of information, or are they containers for moving information within the system?[111] Implicit in this question is a second question: How does the user of a TUI understand the relationship between the form and function of the object they are manipulating and the opportunities for action it affords them within the digital environment to which it is coupled? Unlike the hammer belonging to Heidegger's carpenter, in the world of TUIs, the relationship between a tool's physical "handle" and its digital "tip" can be reprogrammed, often in real time.

One possible solution is to draw on Kaptelinin and Nardi's concept of instrumental technology affordances, previously discussed, and to think of the handling affordances of the tangible being coupled to the end-effector affordances of the mapped behavior of its digital representation.[112] Rather than using a computer mouse to control a virtual paint brush, for example, one might consider augmenting a physical paint brush handle with sensors to track its movement and then coupling this to a virtual representation of its bristle configuration. The handling affordances of the brush handle might be more likely to elicit painting-like gestures from the hand of the user than a mouse. In this way, the metaphor of the TUI—easel, workbench, musical instrument, architectural model—becomes the mediating layer that provides the context for both the form of the tangible object as a handle and the behavior of its digital extension in facilitating the manipulation of the underlying digital information. But even Kaptelinin and Nardi fail to address the question of malleability: If the relationship between a tool handle and its tip is programmable, how does the user of such a system understand, moment to moment, what kind of tool they are attempting to incorporate into their body schema and how their coupling to the environment is changing?

In discussing Heidegger's example of the hammer, I noted that if the goal of a physically augmented system, such as a TUI, is to build on people's experience of engaging with physical objects, then the task of the designer is one of communicating to the user both what a given object is and what it is for. Once a user has picked up an object "in order to" accomplish a task, the object becomes "ready to hand." When users of TUIs must continually move between modes of interaction with objects and modes of inspection of the state of a system conveyed by those same objects, they are essentially transitioning between modes of interaction where objects are ready to hand and those where objects are present at hand. Such interactions are predicated on precisely those moments of breakdown that Heidegger (and later Dourish) point to as being problematic for users of tools.[113]

It is therefore no surprise that TUIs often fail to leverage their users' experience of interacting with physical objects in real environments to bootstrap rich, physically mediated interactions with digital environments.

Hornecker and Dünser cite the example of AR-Jam books (depicted in figure 6.5), an effort to determine whether interacting with augmented physical books would improve young children's motivation to read.[114] The authors note that although children quickly grasped how the system worked, they often attempted interactions not recognized by the system. For example, a story required them to crack eggs to release baby chicks, which they tried to

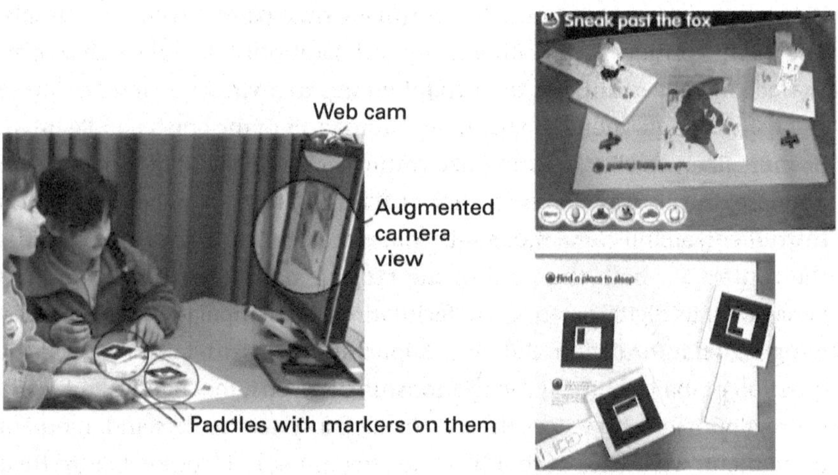

Figure 6.5
Two children interact with tagged physical objects that are a part of an AR-Jam book. On the computer screen, an augmented live video allows the children to control the story's main characters. *Source:* Hornecker, 2012, figure 1. Used with permission from Eva Hornecker. Copyright © 2012 by ACM.

do by hitting paddles off random objects in the room. The children's actions demonstrated that the system could support playful exploration, but they quickly became frustrated that the system did not perform as they expected. Hornecker noted a number of challenges for the children in using this system. She determined that children were frustrated because the physical objects with which they interacted had an infinite variety of affordances, opportunities for action that were not reflected in the more constrained behaviors of the digital objects to which they were coupled. The paddles could be moved freely in any direction in the real environment, and the objects they controlled were rendered as 3D images on-screen. But the motion of the paddles was mapped only to X and Y coordinates, so the on-screen characters could not mirror the movements of the paddles toward or away from the screen. More importantly, the paddles suggest causal actions (i.e., hitting something with a paddle implies force, which should cause an egg to crack, and tilting a paddle with an egg on it should cause the egg to fall off, but these things did not happen). Finally, certain scenes seemed to suggest context-specific actions.

In scenes where paddles could be moved over or around other objects, children expected to be able to interact with all the digitally represented objects in the scene, not just those coupled to the paddles. Hornecker concluded that designers need to explicitly design affordances of hybrid systems such as TUIs because, by default, the affordances of physical objects will always be employed in service of the user's action goals, which cannot be fully controlled or even predicted by designers. This is not necessarily a bad thing, as the rich sensory experience arising from such exploration allows users to creatively and expressively repurpose objects within and beyond the boundaries of the TUI to serve their new exploratory action goals. Trying to leverage a user's experience of interacting with objects in the real world may lead users to believe that such hybrid systems behave like the real world when they do not. Hornecker therefore recommends that designers of TUIs move away from the idea of immediacy and toward a practice that acknowledges the flawed or disjunctive nature of mapping physical manipulations of physical objects onto the physical manipulation of digital representations.[115]

In suggesting that designers turn their attention to the flawed nature of tangible interaction, Hornecker stops short of acknowledging that the source of this disjunction might be one of the central claims for the utility of tangible interaction itself—namely, that TUIs aim to make visible, through the spatial arrangement of their physical objects, the state of the underlying digital system to which they are coupled.[116] At the same time, the tangible objects aim to leverage users' actions in manipulating these same digital representations. The conflict between viewing the objects of inspection and simultaneously grappling with those objects in use is, I argue, the real source of the seamful nature of many tangible user interface designs.

As noted in discussing the work of Singapogu et al. on estimating the lengths of virtual rods, in section 6.5.1, in situations where the available specifying information is ambiguous, users may inadvertently attune to non-specifying information that is coincident but may not be correlated with the opportunities for action available in the rendered environment.[117] Such was the case when the children used their paddles to hit the objects in their immediate surroundings in an effort to crack open eggs to free chicks. Perhaps what is needed is not an acknowledgment of the flawed nature of TUIs but rather an acknowledgment that much of the literature on their usability reports on the experiences of novices encountering systems for the first time. As Singapogu et al. suggest, taking time to help participants attune to the sources of information that specify a system's affordances, even if they are nonideal, will result in more skillful users, and this, in turn, will lead to more expert interactions with augmented and virtual objects. By designing for truly embodied interaction, it is possible to work with, not against, participants as they seek to develop the sense-making strategies to skillfully bridge the gap between physical and digital worlds.

6.6.4 Haptic Proxys, a Tangible Footnote

As we have noted at several points throughout this book, simulating tangible objects in virtual reality is challenging because the mechanoreceptors that form the basis of our cutaneous and kinesthetic senses are distributed widely throughout our skin, muscles, and joints. These mechanoreceptors, in turn, respond to a wide range of object properties associated with global shape, surface texture, weight, center of mass, material density, and so on. In addition, receptors associated with our thermal sense inform us about the thermal properties of materials with which we come into contact. In short, there is no simple mapping between types of receptors and sensed properties of objects. For all these reasons, we are a long way from being able to simulate objects with combined haptic properties that are even close to the quality of simulations of their visual and auditory properties. We are therefore faced with a trade-off:

1. We can retain the flexibility of real-time simulation of a limited number of haptic object properties at the expense of the richness of interaction with material objects.
2. We can incorporate material objects, "haptic proxies," into virtual and augmented environments at the expense of being able to change any of their haptic attributes.
3. We can choose not to simulate the haptic properties of objects at all, instead using visual feedback to provide pseudo-haptic information about, for example, the weight of objects (see section 6.5.1) or contact sounds to indicate collisions and even material attributes of objects (hard or soft, hollow or solid, and so on).

Nilsson et al. define haptic proxies as physical props that serve as proxies for virtual objects.[118] In developing a taxonomy to help designers assess the potential of haptic proxies for given applications, they outline three criteria that should be fulfilled to ensure success:

1. Sufficient similarity: Any haptic proxies with which a user comes into contact should be similar in their haptic properties—for example, in size, weight, and shape—to their virtual counterparts.
2. Complete colocation: When touched and moved, a virtual object should be precisely colocated and aligned with a haptic proxy, mirroring any transformations resulting from those of the proxy.
3. Compelling contact forces: If the user touches a virtual object that is subject to contact forces originating within the virtual environment, such as those arising from collision, then relevant cues should be provided to the hand via the proxy.

The authors note that the first two criteria (sufficient similarity and complete colocation) are orthogonal—that is, it is possible to satisfy one without satisfying the other. For example, a haptic proxy could be completely colocated with a virtual object of a different shape, as when a wooden stick "stands in" for a sword (fulfilling the second criteria but not the first). Or a virtual copy of a haptic proxy could be present but not colocated with the real object, as when movement of a physical handle controls the movement of a virtual handle in the virtual environment that may not be adjacent to the user's hand (fulfilling the first criteria but not the second). But the third criterion, that of compelling contact forces, is not independent. If haptic proxies satisfy both sufficient similarity and complete colocation criteria, such as when the virtual environment is a replica of the real environment, then interactions between virtual objects will coincide with the contact forces arising from the interactions between their real-world haptic proxies.

As a case in point, let us consider the challenge faced by many museums and galleries to find ways for people to interact with rare and fragile objects. Increasingly, haptic proxies are being used to give tactile form to things that are otherwise out of reach.[119] When these objects are rare musical instruments, the goal would ideally be to provide ways for people not just to touch them but also to play them. In recent work, Granzow and Çamcı demonstrate how, by combining 3D scanning, sound synthesis using physical modeling techniques, and a haptic proxy, it is possible to make available a high-fidelity interactive virtual model of a rare instrument (in this case, a Japanese Tsuridaiko) that provides both sound and haptic feedback when it is played.[120] As illustrated in figure 6.6, a circular laser-cut structure that is a proxy for the drum's head is mounted onto the hand controller for an Oculus VR system by a 3D-printed bracket so that its position can be accurately tracked. A secondhand controller is augmented with an oak dowel to serve as a mallet. When struck, the strike point on the haptic proxy determines which resonant mode of the drum will be excited, whereas the strike force determines the amplitude of the resulting sound.

For the player of this virtual instrument, the proxy objects couple the sound and feel of the instrument to enable fine-grain control over timbre, amplitude, and timing. For experienced players, haptic cues may promote transfer of skill in a way that would be impossible if the virtual object could only be seen and heard. Moreover, such skill transfer requires that all three of Nilsson's criteria be met: that there be sufficient similarity between the modeled drum and the proxy's surface and mallet, that these be colocated with the virtual object, and that, as much as possible, the contact forces between the mallet and the wooden proxy should "feel" like interacting with the vibrating head of the drum. In addition, for this sounding object, there is an added requirement that the position and timing of mallet strikes be captured as precisely as possible to ensure that an appropriate source signal is generated to excite the sound synthesis model.

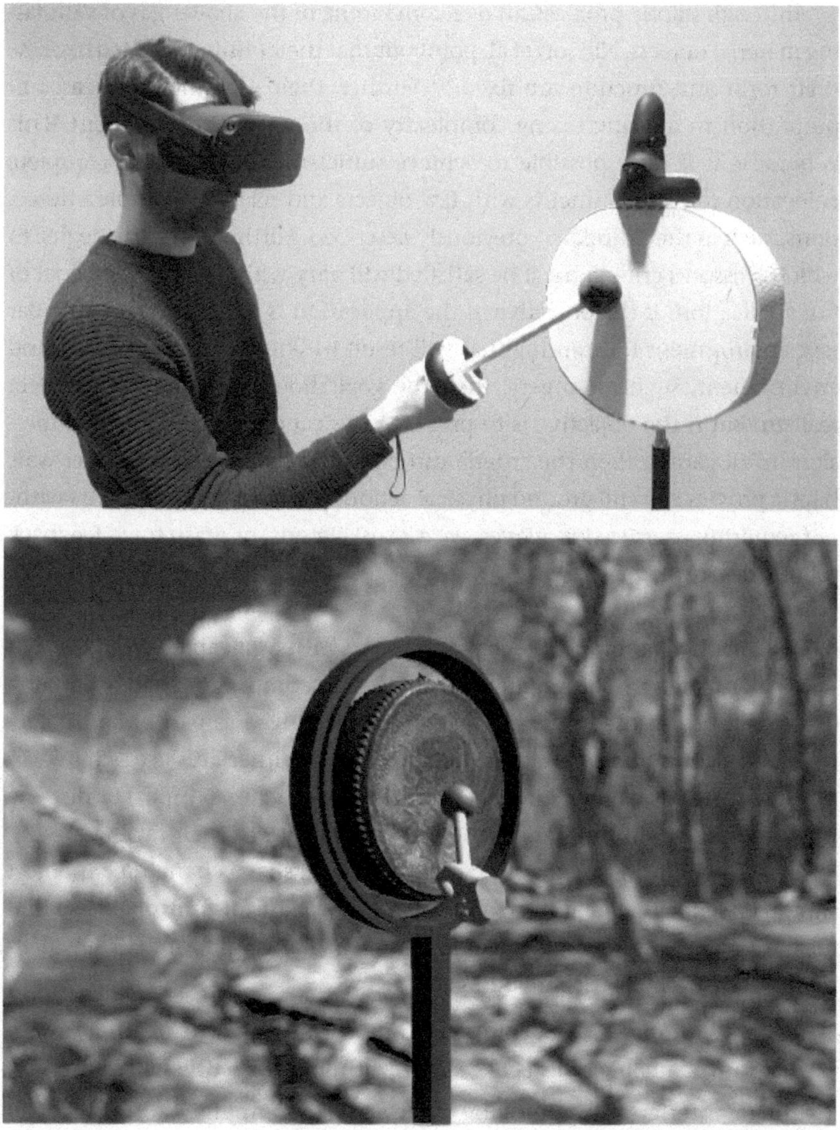

Figure 6.6

Haptic proxy for Tsuridaiko drum. The top image shows a physical drum proxy and Oculus controller augmented with an oak dowel representing a drum mallet. The bottom image shows a virtual representation of the Tsuridaiko drum and mallet. *Source:* Granzow and Çamcı, 2020, figure 5. Reprinted with permission from Anıl Çamcı.

Although haptic proxies can overcome some of the challenges of simulating material objects, Nilsson et al. point out that their utility is limited because their form and function are fixed.[121] Further, their usefulness decreases in proportion to any increasing complexity of the virtual environment. This is because it is only possible to achieve sufficient similarity and complete colocation for environments with few objects and relatively simple interactions, such as the Tsuridaiko previously described. Furthermore, the degree to which Nilsson's criteria must be satisfied will vary with the intended goal of the application. If the objective of the application is to simulate a particular task environment for training that will result in skill transfer to a real-world environment, such as surgery, then the goal should be to achieve perfect realism. But if the objective is to provide physical props to improve immersion in VR games, then the criteria can be considerably relaxed. Either way, haptic proxies can still ground physical actions in virtual environments without requiring users to don gloves, wear exoskeletons, or grasp force-feedback robotic devices.

6.6.5 Shape-Changing Objects: Interacting with Malleable Worlds

A further category of augmented object encompasses shape-changing objects—those whose form can change or be changed as part of an interaction with a computational system. Shape-changing interfaces use a change in physical form as a way of either capturing a user's actions (input) or displaying some change in the state of the system (output). In their review of SCIs, Rasmussen et al. suggest that to be categorized as shape changing, an object should be self-actuated and its self-actuation mechanism controllable so that it can return to its previous state and repeat its change in form.[122] They identify several kinds of SCIs, including:

1. Actuated interfaces (e.g., computer-controlled magnets beneath a surface that move tangible objects like sliders without interacting with a user)
2. Kinetic interfaces (e.g., moving objects under computer control that can apply forces to other objects or to people in their environment)
3. Organic interfaces (devices that record motion from a user's hand and then replay their gestures, capturing the organic nature of their actions)
4. Proactive architecture (incorporating building components that are responsive to the needs of those who live inside them: window blinds that close when it is too sunny, fans that direct air toward parts of a room where people are present, etc.)

The possible transformations that Rasmussen et al. identify as "strong transformations" include changes in form, orientation, volume, and surface texture (with changes in viscosity and permeability included as "weak" transformations). Changes in orientation involve translating or rotating an object while retaining its form,[123] whereas changes in form result in objects whose shape may be substantively altered but whose volume remains the same.[124] In other cases, the overall form of the object may be retained, but its surface texture may be altered or its internal viscosity or compliance may change under program control.[125]

In discussing SCIs, it is easy to be distracted by the reconfigurable nature of these objects and/or the technical means by which this is achieved. The more important questions concern the way in which SCIs mediate the actions of their users and how users make sense of why and how these objects are changing their form. What is the purpose and function of shape changing in a given context? What kinds of interaction is shape changing intended to mediate, and how do the dynamics of the unfolding change in shape impact the possibilities for action afforded by the system?

Saying what SCIs are and what they are for asks us to extend our prior discussion of tangible and augmented objects, as we must confront the fact that the physical tools we "pick up" can be transformed, altering their form and function, often in real time. If, as suggested above, TUIs must often be encountered in ready-to-hand and present-at-hand modes at the same time, how much truer will this be when objects can change form and possibly function while in use? Our carpenter, grappling with nails and walls, can at least rely on a hammer's physical form persisting, whether she uses the head to insert the nail or the claw to withdraw it. Shape-changing objects go beyond functionality that can be altered by reorientation to address directly, through transformations of their physical form, the mediation of human actions with computationally enhanced environments. A physical cylindrical handle, for example, could support the turning and tightening of a virtual screw, and then it could split into two parts with an interior spring to become the handle of pliers for gripping a virtual nut.

SCIs also go beyond TUIs in their capacity to realize Kaptelinin and Nardi's concept of instrumental technology affordances. Not only can the relationship between handling and end-effector affordances be programmatically determined, but this relationship can also be redefined, on the fly, while the tool is in use. In such a paradigm, the question of what such objects are and what they are for becomes mutable.

Much research in the field has endeavored to frame the space of mediated perception–action couplings that arise for such malleable interactions, for without appropriate contextual framing, it is argued, it will be impossible for users to make sense of interactions with objects whose physical form is continually changing. Dynamic affordances have emerged as a theoretical framework that accounts for the relationship between the actions of a user and the shape-changing response of a system. As with the interpretation of affordance in HCI in general, dynamic affordance has been understood from different theoretical starting points. Rasmussen et al., for example, consider the term from the perspective of the user seeking to discover possibilities for action within their computationally mediated environment. They define "dynamic affordance" as "perceived action possibilities that change with changes in shape."[126]

Follmer et al. consider dynamic affordances in terms of "perceived affordances," opportunities for action encapsulated in the design of the system.[127] For them, dynamic affordances are programmatically determined and can, for example, make available physical buttons and sliders, on demand, as the interface to an application changes. Or these buttons and sliders can even arise as dynamic handles available within datasets for users to manipulate data directly. Acknowledging that dynamic affordances place an additional burden on the user, who must track "what they are" as they change, as well as understand the relationship between what they represent and what they afford (i.e., what is present at hand alongside what functionality is ready to hand), Ishii et al. further suggest that SCIs should satisfy three requirements.[128] First, they should "transform" to reflect both changes in the state of the underlying system and changes arising from user input. Second, SCIs should "conform" to constraints imposed by the environment and the available action capabilities of their users. Finally, SCIs should "inform" users of their transformational capabilities (i.e., their dynamic affordances). In this regard, they implicitly acknowledge the challenge of trying to manage the continual need, on the part of the user, to move between encountering the object as a tool in use (*zuhanden*) and encountering the same object as an observer (*vorhanden*).

Sean Follmer, who has systematically studied users' understandings of dynamic affordances for SCIs, concludes that it is the fact that the object or interface changes, rather than how it changes, that is most salient for users.[129] If we are to understand what the dynamic affordances of these interfaces are—that is, what features of a shape-changing object correspond to the needs and capabilities of a user who wants to interact with them to accomplish a task, it is necessary to understand both what is changing and how it is changing. In terms of what can change, Follmer points out that in the coupled environment formed by the body of the user and the interactive system, there are many points where the dynamics of change arise—within the body of the user, within the handling or end-effector affordances of the tool, and within the digitally rendered information being manipulated. (Recall, for example, the points of disjunction in figure 6.3.)

With regard to how shape change can occur, Follmer suggests that the mechanism generating shape change can be a system controlling an actuated object, the user appropriating an object to incorporate into the interface, a user constructing a shape from parts (such as blocks), or the user's actions causing a malleable object to be deformed in some way. Whatever the mechanism, users can ultimately manage the dynamics of a shape-changing system only by actively discovering the opportunities for action that it makes possible and the characteristics of its behavior, such as its temporal unfolding and the stability of its changing and restoring capabilities.

In addition to objects of interaction changing their form, the context in which they are embedded may also change its topology. In the example of dynamic physical buttons/menus in Follmer's data-rendering scenario above, the form and function of the whole interface can be dynamically reconfigured. For Heidegger's carpenter, this would be like continually reconfiguring her workbench and continually recontextualizing her material and the cultural context of her work. It is unclear, at this point, how the high degree of fluidity that shape-changing objects and environments make possible will impact, for example, the carpenter's ability to acquire the domain-specific embodied knowledge to skillfully execute the tasks associated with her trade.

If, as our discussion of Heidegger and Merleau-Ponty suggests, the tools we use shape how we encounter the world, to what extent can we take on the shape-changing world that shape-changing tools bring forth? Can a hammer that continually changes its form really withdraw from the carpenter's awareness, or will it continually force her to move between the hammer as an object in use and the hammer as an object of inspection? If the hammer never withdraws from her awareness, will she be able to incorporate it into her body schema, to extend her perceptual awareness to the point where the tool can take on its role as a structural coupling to her environment?

Although our ability to incorporate technologies in use into our body schema is polymorphic, it is not clear whether such polymorphism can succeed if we are not the agents of this change. Neither is it clear whether the kind of malleability in our body schema that allows us to pick up and put down tools as required by the tasks in which we are engaged will persist if we are not the ones actively choosing the tools. If we cannot wield the hammer as we pick it up or discover the behaviors of buttons and sliders by actively pushing on them, can we come to know how these tools will shape our actions? If the tools we are using keep changing, can we ever acquire the kind of embodied knowledge that allows us to become skilled in their use?

As previously noted, the more skilled we become, the more robust our handling of tools in the face of perturbations in our task environment. But if our tools and our task environment are malleable, what kind of embodied knowledge can we bring to the task? To recall the work of Lambros Malafouris, what does it mean to be co-constituents of a task environment whose agency for change is not clear? It is as if, in the dance of agency between the potter and the clay, or the carpenter and the wall, we must account not only for a digital partner that augments the behavior of our physical tools but also for a physical agent that can reconfigure the equipment that constitutes the task environment. As designers of interactive SCIs, we must weigh their potential for providing highly configurable physical handles to digitally rendered task environments against the need to ensure enough stability in the body–environment coupling in our systems so that users can predict how they will respond to their actions.

Recalling Merleau-Ponty's blind man with his cane, one can only imagine how difficult it would be for him to perceive the topography of his world if the length of the cane continually changed. His ability to skillfully navigate through the world relies on at least this part of the tool–environment coupling being predictable. In designing systems that use SCIs, it is likely that there will be some attributes of objects that are mutable under specific circumstances, such as the handle of the screwdriver changing into the handle of pliers if the task goal within the environment calls for a tool with different affordances (turning vs. grasping). But the decision to invoke such a change in shape is also likely to be most successful when it is under the control of the user or very clearly associated with a change in some boundary condition within the environment, such as a transition to a new task.

6.6.6 Enactive Encounters with Physical Proxys: Designing PebbleBox

One of the most enduring challenges for virtual reality is in simulating the material properties of objects that can support our felt experience of interacting with the things around us. In discussing "haptic proxies" in section 6.6.4, I noted that the designer relaying objects' haptic properties in augmented and virtual reality must choose between haptic simulations that are programmable but limited (e.g., relaying reaction forces to specific parts of the body) and physical proxy objects that can stand in for virtual objects but can't usually be programmed to change their shape or material properties. Considered through the lens of the enactive approach to perception, haptic proxies are interesting because they suggest that there may be more flexibility between the feel of a physical object and the look of its calibrated virtual twin than we might expect. After all, when we pick up a ceramic cup, we expect it to remain ceramic and not suddenly turn into a paper cup of a different size. We are, however, familiar with how the weight and center of mass of that cup change according to how much liquid it contains. More importantly, we will modify how we handle a cup of liquid on the basis of how full it is—in enactive terms, our action of picking up the cup is modified or guided by what it feels like to lift it.

One important question then arises: If we can modify our actions on the basis of the felt behavior of real cups of liquid as they change weight, might there be a way to simulate changes in other material properties of objects—changes that are discoverable only through active exploration? In response to this question, I developed the artifact PebbleBox in collaboration with Georg Essl, as part of the European Union's Enactive network. The goal was to apply the principles underlying the enactive approach to perception and action to the design of human–computer interfaces. This meant finding ways in which a user's actions, mediated by technology, could "bring forth" the world of an application or interactive artwork, much as a ski mediates the interaction between a skier and a snowy mountain and brings forth a skiable world (see chapter 2). In the enactive approach, knowledge about the world does not reside solely in the mind but rather is acquired through interaction with the world. The question posed by PebbleBox was whether such knowledge, acquired in a real-world context, could be repurposed for the control of a digitally mediated experience.

The starting point was the observation that physical interaction with objects is a richly multisensory experience. If we cast a pebble into a pond, we see the ripples as the disturbance of water and we hear the impact sound as a disturbance of air.[130] More importantly, the interaction between the stone and the water carries a lot of information. The volume of the splash that you hear is correlated with the size of the stone and the force with which it is thrown, whereas the pitch of the splash is an indicator of how deep the pond is. The physical laws that govern a stone being thrown into a pond generate information we take in via multiple sense modalities, each attuned to invariant structures concerning stones and ponds in the ambient energy array. In developing what we termed an "enactive approach" to the design of digital musical instruments, we sought to build upon this observation by designing interactive artifacts that embodied action–sound couplings motivated by physical systems in which they naturally co-occur. We then extended these to discover new possibilities where the physical behavior of manipulated objects would shape their sonic behavior in new and flexible ways. The most successful of these artifacts was PebbleBox.

PebbleBox (shown in figure 6.7) is an augmented wooden box that challenges us to explore the relationship between the way things feel and the way they sound. The box contains a shallow, velvet-lined tray of polished pebbles. When moved, the pebbles collide, and their collisions are captured and used to create collision sounds that are drawn from a library of sound "grains." The grains can be the sound of a water droplet, the clink of an ice cube in a glass, or indeed any collision sound. They can even be sounds such as the "collision" of a footstep on gravel. Technically, this is achieved by embedding a microphone in the box's lining and extracting event timing, event amplitude, and event pitch from the audio stream. These parameters then control a granular synthesis algorithm to modify the pitch, timing, and amplitude of the chosen sound sample or grain.

The pebbles invite anyone approaching the box to reach in and move their hand through them or pick them up. When they do, they feel pebbles but hear the sound of randomly assigned grains (e.g., water, ice cubes, etc.). Each time the PebbleBox lid is raised, the movement of the pebbles is associated with a different sound randomly selected from the bank of collision-based sounds. Because the collision of pebbles directly maps to the collision of sound grains, the nuanced movements of the hands are apparent in the resulting sound. Swishing the hands through the pebbles will result in the

Figure 6.7
PebbleBox. When pebbles within the PebbleBox collide, their collision is picked up by an embedded microphone, which then controls a granular synthesis algorithm. The design of PebbleBox was for the exhibition "Touch Me: Design and Sensation," Victoria and Albert Museum. *Source:* Essl and O'Modhrain, 2006, figure 9. Adapted from a photograph by Peter Kelleher, V&A Photo Studio.

sound of the hand swishing, perhaps, through water. Tapping pebbles against each other will sound like individual ice cubes clinking or like water droplets or footsteps.

In designing PebbleBox, we asked ourselves how far we could stretch the relationship between what was being heard and what was being felt. It is one thing to map the collision of pebbles to that of other objects; it is a different matter to map their collision to feet walking or hands moving through water. In watching people interact with PebbleBox, we noticed that their hand movements often changed in response to the sound they heard. If they heard water, they would swish their hands through the pebbles, whereas the sound of coins often prompted people to lift up handfuls of pebbles and let them fall through their fingers. Often, these effects were apparent only for a short period, but we questioned whether the relationship between the feel and the sound of objects might be more open or malleable than we expected. The feel of the pebbles never changed. So, we hypothesized that the way people interacted with these tangible objects could be altered by the behavior of the digital algorithms to which they were coupled. Changing the *sound* in PebbleBox suggested new opportunities for action, even though the feel and form of the objects did not change.

In an unpublished study, Johann Issartle and I tested this hypothesis. We made video recordings of the hand movements of participants manipulating four physical materials—water, large pebbles, fine gravel, and coins (control conditions)—and compared these with their hand movements when they manipulated physical pebbles in the PebbleBox while hearing the sounds of water, pebbles, gravel, or coins (AR conditions). Hand movements were classified according to the exploration strategies participants used in each condition, using a modified and expanded protocol for categorizing EPs.[131] In this case, participants were manipulating multiple objects (and a liquid), so new strategies such as shaking and scooping emerged in both real and augmented conditions. Because our informal observations had suggested that people might initially use hand movements that aligned with what they heard rather than what they felt, we also looked at how exploration strategies changed over time.

Ultimately, our findings demonstrated that although specific strategies emerged for the manipulation of objects according to their material properties (gravel, coins, water), no specific hand movement profiles emerged for AR conditions where the haptic feedback was constant and only auditory feedback was varied. In such cases, hand movements were congruent with those observed for the manipulation of pebbles in the control condition. In short, in conditions where the feel and sound of objects were no longer congruent, the tangible properties of the felt objects outweighed the intangible properties of the sounds they produced. It is as if the pebbles came to shape the coupling between hands and collision sounds not as tools or as handles but rather as material proxies for the objects themselves, bringing forth a world you could reach into and directly impact with your hands.

We lack the technology to render, adequately and simultaneously, multiple material properties of simulated objects in haptic interactions. PebbleBox suggests that there is a space between mediated interaction through tools and direct interaction with hands. Material proxies could couple our actions to a dynamic and malleable world of simulated sensory experience. Actions would be guided by information about the environment picked up through multiple sense modalities, each attuned to some aspect of how user actions relate system responses to structurally or transformationally invariant patterns in the global energy array. These patterns are not sense specific or even technology specific. They are specified by the goal of the interaction between the user and the environment. In this instance, the goal of PebbleBox was to playfully engage users with the behaviors and material properties of colliding objects. We also explored interactions such as crumbling and friction, where material proxies might function in similar ways.[132] By letting go of the idea that sensorimotor couplings to a simulated environment had to be defined by a particular interface technology, we were free to design interactions based on the sensorimotor experiences themselves. In doing so, we discovered a design space where material proxies encapsulating dynamic behaviors (friction, collision, breaking, crumbling) could "stand in" for whole classes of objects, even if other material properties (shape, texture, size, etc.) were not congruent with the simulated material being manipulated. In sum, material proxies offer a way to engage with the material properties of simulated worlds in ways that transcend notions of physical handles for manipulating digital information. Material proxies allow us to physically engage with the digital fabric that is a material part of our increasingly hybrid world.

For me, PebbleBox was more than a project to design an interactive artifact. Instead, it became the starting point for an enactive approach to interaction design centered on the premise that our actions bring forth the world we perceive, even when part of that world is simulated. Moreover, the world that is revealed comes into being only because we actively engage with it. In this view, the role of the designer is less concerned with the construction of a static interaction space than with the dynamic unfolding of an interaction experience that is discoverable by users through their ongoing engagement with an artifact or environment. A large part of the appeal of PebbleBox was in what was not immediately apparent to those who approached it; it revealed its secrets only once they committed to physically engaging with the artifact, once they literally got their hands into it. Designing their enactive encounter required that users' actions be continually rewarded by responses from the object—in this case, sounds that stretched the bounds of plausibility such that people were compelled to continue moving to discover what else their actions might reveal.

As a hybrid object, PebbleBox made no claims to having any purpose other than being an intriguing artifact. But the enactive principles underpinning its design have, in the hands of others, become the foundation for a new methodology for the design of hybrid environments that embody the needs and experiences of the individuals who will inhabit them.

6.6.7 Hybrid Objects: Designing for Embodied Being-in-the-World

So far, our discussion of virtual, augmented, and tangible objects has centered on the ways in which they couple the actions of a person in the physical world to the manipulation of information encoded in an omnipresent digital space. Depending on the technology mediating our interaction with these "hybrid" objects, the boundary between digital and physical representations can lie anywhere between the boundary of the body and a digital artifact. At one end of this spectrum, in the case of fully simulated objects such as visuo-haptic holograms of artifacts in a virtual environment, the boundary is between the hand and the virtual world. At the other end of the spectrum, in the case of a tangible object such as a block on a table coupled, via sensors, to a digital representation of the handle of a slider, the boundary lies between the manipulated block and the digital representation.

As van Dijk points out, mixing physical and digital forms in this way has brought into our lived environments new kinds of hybrid materials that we are not readily equipped to understand.[133] This became apparent, for example, in our discussion of AR-Jam books, where children struggled to understand the differences in the affordances and constraints of the blended physical and digital representations within the world of the interactive book. Van Dijk's observation further supports the point raised by Baranyi et al. that our current lack of common and pervasive modes of interaction, combined with fragmented modes of representation, means we are not yet able to manipulate and organize information in ways that are independent of the digital medium we are using.[134] And, like Baranyi and colleagues, van Dijk suggests that getting beyond this fragmented augmented space will involve rethinking the ways in which we encounter digitally augmented artifacts within the lifeworlds we inhabit.

In addressing the challenge of designing for this new form of hybrid materiality, van Dijk and Hummels argue for a radical shift in thinking, one that no longer considers interactive artifacts as having special privilege but rather repositions the hybrid as a part of the fabric of the lived environment of its user.[135] In this approach, which they term "design for embodied being-in-the-world," hybrids are not simply augmented objects or systems that coexist alongside us but are "conceived as participating in socially situated, sensorimotor couplings that govern the way the lived body operates in the lifeworld."[136] In chapter 2, I drew attention to the ways in which meaning, for living organisms, is brought forth through their active involvement and coordination with their lived environment. I also noted that in environments that happen to contain interactive artifacts, the kinds of environments "brought forth" for the individual are mediated by these technologies. In laying out the framework for D4EB, van Dijk poses the same challenge in a far more elegant way: "The question for designers of interactive technologies is how particular design choices impact the dynamics of this embodied and situated self-organizing sense-making network."[137]

For van Dijk, this means that designers of hybrid systems must let go of the idea of boundaries between physical and digital modes of representation and between the inner mind and the outer world. In drawing upon the kinds of embodied knowledge people bring to interactions with physical artifacts in their environments, systems employing tangible user interfaces begin to dissolve those boundaries. This enables users to employ their skills and experience in thinking with and through physical objects to solve problems. They distribute the task of processing information across the network of resources consisting of their mind, their body, and the physically augmented digital artifacts of the TUI. In the case of Illuminating Clay, discussed earlier, our hypothetical geology students used their hands, guided by their thought processes, to reshape the physically present sand in order to manipulate the digital representation (the topographic map). Only by combining all three resources could they discover, through their actions, how reshaping a topography might alter a watershed.

But designers of TUIs insist on preserving a boundary between physical and digital representations of information. The distinction between their content and their physical form is retained.[138] In contrast to the embodied design approach characteristic of systems such as TUIs, D4EB takes as its starting point Merleau-Ponty's assertion that our individual lived experience is constituted by the way in which our bodies inhabit our lifeworld.[139] Everything, they argue, begins with our phenomenal experience of "being-in-the-world." Key to the concept of the lifeworld is the understanding that the same physical environment will constitute a different lifeworld for each of us. This same environment will show up for each individual, depending upon the activities they need to accomplish within it: "Kitchen cupboards, windowsills, and the spaces underneath chairs have different significances for cats and people, as do balls of yarn, upholstery, television sets, and other cats."[140] Technologies that fail to acknowledge the unique lifeworlds of their users, van Dijk argues, will not establish the kind of contextual framing that situates an individual within the lifeworld of their skills and their sociocultural experience.

In "designing for embodied being-in-the-world," van Dijk radically reframes the process of creating hybrid artifacts.[141] This approach assumes that every user has a unique lifeworld. And designed artifacts must take their place within the user's lifeworld alongside all other artifacts in their environment. Augmented artifacts are not privileged. They are there to augment the structure of the individual's environment so they can accomplish their goals. To illustrate how D4EB is used in the design of situated hybrids, van Dijk and Hummels report on three case studies.[142] I will highlight one of these, MyDaylight, to draw out some of the unique contributions of their approach.

MyDaylight, depicted in figure 6.8, is a collection of augmented, connected lamps designed to support those living with autism spectrum disorder in managing activities within their home. By allocating different tasks and events to different lamp colors, arranging the placement of lamps, and scheduling when they turn on, the user of the system can augment their environment with reminders relevant to the time and space for an intended activity. Lamps can also coordinate situated social activities. In planning the visit of a friend, a user might place a lamp where the visitor will sit and might choose its light color to enrich the experience of the anticipated interaction: "For Sonja I choose a nice violet color for the couch-lamp, so I'll put her in a beautiful light when she sits there tomorrow."[143]

In their D4EB framework, van Dijk and Hummels identify seven principles that can serve as guides for designers. They emphasize that the design process should integrate these principles to form a fully situated system, which, as with MyDaylight, takes on meaning when embedded in the lifeworld of the

Figure 6.8
MyDaylight. The lamp is controlled through an iPad app, where users can set its colors to represent different tasks and events in their calendar. The device is intended to support those with autism spectrum disorder through color changes that act as visual reminders. *Source:* van Dijk and Hummels, 2017, figure 2. Reprinted with permission from Jelle van Dijk. Copyright © 2017 by ACM.

user. The first three principles—sustaining skills, scaffolding social coordination, and embedding in the lifeworld—concern ways in which a D4EB design can support activities in the here and now. MyDaylight, for example, does not ask its user to consult a calendar of events constructed in the past and projecting forward into an abstract future. Rather, through objects situated in the users' immediate environment, MyDaylight draws attention to what needs to be accomplished at a given place and time. A user can develop situated ways of negotiating the planning and execution of actions and routines as a reflection of desired activities rather than as a representation of planned tasks. In meetings with the participant and his caregivers, researchers noticed that discussions of how to program MyDaylight brought together the user's experience and the professional perspective of the caregivers. This resulted in a move away from a task-based representation of events to be managed to a representation that reflected the current system state in a way that allowed them to collaborate. Since the things embedded in our lifeworld are already part of our skills and routines, they are already endowed with meaning and frequently show up for us when they are relevant to our current task. Similarly, the lamps that are part of MyDaylight convey information to a user only when they are embedded in his lived environment. Taken out of this space, their behavior loses meaning for him.

The next three design principles of van Dijk and Hummels's framework—forming and expressing identity, social relationships, and transforming the lifeworld—are intended to support the development of skills and knowledge over a longer period. Forming and expressing identity concerns the way in which a system can be designed to serve as what Heidegger calls equipment to support a user in developing their skills, situating themself in a sociocultural space, and "growing" their lifeworld.[144] MyDaylight provided its user with a way to take charge of his environment so that managing his daily routine became something he was doing rather than yet another task imposed by his caregivers. As we noted, MyDaylight also gave its user a sense of control over planning a social event. In acknowledging that an individual's lifeworld transforms in response to their changing needs and goals, the principle of social relationships acknowledges that the function of a given hybrid within the user's lifeworld will also transform over time. The user of MyDaylight might develop new ways of deploying the lights to support new tasks and routines as he becomes more skilled in managing his time and as the objects take on more meaning for him as constituents of his lifeworld.

Van Dijk and Hummels's final principle, reflection in action, offers designers a way to step back from their work and avoid becoming entrenched in one mode of thinking. In the case of MyDaylight, reflection in action enabled the design team to see how the planner-based view of the system for programming the lamp did not afford collaborative meaning making. Going back to the drawing board, they developed an interface that reflected the system's current state rather than focusing on its future timeline, and they thereby facilitated an entirely new mode of situated, collaborative interaction.

In short, D4EB serves as a way for designers to move beyond notions of digital artifacts as physical representations of digital information to consider them as hybrid materials, new ways of augmenting structures within the lifeworlds of their users, co-constructed by users' own actions and in service of their own task goals. Is it possible that D4EB, in seeking to dissolve the boundaries between the inner mind and the outer world, between physical and digital modes of representation, could ultimately allow us to manipulate and organize information independently of the digital medium of encounter? Could it bring us to the point where Baranya's orthogonal axes of the real–virtual and the physical–digital collapse because they are no longer meaningful?

The D4EB framework repositions hybrids as augmented artifacts that participate as situated sensorimotor couplings, shaping how we operate in a digitally augmented lifeworld. They demonstrate how it is possible to refocus the design of augmented environments away from the fragmented technologies of today and toward a more embodied understanding of the fabric of a truly hybrid world.

6.7 Conclusion

Interactions with augmented and virtual objects are, like those with real objects, framed by the contexts in which they occur. Perceptual experiences of such encounters are shaped by the goals such contexts define. But in AR/VR, the objects we encounter and their contexts of use are malleable. How do we, then, understand what objects actually are and what they are for? The first thing to understand is how users of AR and VR determine what kind of information is perceptually salient for them, given the modes of interaction available and the context that is given by a partly or wholly rendered task environment. As in the real world, it turns out that this process of attunement in an AR/VR world occurs through action: We seek to learn how the environment changes in response to our movements and to learn which items in the environment contain higher-order information-specifying invariant properties. With time and perhaps training, we can become skilled users of environments and the objects they contain.

Given that we can now create objects whose properties are mutable, what are the invariant structural and transformational properties within our designed environments? What provides the information users seek to accomplish task goals? Only certain properties of our lived environment "show up" for us as we go about daily tasks. Our sense organs respond to certain environmental information, but only some of that information is relevant. This is also true in augmented and virtual environments. But this is crucial: Movement does not constitute an invariant. Movement is instrumental for users in discovering task-relevant information, specified by the properties of the environment, that is structurally or transformationally invariant in the presence of the otherwise fluctuating energy in the global array.

In their work with rugby goalies, Dessing and Craig showed that they tracked not the color of the ball or its absolute size but rather the rate of change of the angle of expansion of the ball's image on the retina.[145] When this was reproduced in a virtual environment, they could transfer their skillful recognition of this invariant into the virtually rendered world. In rendering real-world task environments such as rugby, attunement as an approach demonstrates the possibility that more opportunities for action will show up for users. Dessing and Craig were able to present their goalies with balls arriving from distances and angles they had not encountered in their virtual environment training, thus demonstrating that, having attuned to the relevant task variables as rendered, they could generalize this learning to adapt to a broader range of encounters with incoming shots on goal.

It seems, then, that the designer of an augmented or virtual world has to manage several things:

1. The constant properties—visual, auditory, haptic—of objects
2. The ways in which the presence of these constant properties structure the partly or wholly simulated global energy array for the user of their system (the structural and transformational invariants to which they can become attuned)
3. The way the system behaves in response to the actions of the user (the lawful relationship between their actions and the consequences of those actions in the augmented or virtual world)

As long as there is a predictable, stable relationship between users' actions and the behavior of the augmented or virtual environment, users can become attuned, skillful practitioners, mastering the tools and task environments that designers create. Furthermore, as noted in the case of judging the lengths of virtual rods, it is possible to support users in attuning to invariant properties of such interactions, even when the available technologies cannot render the most perceptually salient higher-order invariants that would typically support participants' interactions with physical tools.

At this moment of disjunction, where the world of atoms is increasingly entangled with the world of bits, we must confront the fact that such entanglement has consequences for every aspect of our lives, for our very way of being in the world. Sometimes, the consequences of our interactions with objects are situated in the material environment; at others, they are realized in a digital world. In the case of hybrid objects, our actions can have consequences in both. To truly inhabit such hybrid or blended worlds, we will need to move beyond notions of digital artifacts as physical representations of digital information. We will need to learn to understand them as augmented structures that support us in skillfully negotiating our individual lifeworlds. Like Molyneux's newly sighted patient encountering a new form of sensory stimulation for which he had yet to develop sensorimotor contingencies, we will need to develop methods of attunement and calibration to help us understand how to discover invariant properties of an extended, blended energy array—structured encounters with appropriate feedback that can support us in negotiating the entangled hybrid lifeworld that our actions now bring forth.

7 Designing Interactions with Bodies in Mind

7.1 Designing for Current Bodies and for Future Entangled Worlds?

When I started working on this book in the fall of 2017, the Oculus Rift and HTC Vive had been available to consumers for just about eighteen months.[1] With their arrival, fully immersive VR moved swiftly from the lab to the home, and there was a rapid increase in the release of games that relied on full-body motion tracking. The experience of moving between real and simulated worlds had quickly become available to many more people. The shift in the relationship between perception and action that accompanied the development of immersive VR has few parallels in our evolution. We certainly received a perceptual jolt when steam trains suddenly allowed us to move so much faster through our environments, and as James Gibson says, we had to acquire a whole new array of perceptual and motor skills once we were able to fly planes.

In this book, I have argued that designing for this new entangled world, where the boundaries between digital and physical reality are increasingly porous, involves acknowledging that our bodies are in continuous dialogue with the blended materiality of a digitally augmented lived environment that is something quite new.

The dialogue of the senses that gives this book its name is not a dialogue among sense modalities, in which vision or hearing or touch must make a case for our attention. Rather, it is a dialogue between our bodies and the world. It is a dialogue between exploring action possibilities and focusing on those actions that successfully support us in reaching our goals. With the arrival of digitally mediated experiences, the dialogue of the senses has also become a dialogue between our bodies and an existence that newly entangles physical and digital realities. Like a toddler exploring the affordances of everything within their reach, we need space in this new world to push and pull, to climb up and fall down, to navigate all the gnarly byways of our digitally augmented world so we can learn what is possible.

The claim that lies at the center of this book is that our ability to act in and to understand our environment is dependent on the ongoing interaction between our moving body and the world. More particularly, our knowledge of the world is dependent on the relationship between the structure of our body and the structure of the environment. This relationship shapes both the actions we perform and the reactions we can observe. It is by means of this dynamic coupling between our body and the environment that we learn to adaptively cope in an ever-changing landscape of sensory information. Through active exploration, we discover the opportunities for action the environment affords us, and through repeated encounters, we attune to the sources of energy that can most usefully guide our actions.

7.2 Allowing Bodies to Bring Forth Worlds

The goal of this book has been to argue that the designer of virtual worlds should start with the seeing, feeling, moving body that will accomplish a task rather than with the performance capabilities of the machine. In other words, the designer should acknowledge the role of the user in bringing forth their individual experience of using the artifacts and systems we build. The four aspects of the relationship between our bodies and the world that I have drawn particular attention to in this book are:

1. Movement brings forth experience (chapter 3)
2. Actions bring forth time (chapter 4)
3. Bodies bring forth space (chapter 5)
4. New interactive objects bring forth new blended worlds (chapter 6)

Combined, these aspects represent dimensions of an embodied approach to interaction design founded on the premise that we cannot design the experiences of our users. At best, we can put in place the conditions under which users can bring forth, through their actions, an embodied experience of the world that also encompasses their experience of the things we build. Through their movements, people discover the opportunities for action available to them in their environments, and through the temporal unfolding of their actions, they learn to skillfully negotiate the relationship between their bodies and the world. It is also by moving through the world that people and animals discover its spatial attributes, such as the size of gaps that can be passed through or whether approaching objects will intercept their paths. Through the objects and artifacts we design, we also make available opportunities for people to discover, through their actions, entirely new blended worlds.

7.2.1 Movement Brings Forth Experience

The world we experience is the world we bring forth through our movements and our interactions with our environments. Like toddlers, we use active exploration to discover the opportunities for action our environment affords us as well as the sources of information within our bodies and in the external world that can guide us toward success. We learn to adapt when the environment does not respond as we expect and to cope with perturbations in the unfolding relationship among our body, our actions, and the world.

In chapter 2, I drew attention to the many sources of information that guide people as they move about in the world, suggesting that, in any successful interaction design, this information must be available for people to discover. The chapter provided an overview of current research on action, perception, and embodied cognition in order to guide designers in determining how to provide adequate and appropriate perceptual cues for users of their systems. A central argument was that designers cannot design the structural couplings between users' bodies and the augmented artifacts or systems we design. Rather, we can only create the conditions under which users can discover opportunities for action and the associated structural couplings appropriate for the tasks they are performing.

On a practical level, chapter 3 reviewed the many sources of variability in sensor data from motion-capture systems, pointing out that this data often contains useful material on which to build a richly interactive system. Capturing the ancillary body movements of musicians, for example, can allow one to observe how their body movements change as they acquire skill. I also introduced a number of design methodologies that consider movement as a design material, just like the hardware and software that will be used in realizing an interactive system. In these approaches, designers are encouraged to become experts in moving themselves to access a greater level of awareness of the many ways in which bodies move through the world.

Understanding the process of designing movement from a user's first-person felt perspective, I argue, begins with turning your awareness toward your own felt experience of moving. The first step is to engage deeply with an established movement practice, to feel what it is like to skillfully move one's own body, and to inhabit the dynamically evolving sensation of being physically engaged in moving through the world. Beyond this, these practices serve as scaffolds for turning felt experience into designed products or artifacts. Design movement provides the concept of choreographies of interaction, observations of movements, and their qualities in real-world contexts that can shape the qualities of interactions with digital artifacts. Choreographies of interaction reflect not only which parts of the body are moving but also aspects of how a person approaches an interaction—joyfully, greedily, carefully, and so on. To truly inhabit the shoes of future users, you have to reposition yourself to experience your designs as they will encounter them. You must raise your awareness of the different bodies and life experiences people will bring to encounters with your products.

Many of the design approaches introduced in this book provide useful starting points for a practice that places the body at the center of the design process. Design movement emphasizes designing action before product—that is, developing the choreography of an interaction before determining how it will be realized in the design of an interactive digital artifact or system.[2] Soma design emphasizes the designer's need to train their body to be more aware of the felt experience of movement in order to imagine, through their own bodily sensations, the felt experience of those who will ultimately use what they design.[3] Loke and Robertson's practice of "moving and making strange" encourages designers to step out of their comfort zones and into the shoes of potential users, whose experiences may be very different from their own.[4] Loke and Robertson argue that by becoming accountable to the first-person perspective, the designer learns to acknowledge the body as a source of knowledge about movement, the place where physical experience is situated and physical skills are developed.

In the end, design movement, somaesthetics, and the work of Loke and Robertson all enrich interaction designers' understanding of bodily movement, with the shared goal of deepening their appreciation for the richness of movement as a design material. Most importantly, they all focus on deeply engaging with the first-person felt perspectives of those who will use the products.

7.2.2 Actions Bring Forth Time

The movements that bring forth our experience of the world unfold in time. Our actions shape the time of our lived experience. Our interactions with the environment are planned within the context of what we have just experienced, in awareness of the present, and in anticipation of how the future might unfold as we direct our actions toward accomplishing a given task goal.

In chapter 4, I introduced the idea that the time of our experience is shaped by the time we can perceive and the time in which our bodies can act. There are events that unfold at a timescale faster than our sense organs can pick up, and others that unfold so slowly that we do not perceive the rate at which they change. Similarly, there are situations in which we know that we can act quickly enough to avoid a collision, or slowly enough to control a fine-grain action (e.g., threading a needle). This is because, from experience, we know what our individual bodies are capable of accomplishing. Our lifetime of being in our bodies, therefore, shapes the time of our lived experience.

The central argument of chapter 4 was that interaction with computational artifacts, like interaction with any other object in our environment, unfolds as a temporal arc, where we plan actions in the context of what has just happened, in awareness of the present, and in anticipation of what is to come. I further argued that since we lack a theoretical framework in the field of HCI to account for time in interaction this way, we can adopt the theoretical framing of tau-guided movement. Using the example of tau for guiding movement in simulated dynamic environments such as virtual soccer and rugby games, I suggested that tau theory is relevant because it is sensitive both to the context in which movements are executed and to the bodily capabilities that a user brings to that context. As such, tau-guided movement as the basis for interaction design provides a practical methodology for delivering ecologically grounded, context-specific perceptual information to support users in planning and executing action.

As designers of interactive systems, we do not typically think of an unfolding temporal arc as shaping people's encounters with digital artifacts and environments. It is also easy to forget that these interactions are embedded within the wider context of people's encounters with everything else in their surroundings. The applications that populate our digital devices are designed to grab our attention, and in doing so, they also take our time. By interrupting the temporal arc of a current task, their notifications and alarms force us to redirect our actions, postponing or replacing the anticipated accomplishment of current task goals.

As we move away from the event-driven interfaces that characterized first- and second-wave HCI and move toward embodied, situated interaction design, we face the challenge of determining how to support users in discovering, through their actions, what new affordances our artifacts and devices make available. The more nuanced account of time consciousness set forth by Gallagher and Zahavi, in which our experience of the present is brought forth by our active engagement with our environment, provides a conceptual framing for interaction design in the third wave.[5] Only by reorienting our thinking toward the temporal arc of interaction design in this way can we move beyond a world where we interact with our digital stuff in a notional parallel cloud toward a world where our digital artifacts are situated in our environment and contribute to bringing forth the time of our lived experience. If we can remove our digital devices from the pedestal they occupy and learn to think of them as physical artifacts that just happen to be augmented with digital capabilities, then we may be able to cast new light on their role in shaping the temporal arc that constitutes the planning and execution of our interaction with the world around us. This more nuanced account of time as being brought forth through active engagement with the environment will require us, as designers, to become aware of how people's everyday experiences both shape and are shaped by the interactive artifacts we create.

7.2.3 Bodies Bring Forth Space

The space we inhabit, whether real or simulated, turns up in our lived experience not as abstract geometry but rather as a series of relationships among our bodily capabilities, our action goals, and the ability of the objects and surfaces around us to support us as we go about our daily lives. The structure of space is brought forth for us by our actions and scaled by our bodily dimensions and our physical capabilities. We perceive objects and structures in our environment not in terms of their size or their distance from us in metric units but in terms of whether they can help us accomplish a given task goal: Is that object reachable? Can I pass through that gap? and so on. The sources of information on which people base these judgments may be present in their environment as optic, acoustic, and/or mechanical energy. What is important is that people learn to attend to those sources of information that are most reliable in guiding their actions.

In chapter 5, I argued that the most relevant information that an individual requires as they move and act in space is the information that directly pertains to their current action goal, specified in terms of their current action capabilities. An individual, therefore, does not need to know how to perceive, for example, the size of a gap in metric units. They just need to know whether the gap affords them the ability to pass through. Whether interacting with static objects or intercepting moving things, what matters for an individual is what will happen in their "current future," assuming current conditions prevail. In negotiating space, perception of one's current future is as important as the perception of properties such as the size or form of objects.

For designers of digitally augmented artifacts and environments, this poses a challenge because it is difficult to predict which source of information an individual user might find most relevant within a given task context. The solution proposed in this book is that the rendered or partly rendered environment should contain a sufficiently rich array of sensory stimuli to support participants in discovering and attuning to sources of task-relevant information that can guide them toward their goal. In designing partly or wholly rendered interactive spaces, we should focus less on rendering realistic, metrically accurate geometries. Instead, we need to pay attention to the sources of information that individuals attune to as they plan and execute actions directed toward the objects and surfaces around them. As I demonstrated in chapter 5, the spatial properties of the environment, such as whether gaps are passable or objects are reachable, are specified for an individual on the basis of discoverable opportunities for action.

This is no less true for augmented and virtual environments. Think again of the player of a virtual rugby match who has attuned to the cues defining the tau of a closing gap between the avatars of two members of the opposing team in order to determine whether he can run past them. Similarly, think of the blind cane user navigating a virtual world who has attuned to the changing spectra of sounds created by the tapping of her cane to judge the distance to nearby objects. As designers, we cannot predetermine the sources of information in our partly or wholly simulated global energy array that users will attune to in order to guide their actions. But our goal should be to incorporate this information so that it is available to be discovered. For example, when we are rendering an approaching ball that we intend a user to intercept, we can incorporate changes in the shape of its image that reflect the angle and curvature of its trajectory for the catcher, who is attuned to the rate of change of the angle of expansion of this image on their retina as a perceptual invariant defining the time-to-contact with the ball. The more we can support people by making available the information necessary for them to negotiate the augmented and virtual spaces we create, the more skillful they can become in moving about in the worlds we design.

7.2.4 New Interactive Objects Bring Forth New Blended Worlds

As Merleau-Ponty asserted, our individual lifeworld is brought forth through our lived experience. The same physical environment constitutes a different lifeworld for each individual. Moreover, that environment will show up for each individual in different ways, depending on what they want to do. If a toddler needs support to stand, they grab onto a chair. If they want to reach something on a countertop, they climb on that chair. The meaning of the form of the chair is brought forth by coordination between the toddler's action goal and their understanding of the structure and capabilities of their body. No doubt the designer of the chair was given a brief describing its context of use, including the average height and weight of those who would sit on it. But were they asked to consider the lifeworlds of toddlers?

It is easy to understand how chairs might show up in different ways for people with different bodily capabilities and action goals. When we design artifacts or systems that include digitally mediated functionality, it is easy to forget that this functionality will show up in the lifeworlds of users in many different ways, depending on their individual bodily skills and abilities and the variety of tasks they wish to accomplish. Digital artifacts or encounters with rendered, augmented, and virtual worlds are not privileged. Like encounters with physical objects, the worlds of possibilities they make available are still brought forth through the actions of those who use them, and those users' experiences will still be shaped by what they wish to do.

In chapter 6, I used the metaphorical lens of interaction with augmented and virtual objects to propose that the choices designers make ultimately impact the relationship between an individual user's bodily capabilities and experiences and their ability to make sense of and act within digitally augmented lifeworlds. This is because augmented and virtual objects fundamentally challenge traditional notions of materiality by blurring the boundary between the physical world, in which we move and act, and the digital world, where the consequences of our actions unfold. In the case of hybrid artifacts such as "smart" objects (e.g., networked doorbells), our actions can have consequences in both our physical and digital environments—we can see the delivery driver at our door and perhaps remotely open a lock to allow them to leave a package. Conversely, as with the use of lamps in the MyDaylight project, physical objects can be repurposed to make explicit the underlying functionality of a digital system—in this case, a calendar.[6]

My assertion here is that to design the objects that will mediate the actions of people in hybrid or blended worlds, we will need to move beyond notions of digital artifacts as physical representations of digital information. In this new entangled world, the artifacts we design are hybrid materials. They are new ways of augmenting structures within the lifeworlds of their users. Through active engagement with the hybrid artifacts we build, people will discover the appropriate structural invariants in the partly or wholly simulated global energy array—the patterns of light, sound, or mechanical feedback that will support them in skillfully accomplishing their task goals.

Designing interactions with blended worlds is ultimately about designing with the body in mind. It is about dissolving the boundaries between the inner mind and the outer world by allowing people to discover, through their movements and actions, the possibilities that blended environments afford. Like Molyneaux's newly sighted man, our users need to discover how their actions shape the hybrid sensory milieu they are encountering and to learn new sensorimotor contingencies to help them navigate this new entangled life. And although we cannot design new sensorimotor contingencies, our job is to design the conditions under which users can discover appropriate strategies for themselves. This does not mean that we, as designers, need to change what we do. But it does mean we need to change how we *think about* what we do. Like van Dijk and colleagues designing for embodied being-in-the-world, we need to have the confidence to allow the context for which we are designing to tell us what the technology needs to be—if, indeed, there needs to be technology at all.

7.3 Future Dialogues, Future Directions

In this book, my aim has been to present a rich exploration of the theories, principles, and research findings related to the fields of embodied cognition and embodied interaction design for students and professionals interested in situating the human body at the center of their design practice. Along the way, I have had to make many choices about what to include, as well as what to leave out, in order to engage with the broadest possible audience. There are, for example, whole areas of practice where the body has always been the starting point for interaction design, including interactive dance performance, digital musical instrument design, and the design of interactive art installations. In addition, I fully acknowledge the normative bias that informs this book, arising from the fact that the work it cites has prioritized the experiences and bodily capabilities primarily of able-bodied white individuals. This is not my own experience, nor is it the experience of many of the people who ultimately use the things we design. In closing, I therefore wish to offer some thoughts about possible ways of expanding this current work, in the hope that others can pick up some of the threads in the fabric of this project that I have left unwoven.

Below, I draw attention to three such threads: the normative nature of embodied cognition and its impact on embodied interaction design, the particular case for users with disabilities, and the potential for embodied practices in interactive music performance. These areas provide a lens through which we can reflect on interaction design that truly centers the specific needs and lifeworlds of the individuals with whom we work. It is well beyond the scope of this chapter to address these topics in detail, but my goal is to draw attention to the future directions in which they might take the field of embodied interaction design.

7.3.1 Designing Interaction with Whose Bodies in Mind?

The embodied approach to cognition that has grounded the theory and practice presented in this book centers the role of the body in the active process of "sense-making," of discovering the possibilities for action the environment affords an individual in negotiating their lifeworld. As De Jaegher and Di Paolo state, "Sense-making is an inherently active concept. Organisms . . . actively participate in the generation of meaning in what matters to them; they enact a world."[7] As I have noted throughout this book, scholars of the enactive approach to cognition posit that the world that is brought forth for an individual is shaped by their actions and is constituted by the structural coupling between their body and the environment. Thus, the structure of the feet of a lizard affords them the ability to climb walls and walk across ceilings. Humans, like many other animals, have figured out how to co-opt or design tools to extend the ways in which their bodies can be structurally coupled to their worlds—tools such as hammers for constructing shelters and skis for moving through snowy mountain landscapes.

Implicit in the enactive approach is the idea that an individual's encounter with their world is particular to their bodily resources and their lived experience. But enactivists rarely consider the range of bodies and lived experiences that might exist for any given individual within a species. The assumption that each "niche" is occupied by a group of individuals with generic bodies has therefore led to a justified critique of the enactive approach to cognition. This assumption on the part of enactivists is unfortunate because, as van Grunsven notes, enactivists thereby fundamentally miss the power of the enactive approach to account for the infinite number of ways in which embodied beings make sense of an infinite variety of individual lifeworlds.[8]

In her book *Invisible Women*, Caroline Criado Perez points out the many ways in which normative assumptions about the bodies of those who will use the artifacts we build creep into every aspect of design. For example, she notes that the average handheld phone has a screen size of about 5.5 in., a format that fits comfortably within the hand of an average Caucasian man. Although smaller phones have been made that were better suited to smaller hands, these were typically underpowered when compared with larger phones of the same generation. People with smaller handspans, it seems, have to make the choice between functionality and ease of use.[9] Similarly, the head-mounted displays and handheld controllers used in today's VR systems do not accommodate the hand size or interpupillary distance of the bodies of women and children. As Peck et al. note, this is not surprising, since most of the designers who develop these devices come from the roughly 3 percent of the population who can be described as male, white, educated, industrialized, rich, and Democratic (M-WEIRD).[10] As Munafo and colleagues discovered, one impact of this bias in the design of VR hardware and software is that women are more likely to suffer from the effects of VR sickness than men.[11]

Developers of technologies such as phones and fitness trackers will defend such a one-size-fits-men approach by saying that it is difficult and expensive to cater to the huge variety of bodies and contexts of use their devices will meet out in the world. But researchers such as Katta Spiel point out that this approach only further entrenches the impact technologies have on bodies that don't conform to manufacturers' assumptions. In a discourse on the impact of fitness trackers, Spiel points out that their very form factor excludes people with broader arms because their wristbands just don't fit and their calorie-oriented goal structure prioritizes slimming norms, enabling disordered eating and high anxiety in many of those who use them.[12] Spiel's argument is that unless researchers pay attention to how the tools they design shape the lives of those who use them, we cannot break the cycle of normative design.

In a 2021 analysis of forty-five papers published in the proceedings of the Conference on Tangible and Embodied Computing, Spiel examines how normative assumptions relating to users' bodies pervade the field. Spiel has found that when the tangible embedded and embodied interaction community talks about bodies, they are idolized, disembodied beings whose lifeworlds are sanitized and will, therefore, never reflect the messy reality of users' lived experiences. Instead, researchers' approach to establishing bodily norms is not to specify them but rather to leave them unmarked, naming only what is considered "other." The challenge with this strategy, Spiel argues, is that unspecified norms remain unidentified and out of reach.[13]

To address this problem, Spiel presents the community with four design "provocations":

1. Check privileges: Reflect on how our own embodiments impact our designs. If we are young, slim, and fit, the body we bring to the design process is a minority, privileged body with access to health care and probably time and space for recreational activities.
2. Acknowledge imperfections: When made aware of a shortcoming in one's work as a designer, acknowledge failure gracefully. As the author of this book, for example, I acknowledge that I have often failed to confront the normative biases in the scholarly work I have cited and may have also paid these views forward. I strongly believe that we, as a community, can turn this corner, prompted by the critical stance that is now a central part of work in our field.
3. Name norms: Spiel points out that it is only by naming norms and assumptions in our design methodologies and practices that we can confront them. This may be uncomfortable for those of us who represent normative populations. It is on us to recognize such positionality and to work toward confronting our own biases and assumptions (e.g., racism and ableism).
4. Embrace messes: Remember that living bodies are messy. They stumble, swear, and sweat; they acquire injuries and grow old. In other words,

> bodies are always in transition, and their lifeworlds continually evolve. So, an artifact we design must respect the plurality of bodies and circumstances in which it will be encountered.[14]

This is all very well, you might say, but how can these "provocations" be incorporated into a design practice? Let's return to the case of the phone. In her book, Criado Perez recounts a moment when she watched her mother struggling with the voice-recognition system in her car as she tried to call her sister. After a number of failed attempts, her mother tried lowering the pitch of her voice, whereupon the system instantly recognized her. The normative assumption here was that the driver of this car, a Volvo, would most likely be male; the designers failed to include adequate variance in pitch range in their voice-recognition model. This simple assumption resulted in very real stress for an individual driving a car, and yet it could have been completely avoided if the system designers had been open to questioning their own normative design biases. Such biases creep in early in the design process and can only be addressed if they are questioned at every stage.

The first thread that I leave unwoven at the end of this book is, therefore, a need for design methodologies that start by questioning normative assumptions in embodied interaction design, even before hardware systems and software algorithms are considered. Some promising routes forward, as I note in chapters 3 and 6, are body-centered, movement-first design practices such as that proposed by Höök and the practice of designing for embodied being-in-the-world proposed by Hummels and van Dijk.[15] By building outward from a deep understanding of the lifeworlds of the people who will use their designed artifacts, D4EB practitioners are, I argue, much less likely to overlook the nature of the particular bodies, minds, and worlds for which they are designing.

7.3.2 Embodiment, Disability, and Inclusive Design

A second unwoven thread in this book concerns the representation of the bodies of people with disabilities within the practice of embodied interaction design. I have already noted how people with different bodily abilities discover specific invariant properties within the global energy array that can guide their actions. Merleau-Ponty, for example, noted how the cane of a blind person provides a structural coupling to the environment that extends their reach to the tip of their cane, so that they can detect the presence of objects beyond arm's length. Over time, Merleau-Ponty suggests, the person incorporates the extent of their cane into their body schema so that they come to understand the distance to their cane tip with respect to their walking speed.[16] Similarly, acoustic information about the presence of obstacles and openings can be revealed by a cane's tapping to guide the actions of a blind individual in estimating the distance to a wall or the size of passable gaps.[17]

Of course, the sources of energy that blind individuals are attuned to in these cases are present in the global energy array for everyone. It is just that in the absence of visual information to guide their actions, these tactile and auditory cues become more salient in bringing forth the world of the blind traveler. It is not that the lifeworld of a blind individual is "deprived" of visual cues. Rather, their lifeworld is brought forth, fully featured, in response to nonvisual sources of information. As van Grunsven points out, "Enactivism is committed to foregrounding that, even when someone's world is a harder one to navigate, it is still their world with perceptual saliences that might matter deeply to them."[18]

As I have emphasized throughout this book, the strength of the enactive approach as a starting point for designing digitally augmented artifacts and environments lies precisely in its emphasis on the role that the individual's embodied mind plays in bringing forth their world, even when that world is partly or wholly simulated. The work of van Dijk, Hummels, and colleagues on projects such as MyDaylight introduces a design methodology that seeks to understand the lifeworld of individuals. It is a methodology that, by design, has to account for their unique embodied knowledge and experience and their unique ways of making sense of the world. Rather than defining general design guidelines to accommodate all those whom designers of technology determine to have specific accessibility needs, this approach imagines the embodied lives and sensory, motor, and cognitive milieus of all participants and potential end users. In van Grunsven's words, "I see it as a central commitment of enactive 4E Cognition that it adopts a stance towards a person's embodied life, disabled or otherwise, as anchoring them to a meaningful world through their particular history of ongoing self-constitution, which is reflected in their ways of moving and coping with their environment and their styles of embodied expressivity."[19] Writing about the experiences of autistic people, De Jaegher further expands van Grunsven's observation, noting, "If autistic embodiment is intrinsically linked with autistic sense-making, we can hypothesize that many autistic people will find joy or significance in behaviors and embodied styles of sense-making that are considered 'autistic.'"[20] For the person for whom MyDaylight was designed, there was clearly pleasure in selecting the color and placement of the light that represented the anticipated visit of his friend.

Notably, the rapidly growing discipline of crip technoscience, the practice of disabled users of technology critiquing, altering, and reinventing their material-discursive world, is sharply critical of approaches such as inclusive design (ID) and accessible design (AD)—approaches that seek to encompass the needs of most potential users of technologies most of the time.[21] The aim of AD, for example, as stated in ISO guide 71, is to extend "standard design" to encompass the needs of "people with some type of performance limitation to maximize the number of potential customers who can readily use a product, building or service." The guidelines state that this may be accomplished by:

1. Designing products, services, and environments that are readily usable by most users without any modification
2. Making products or services adaptable to different users (adapting user interfaces)
3. Having standardized interfaces to be compatible with special products for persons with disabilities[22]

The problem is that such so-called accessible design is often achieved at the expense of the roughly 10 percent at the margins whose individual needs and experiences are not encompassed by the AD umbrella.[23] From the perspective of the embodied approach to interaction design set forth in this book, AD can also result in designers overlooking the variety and richness of the embodied experiences of disabled people. Hamraie and Fritsche, in their "Crip Technoscience Manifesto," point out that such design practices often fail to commit to disability as a difference but instead incorporate regulatory compliance as an inconvenient set of rules that must be adhered to.[24]

Like others before them, Hamraie and Fritsche emphasize the fact that people with disabilities are expert problem solvers, skilled at deploying the resources in their environment to overcome the mismatches between normative design practices and their particular physical and cognitive abilities. Indeed, the first sidewalk curb cuts, the environmental modification for wheelchair users that is often held up as the poster child of universal design, came into being only because of the persistent activism of wheelchair users such as Jack Fisher and Ed Roberts. Roberts and his friends went so far as to reshape sidewalks and pour asphalt under the cover of darkness so that Ed could get to and from his classes at the University of California at Berkeley.[25] Then, it turned out that curb cuts were useful for everyone, whether riding a bicycle, pushing a baby in a stroller, or wheeling a cart. So, they are now ubiquitous features of our urban streetscapes—and for this very reason, we often forget their presence was hard won and that their design was the outcome of the kind of problem-solving expertise developed by the disabled people who needed a way to negotiate curbs.

The phrase "Nothing about us without us," first coined by James Charlton, is as much a call for a practical reorienting of time and resources toward supporting disabled designers as it is a call for disabled people to become active and activist participants in ensuring that we are not locked out of a technologically mediated future.[26] Responding to Charlton's call, researchers with disabilities in the field of HCI conducted a workshop in 2020 to specifically address the challenges of including their voices in the design of access technologies, with the goal of establishing a research community to look at the field of HCI through the critical lens of disability studies and to meaningfully engage people with disabilities in making information and interactive systems accessible.[27] Like Charlton, Spiel and colleagues acknowledge that not only are disabled people design experts, they are also epistemic activists. In the words of Hamraie and Fritsche, "Technoscience recognizes that disabled people's world-dismantling and world-building labors stem from situated experiences of 'misfitting' in the world."[28]

Increasingly, funding bodies such as the US National Science Foundation are responding to pressure from researchers with disabilities.[29] They are asking that, where applicable, research teams include people with disabilities as primary or coinvestigators on grants, as project officers, and on review panels so that the lived experiences of disabled people are centered at every stage of the research process.[30] Like van Grunsven, my argument here is that the embodied approach to interaction necessarily incorporates, quite literally, the lived experiences of those for whom we are designing. If we are committed to this starting point, then the goals of approaches such as accessible design will be an outcome, not a driver, of our design process. Janna van Grunsven's paper titled "Disabled Body-Minds in Hostile Environments" argues that as we think toward a world where more and more people are displaced by climate change and forced to live in environments that are less hospitable and less navigable than those to which we in the Western world are accustomed, we will also be forced to embrace a much broader range of use cases and thus a much larger variety of bodies and lived experiences for which we must design.[31]

Rather than continuing to think of people with disabilities as non-normative edge cases, it is time to embrace their lived experiences as a resource, to build on their design expertise and their continual capacity to adaptively cope with environments that do not accommodate their needs. At the limit, van Grunsven notes, none of us is equipped to survive the lifeworlds we will encounter if we ever leave this planet.

In this book, I have argued that design approaches such as D4EB and soma design already point toward a more encompassing design future.[32] Now all we need is for the design community to commit to engaging with the bodies and minds of the people with disabilities who can bring this future into being.

7.3.3 Embodied Interaction Design Through the Lens of Interactive Music Performance

The final thread I wish to pick up here grows out of my own work on what Jensenius has recently termed "embodied music technology."[33] It concerns the role that practices such as interactive music performance have played in setting the stage for a person-first approach to embodied interaction design. For every new technology that humanity has developed to transduce human movement into another form of energy (mechanical, electrical, digital), musicians have been quick to grasp its potential as a means of amplifying or translating expressive bodily movements to control sound. The mechanical tracker organ developed in the seventeenth century, for example, made it possible for church musicians to play pipes located throughout large cathedrals so that churchgoers found themselves fully immersed within the body of the distributed instrument. In 1920, Leon Theremin discovered how to exploit the body's ability to absorb electrical charge to create a contactless and apparently magical instrument where players could control sound by moving their hands through the air.[34] Indeed, Jensenius points out that the Theremin is truly an embodied instrument, as it requires the presence of the player's body to both create and control sound.[35]

With the advent of digital sensors that can capture bodily gestures and digital methods to map sensor signals to control algorithmically generated synthesized sound, musicians have again become some of the earliest adopters of these new technologies, often pushing their boundaries long before mainstream applications recognized their potential. In 1982, for example, David Rokeby's "Very Nervous System" used a home-built box containing sixty-four light sensors to track the movements of a dancer. The sensor signals were then mapped in software so that the dancer could control individual synthesized musical instruments within a digital ensemble.[36] At around the same time, the performer Michel Waisvisz developed his "hands" system: low-latency, multi-axis motion-capture gloves that he used to control sound.[37]

In short, performers have been working with interactive systems that capture movements of diverse bodies, for diverse purposes, for far longer than designers working in mainstream HCI. Moreover, the temporal and spatial demands of live performance have pushed the boundaries of real-time responsiveness, for both hardware and software, which have historically far exceeded those for general purpose interactive computing, at least until the advent of high-powered consumer-level interactive gaming systems in the early 2000s.

My proposal here is that if we are looking for design methodologies and practices that can accommodate a diversity of bodies and ways of moving, we could learn much from practices such as interaction design for music performance.

In my own work, I have found this to be an extremely fruitful practice. In PebbleBox, for instance, there is no constraint on how people interact with pebbles; the signature of hand movements is transferred directly through to sculpt the output sound with minimal mapping steps (movement amplitude controls sound amplitude, and movement timing controls the timing of pebble collisions).[38] The result is an interactive experience that has proven to be engaging for young and old, musician and non-musician alike. Although this experience is compelling, the interaction does not support the development of any deep musical skill. Learning researchers such as Mitch Resnick would describe this interaction as having a low floor and low ceiling—that is to say, it is easy to figure out how the interaction works, but there is little room to become an expert user.[39]

Working with PebbleBox led us to consider how one could design a sensor-based motion-capture system for interactive musical instruments that would support a player in developing musical skill and expertise while at the same time being scalable. This ultimately led to Nick Gillian's work on the Gesture Recognition Toolkit (GRT).[40] Gillian recognized that, for music performance, it is not necessary to recognize the movements of most people most of the time. What is more valuable for a performer is that a system should recognize one musician's movements all of the time. This realization allowed Gillian to build into the GRT the ability for performers not only to tune the response of machine learning algorithms during rehearsal but also to lock down an algorithm's behavior for a given performance.[41] As Gillian notes, this is critical when sensor systems are being used in different performance situations (e.g., in different venues) because players can then retune their instruments to allow for different lighting conditions or even the presence of different amounts of electrical interference on stage. In the case where a digital musical instrument of this kind is passed on to a different performer, there is built-in flexibility at the training/rehearsal phase for them to attune its response to their own body and to meet the goals of their own performance requirements.

We are not the only digital musical instrument designers to recognize the power of machine learning to provide a means of balancing the needs of individual performers with the desire to build scalable interactive systems that track movement. Tools such as the Wekinator and, more recently, AI point toward very practical ways to implement systems that can account for individual bodies while still being highly configurable and deployable across a range of hardware controllers, software environments, and physical spaces.[42] The challenge is now to develop ways for people to train their devices to recognize and flexibly map their movements onto the functionality of their devices, much as phones and digital home assistants now learn to understand the voices and speech patterns of their users.

This is no easy feat, but it is, I believe, the practical step we will need to take if we are to realize the goals of embodied interaction design: meeting the needs of the bodies of all users, designing systems and artifacts that they will want to bring into their individual lifeworlds. In this book, I have proposed that meeting these design goals requires us to keep in mind the following three considerations:

1. To consider bodily movement as the starting point for any interaction design
2. To consider how the interactive systems we build will be integrated into the vastly varied lives of the people we build for—in particular, how our

designs will build on the perceptual and motor skills of people to support them in making sense of the artifacts and environments we build
3. To consider how these artifacts and environments support people's sense of being-in-the-world, when the world we are designing is partly or wholly simulated, so that we are truly enriching their lived experience

7.4 Coda

This book concerns the ways in which individual bodies experience the world, with all the complexity of its newer hybrid and virtual manifestations. I hope this investigation into how the world of an individual's experience is brought forth by their actions can be the springboard for approaching virtual and hybrid projects from the perspective of moving, acting, thinking, feeling bodies, each possessing skills and capabilities that the individual user will bring to bear on our designs. From this perspective, there is no such thing as accounting for individual differences in how we build or evaluate design. To paraphrase the words of Monty Python in *The Life of Brian*, we "are all individuals."[43]

In the foregoing chapters, I have laid out evidence to support a truly embodied approach to design that has roots in phenomenology, ecological approaches to perception and action, and embodied approaches to interaction design. Putting the living, moving body at the center of every aspect of the design process certainly means that technological advancement as a project will be knocked from its pedestal. As Fallman anticipated, doing so will require us to reassess what we mean by "good" design.[44] Throughout this book, I have outlined several candidate approaches to the design of interactive artifacts and of augmented and virtual environments, in particular the work of Loke and Robertson; the Danish design school; and Höök, Hornecker, and van Dijk and their colleagues on D4EB. If we bring this work to the fore, our designs will expand users' lived experiences and enrich their environments with new possibilities for action, arising from coordination between their bodies, their capabilities, and the interactive objects and environments we build. In doing so, we will design technologies for embodied being-in-our-world rather than technologies that insist on bringing us into their worlds.

Notes

Chapter 1

1. Walker, "Curse of a Sunburnt Land."

2. Gibson, *The Senses Considered as Perceptual Systems*, 54.

3. Kay, *Doing with Images Makes Symbols.*

4. As a reviewer of this manuscript pointed out, pre-metric body-scale measures included the foot (12 in., believed to have been derived from the length of the booted foot of King Henry I), the hand (4 in. or the width of a man's hand, used for measuring the height of horses), and the fathom (6 ft. or the span of a man's outstretched arms, used for measuring the depth of water). See Vincent, *Beyond Measure.*

5. Von Helmholtz, *Handbuch der physiologischen Optik.*

6. Simons and Chabris, "Gorillas in Our Midst."

7. McGann, "'Enactive' Cognition."

8. Fajen and Phillips, "Spatial Perception and Action."

9. Souza et al., "Measuring Presence in Virtual Environments."

10. Höök, *Designing with the Body.*

11. Teixeira and Braga, "The State of UX in 2024."

12. Fiske, "Mind the Gap," 123.

13. Dichgans and Brandt, "Visual–Vestibular Interaction."

14. Zadra and Proffitt, "Optic Flow Is Calibrated to Walking Effort."

15. The Nobel Prize in Physiology or Medicine 2004, NobelPrize.org, https://www.nobelprize.org/prizes/medicine/2004/summary/.

16. See Gullstrand (1911), Granit, Keffer, Hartline, and Wald (1967), and Huber and Wiesel (1981) at All Nobel Prizes, Nobelprize.org, June 25, 2018, http://www.nobelprize.org/nobel_prizes/lists/all/.

Chapter 2

1. See, e.g., Goldstein, *Sensation and Perception*, chap. 1.

2. Hurley, "Perception and Action," 3.

3. Carello and Turvey, "The Ecological Approach to Perception."

4. Gibson, *Perception of the Visual World.*

5. Gibson, *The Senses Considered as Perceptual Systems*, 3.

6. See Gibson, *The Senses Considered as Perceptual Systems*, 50.

7. Carello and Turvey, "The Ecological Approach to Perception."

8. Jenison, "On Acoustic Information for Motion."

9. Rosenblum et al., "Relative Effectiveness of Three Stimulus Variables."

10. Gibson, *The Ecological Approach to Visual Perception*, 119.

11. Gibson, *The Ecological Approach to Visual Perception*, 119.

12. Turvey, "Affordances and Prospective Control."

13. Chemero, "An Outline of a Theory of Affordances."

14. Baggs and Chemero, "The Third Sense of Environment."

15. Wilson and Foglia, "Embodied Cognition."

16. See, e.g., the second of the "Meditations on First Philosophy" in Descartes, *Meditations on First Philosophy.*

17. Shapiro, *Embodied Cognition.*

18. Adolph et al., "Crawling Versus Walking Infants' Perception of Affordances."

19. Smith and Thelen, *A Dynamic Systems Approach to Development.*

20. Hurley, "Perception and Action"; McGann, "'Enactive' Cognition"; and McGann, "Perceptual Modalities."

21. Shapiro, *Embodied Cognition.*

22. Transcendental idealism was first defined by Immanuel Kant, who, in his *Critique of Pure Reason*, posited that the self, or "transcendental ego," constructs knowledge from unstructured sensory data onto which it imposes "categories," generalized concepts derived from prior experience that give sense impressions meaning.

23. Heidegger, *Sein und Zeit.*

24. Dreyfus, *What Computers Still Can't Do*; and Kiverstein and Wheeler, *Heidegger and Cognitive Science.*

25. Wheeler, "The Revolution Will Not Be Optimised"; and Dourish, *Where the Action Is.*

26. Winograd and Flores, *Understanding Computers and Cognition*; and Dourish, *Where the Action Is.*

27. Kaufer and Chemero, *Phenomenology*, 51.

28. Kaufer and Chemero, *Phenomenology*, 58.

29. Winograd and Flores, *Understanding Computers and Cognition*; and Dourish, *Where the Action Is.*

30. Merleau-Ponty, *Phenomenology of Perception.*

31. Merleau-Ponty, *Phenomenology of Perception*, 78, cited in Kaufer and Chemero, *Phenomenology*, 100.

32. Kaufer and Chemero, *Phenomenology*, 104.

33. Kaufer and Chemero, *Phenomenology*, 105.

34. Kaufer and Chemero, *Phenomenology*, 111.

35. Menary, "Intentionality, Cognitive Integration and the Continuity Thesis."

36. For an excellent introduction to phenomenology, particularly to those aspects that are most relevant to embodied cognition, see Kaufer and Chemero's book *Phenomenology: An Introduction.*

37. Cutting, "Two Ecological Perspectives."

38. Cutting, "Two Ecological Perspectives," 216, cited in Withagen et al., "Affordances Can Invite Behavior."

39. Withagen et al., "Affordances Can Invite Behavior."

40. Ju and Takayama, "Approachability."

41. Kugler et al., "On the Concept of Coordinative Structures as Dissipative Structures."

42. Kugler et al., "On the Concept of Coordinative Structures as Dissipative Structures."

43. See Kaufer and Chemero, *Phenomenology*, chap. 9, for more detail.

44. Smith and Thelen, *A Dynamic Systems Approach to Development.*

45. Kaufer and Chemero, *Phenomenology*, 449–451.

46. Chemero, *Radical Embodied Cognitive Science.*

47. Maturana and Varela, *Autopoiesis and Cognition.*

48. Varela et al., *The Embodied Mind.*

49. Merriam-Webster.com Dictionary, s.v. "autopoiesis," accessed December 8, 2024, https://www.merriam-webster.com/dictionary/autopoiesis.

50. Kaufer and Chemero, *Phenomenology*, chap. 9.

51. Di Paolo, "Autopoiesis, Adaptivity, Teleology, Agency."

52. Di Paolo, "Autopoiesis, Adaptivity, Teleology, Agency"; Noë, "Neural Plasticity and Consciousness"; O'Regan, *Why Red Doesn't Sound Like a Bell*; and Weber and Varela, "Life after Kant."

53. Piaget, *The Construction of Reality in the Child*, 161, as cited in Kelso, "On the Self-Organizing Origins of Agency."

54. Penny, *Making Sense*, 250.

55. Thompson and Stapleton, "Making Sense of Sense-Making."

56. Stewart et al., "Enaction and Engineering."

57. Bennett et al., "SimpleFlow"; Hurley and Noë, "Neural Plasticity and Consciousness"; O'Regan, *Why Red Doesn't Sound Like a Bell*; and O'Regan and Noë, "A Sensorimotor Account of Vision and Visual Consciousness."

58. O'Regan and Noë, "A Sensorimotor Account of Vision and Visual Consciousness."

59. McGann, "Perceptual Modalities," 83.

60. Bach-y-Rita et al., "Vision Substitution by Tactile Image Projection."

61. Auvray et al., "There Is Something Out There."

62. Smith and Thelen, *A Dynamic Systems Approach to Development.*

63. Smith and Thelen, *A Dynamic Systems Approach to Development.*

64. Piaget, *The Construction of Reality in the Child.*

65. Smith et al., "Knowing in the Context of Acting."

66. Bell and Fox, "The Relations Between Frontal Brain Electrical Activity and Cognitive Development."

67. Smith et al., "Knowing in the Context of Acting," 258.

68. Smith and Thelen, *A Dynamic Systems Approach to Development.*

69. Kirsh, "Embodied Cognition and the Magical Future of Interaction Design."

70. Lakoff and Johnson, *Metaphors We Live By.*

71. See, e.g., Fishkin, "A Taxonomy for and Analysis of Tangible Interfaces."

72. Harrison et al., "The Three Paradigms of HCI."

73. Sheets-Johnstone, *The Primacy of Movement.*

74. Sheets-Johnstone, *The Primacy of Movement*, chap. 3.

75. Sheets-Johnstone, *The Primacy of Movement*, 127.

76. Here, "presence" is an abbreviation of the term "telepresence," which is specific to the phenomenal experience of being immersed in an augmented or virtual environment. For a more detailed definition, see the website of the International Society of Presence Researchers: http://ispr.info/about-presence-2/about-presence/.

77. Thornson et al., "Predicting Presence," 62.

78. Dourish, *Where the Action Is.*

79. Stewart et al., "Enaction and Engineering."

80. Kirsh, "Embodied Cognition and the Magical Future of Interaction Design."

81. Kirsh, "Embodied Cognition and the Magical Future of Interaction Design," 11.

82. Almeida et al., "Unconscious Processing Dissociates Along Categorical Lines," cited in Kirsh, "Embodied Cognition and the Magical Future of Interaction Design."

83. Kirsh, "Embodied Cognition and the Magical Future of Interaction Design," 11.

84. Harrison et al., "The Three Paradigms of HCI."

85. Card et al., *The Psychology of Human–Computer Interaction*; and Norman and Draper, *User Centered System Design.*

86. Norman and Draper, *User Centered System Design.*

87. Fallman, "The New Good."

88. Harrison et al., "The Three Paradigms of HCI."

89. Dourish, *Where the Action Is*; Gaver, "Technology Affordances"; Hornecker and Buur, "Getting a Grip on Tangible Interaction"; and Fallman, "The New Good."

90. Svanæs, "Interaction Design for and with the Lived Body."

91. Winograd and Flores, *Understanding Computers and Cognition.*

92. Suchman, *Human–Machine Reconfigurations.*

93. Ethnomethodology is the study of methods used by a particular sociocultural group in relation to their shared activities. In this case, Suchman used techniques from ethnomethodology to study people working in offices or businesses to observe the common activity of using photocopiers.

94. Suchman, *Plans and Situated Actions*, 76.

95. Robertson and Loke, "Designing Situations."

96. Suchman, *Human–Machine Reconfigurations.*

97. Robertson and Loke, "Designing Situations."

98. Larssen et al., "Understanding Movement as Input for Interaction"; and Loke, "Understanding Movement for Interaction Design."

99. Robertson and Loke, "Designing Situations."

100. Robertson and Loke, "Designing Situations," 7.

101. Harrison et al., "The Three Paradigms of HCI," 7.

102. Harrison et al., "The Three Paradigms of HCI," 7.

103. Gaver et al., "Ambiguity as a Resource for Design."

104. Fallman, "In Romance with the Materials of Mobile Interaction," 107.

105. Gaver et al., "Ambiguity as a Resource for Design."

106. Harrison et al., "The Three Paradigms of HCI."

107. Hummels and Van Dijk, "Seven Principles to Design for Embodied Sensemaking."

108. Hummels and van Dijk, "Seven Principles to Design for Embodied Sensemaking."

109. Harrison et al., "The Three Paradigms of HCI."

110. Harrison et al., "The Three Paradigms of HCI."

111. Hartmann et al., "Implementing Expressive Gesture Synthesis."

112. Dourish, "Epilogue," 2:2.

113. Klemmer et al., "How Bodies Matter."

114. Underkoffler and Ishii, "Illuminating Light."

115. Klemmer et al., "How Bodies Matter."

116. Djajadiningrat et al., "Easy Doesn't Do It."

117. Kirsh, "Embodied Cognition and the Magical Future of Interaction Design."

118. Sellen and Harper, *The Myth of the Paperless Office.*

119. Klemmer et al., "How Bodies Matter."

120. Svanæs, "Interaction Design for and with the Lived Body," 3.

121. Fallman, "The New Good," 1.

122. Fallman, "The New Good," 3.

123. Ihde et al., *Technics and Praxis*, cited in Fallman, "The New Good."

124. Fallman, "In Romance with the Materials of Mobile Interaction."

125. Kirsh, "Embodied Cognition and the Magical Future of Interaction Design."

126. Stewart et al., "Enaction and Engineering."

127. Culkin, "A Schoolman's Guide to Marshall McLuhan," 70.

128. Fallman, "The New Good."

129. van Dijk, "Cognition Is Not What It Used to Be."

130. Gibson, *The Ecological Approach to Visual Perception.*

131. Stewart et al., "Enaction and Engineering."

132. McGann, "Perceptual Modalities."

133. Harrison et al., "The Three Paradigms of HCI."

Chapter 3

1. Sheets-Johnstone, "Maxine Sheets-Johnstone."

2. Loke and Robertson, "Moving and Making Strange."

3. Sheets-Johnstone, "Maxine Sheets-Johnstone," 2.

4. Kay, *Doing with Images Makes Symbols.*

5. For a recent review, the reader is referred to Uchida and Delp's 2021 book, *Biomechanics of Movement.*

6. See, e.g., Prilutsky and Zatsiorsky, "Tendon Action of Two-Joint Muscles."

7. Dimitriou, "Enhanced Muscle Afferent Signals During Motor Learning."

8. Yang et al., "Proprioception Assessment."

9. Cole, *Pride and a Daily Marathon.*

10. Rosenblum et al., "The Supramodal Brain."

11. Auvray et al., "There Is Something Out There."

12. Lashley, "Integrative Functions of the Cerebral Cortex."

13. Bernstein, *The Co-ordination and Regulation of Movements.*

14. Bernstein, "A New Method of Mirror Cyclographie," cited in Latash, *Synergy.*

15. Latash, "The Bliss (Not the Problem) of Motor Abundance (Not Redundancy)."

16. Cited in Turvey et al., "The Bernstein Perspective."

17. Berthoz, *The Brain's Sense of Movement*, 140.

18. Viviani and Flash, "Minimum-Jerk, Two-Thirds Power Law, and Isochrony."

19. Berthoz, *The Brain's Sense of Movement*; and Bernstein, *The Co-ordination and Regulation of Movements*.

20. Viviani and Stucchi, "Biological Movements Look Uniform."

21. Card et al., "Evaluation of Mouse."

22. Fitts, "The Information Capacity of the Human Motor System."

23. Shannon and Weaver, *The Mathematical Theory of Communication*.

24. Card et al., "Evaluation of Mouse."

25. Bootsma et al., "Behind Fitts' Law."

26. Mottet et al., "Two-Handed Performance."

27. Bootsma et al., "Behind Fitts' Law."; and Mottet and Bootsma, "The Dynamics of Goal-Directed Rhythmical Aiming."

28. Mottet et al., "Two-Handed Performance."

29. Lee, "How Movement Is Guided."

30. McBeath et al., "How Baseball Outfielders Determine Where to Run."

31. Craig et al., "Optic Variables Used to Judge Future Ball Arrival Position."

32. Shaffer et al., "How Dogs Navigate to Catch Frisbees."

33. Lee and Reddish, "Plummeting Gannets."

34. Gibson, *The Ecological Approach to Visual Perception*.

35. Lee et al., "Guiding Contact by Coupling the Taus of Gaps."

36. Card et al., "Evaluation of Mouse."

37. Rodger et al., "Qualitative and Quantitive Assessment of a Virtual Bowed String Instrument."

38. Kuo, "The Six Determinants of Gait."

39. Hausdorff, "Gait Dynamics in Parkinson's Disease."

40. Young et al., "Auditory Observation of Stepping Actions."

41. Ojeda and Borenstein, "Non-GPS Navigation."

42. Lederman and Klatzky, "Hand Movements."

43. Zhu and Bingham, "Learning to Perceive the Affordance."

44. Hausdorff et al., "Interaction Between Cognition and Gait."

45. Wanderley et al., "The Musical Significance of Clarinetists' Ancillary Gestures"; and Nusseck and Wanderley, "Music and Motion."

46. Rodger et al., "Temporal Guidance of Musicians' Performance Movement." See also section 4.3.6.

47. Wanderley et al., "The Musical Significance of Clarinetists' Ancillary Gestures."

48. Hajdu, "Embodiment and Disembodiment."

49. Hudlicka, "Affective Game Engines."

50. Hausdorff, "Gait Dynamics in Parkinson's Disease"

51. Goel et al., "GripSense."

52. Ruggieri and Katsnelson, "An Analysis of a Performance."

53. Ruggieri and Katsnelson, "An Analysis of a Performance."

54. Bernstein, *The Co-ordination and Regulation of Movements.*

55. Rodger et al., "Temporal Guidance of Musicians' Performance Movement."

56. Picard, *Affective Computing.*

57. Dittrich et al., "Perception of Emotion from Dynamic Point-Light Displays."

58. Boehner, "How Emotion Is Made and Measured."

59. Nusseck and Wanderley, "Music and Motion."

60. Isbister and Mueller, "Guidelines for the Design of Movement-Based Games."

61. Johansson, "Visual Perception of Biological Motion."

62. Dittrich et al., "Perception of Emotion from Dynamic Point-Light Displays."

63. Runeson and Frykholm, "Visual Perception of Lifted Weight."

64. Michaels and Oudejans, "The Optics and Actions of Catching Fly Balls." See also chapter 5 in this book.

65. Dessing and Craig, "Bending It like Beckham."

66. Jacobs et al., "Learning to Perceive the Relative Mass of Colliding Balls."

67. Benford et al., "Expected, Sensed, and Desired."

68. Fleisig et al., "Kinematic and Kinetic Comparison."

69. Knudson and Morrison, *Qualitative Analysis of Human Movement.*

70. Rodger et al., "Temporal Guidance of Musicians' Performance Movement."

71. Chagnon et al., "On the Use of Laban-Bartenieff Techniques."

72. O'Modhrain et al., "Qualitative and Quantitative Assessment."

73. Schiphorst, "A Case Study of Merce Cunningham's Use."

74. Fdili Alaoui et al., "Strategies for Embodied Design."

75. Hamburg, "Coaching Athletes Using Laban Movement Analysis," 34.

76. Loke et al., "Understanding Movement for Interaction Design."

77. Jensen, "A Physical Approach to Tangible Interaction Design."

78. Ward, *Effortful Interaction.*

79. Harrison et al., "The Three Paradigms of HCI."

80. Klooster et al., "Design Education Moves."

81. Loke and Robertson, "Moving and Making Strange."

82. Fogtmann et al., "Kinesthetic Interaction."

83. Klooster et al., "Design Education Moves."

84. Hummels et al., "Move to Get Moved."

85. van Dijk and Hummels, "Designing for Embodied Being-in-the-World." See also section 6.5.7.

86. Höök et al., "Soma Design and Politics of the Body."

87. Höök, *Designing with the Body.*

88. The Feldenkrais Method, named for Moshé Feldenkrais, is a movement practice that proports to restore connectivity between the motor cortex and the body, enhancing the quality of movement and improving one's sense of well-being. For further details, see Kaetz, *Making Connections.*

89. van Dijk and Hummels, "Designing for Embodied Being-in-the-World."

90. Höök, *Designing with the Body.*

91. Höök et al., "Soma Design and Politics of the Body," 25.

92. Loke and Robertson, "Moving and Making Strange."

93. Loke and Robertson, "Moving and Making Strange."

94. Höök et al., "Soma Design and Politics of the Body," 25.

95. Broughton and Stevens, "Analyzing Expressive Qualities in Movement and Stillness."

96. Djajadiningrat et al., "Easy Doesn't Do It."

97. Benford et al., "Expected, Sensed, and Desired."

98. Sheets-Johnstone, *The Primacy of Movement.*

Chapter 4

1. Heidegger, *Being and Time.*

2. Winograd and Flores, *Understanding Computers and Cognition.*

3. Plater, *Only a Matter of Time.*

4. Knowles, "I'm So Sorry, Says Inventor of Endless Online Scrolling."

5. Zahrai et al., "Either You Control Social Media or Social Media Controls You."

6. Adgate, "Agencies Agree."

7. See, e.g., Baldassarre, "Council Post."

8. Beus, "Why (Almost) Everything You Knew."

9. Jönsson, *Ten Thoughts About Time.*

10. Dawkins, "Why the Universe Seems so Strange."

11. Katsimbri, "The Biology of Normal Bone Remodelling."

12. Gallagher and Zahavi, *The Phenomenological Mind.*

13. Edmund Husserl, "The Phenomenology of Internal Time-Consciousness," 30.

14. Zahavi, "Inner (Time-) Consciousness," 320.

15. Husserl, *Phenomenological Psychology.*

16. Gallagher and Zahavi, "Primal Impression and Enactive Perception," 97.

17. Gallagher and Zahavi, "Primal Impression and Enactive Perception."

18. Husserl, *Phänomenologische Psychologie.*

19. Stetson et al., "Does Time Really Slow Down During a Frightening Event?"

20. Wittmann, "Embodied Time."

21. James, *The Principles of Psychology*, 609.

22. Pockett, "How Long Is 'Now'?"

23. Pockett, "How Long Is 'Now'?"

24. Gallagher and Zahavi, *The Phenomenological Mind*, 79.

25. Pockett, "How Long Is 'Now'?"

26. Gallagher and Zahavi, *The Phenomenological Mind.*

27. Pöppel, *Mindworks,* 92. It is beyond the scope of this chapter to delve into the structure of memory and the ways in which memories are reshaped and reinterpreted as a function of the contexts in which they are recalled. For a review, see Baddeley, *Essentials of Human Memory.*

28. Varela, "The Specious Present."

29. Gallagher, *Action and Interaction.*

30. Varela, "The Specious Present"; and Gallagher, *Action and Interaction.*

31. Gallagher, *Action and Interaction.*

32. Varela, "The Specious Present"; and Gallagher, *Action and Interaction.*

33. See Gallagher, *Action and Interaction.*

34. Gallagher, *Action and Interaction.*

35. Gallagher, *Action and Interaction,* 35.

36. Gallagher and Zahavi, "Primal Impression and Enactive Perception."

37. Engbert and Wohlschläger, "Intentions and Expectations in Temporal Binding."

38. Farrer et al., "The Time Windows of the Sense of Agency."

39. Gallagher and Zahavi, "Primal Impression and Enactive Perception."

40. Gallagher and Zahavi, "Primal Impression and Enactive Perception," 84.

41. Gallagher and Zahavi, "Primal Impression and Enactive Perception," 98.

42. Gallagher and Zahavi, "Primal Impression and Enactive Perception," 85.

43. Gallagher and Zahavi, "Primal Impression and Enactive Perception."

44. Sheets-Johnstone, *The Primacy of Movement,* 133.

45. Sheets-Johnstone, *The Primacy of Movement,* 133.

46. Sheets-Johnstone, *The Primacy of Movement,* 135.

47. Husserl, *Phenomenological Psychology.*

48. Heidegger, *Sein und Zeit.*

49. Seow, *Designing and Engineering Time,* chap. 3.

50. Brebner and Welford, "Introduction: An Historical Background Sketch."

51. Seow, *Designing and Engineering Time*, chap. 3.

52. Seow, "Information Theoretic Models of HCI."

53. Brysbaert, "How Many Words Do We Read per Minute?"

54. Oulasvirta et al., "Interaction in 4-Second Bursts."

55. Seow, *Designing and Engineering Time*, chap. 3.

56. Shneiderman, "Response Time and Display Rate," 266.

57. Miller, "Response Time in Man–Computer Conversational Transactions."

58. Fitts, "The Information Capacity of the Human Motor System."

59. Fallman and Waterworth, "Dealing with User Experience," 1.

60. Fallman, "The New Good." See also the discussion in chapter 2.

61. Shin and Ramanathan, "Real-Time Computing."

62. Alvarez Igarzábal, *Time and Space in Video Games*.

63. Alvarez Igarzábal, *Time and Space in Video Games*, 42.

64. West, "Programming Responsiveness," cited in Alvarez Igarzábal, *Time and Space in Video Games*.

65. Alvarez Igarzábal, *Time and Space in Video Games*, 42.

66. Wessel and Wright, "Problems and Prospects."

67. Kudrle et al., "Fingerprinting for Solving A/V Synchronization Issues."

68. Carôt and Werner, "Fundamentals and Principles of Musical Telepresence."

69. Chew et al., "Musical Interaction at a Distance."

70. Chew et al., "Musical Interaction at a Distance."

71. Carôt and Werner, "Fundamentals and Principles of Musical Telepresence."

72. Chafe, Cáceres, and Gurevich, "Effect of Temporal Separation on Synchronization in Rhythmic Performance."

73. H. Cheston, personal communication with author, June 23, 2022.

74. Chafe, "Tapping into the Internet."

75. Stauffert et al., "Latency and Cybersickness."

76. Chen et al., "Virtual Reality over Wireless Networks."

77. Seow, "Information Theoretic Models of HCI."

78. Cummings and Bailenson, "How Immersive Is Enough?"

79. Van Damme et al., "Human-Centric Quality Management."

80. Wechsung and De Moor, "Quality of Experience Versus User Experience."

81. Gillespie and Cutkosky, "Stable User-Specific Haptic Rendering."

82. Van Damme et al., "Human-centric Quality Management."

83. Kasahara et al., "First Person Omnidirectional Video."

84. Waltemate, "The Impact of Latency on Perceptual Judgments."

85. Waltemate, "The Impact of Latency on Perceptual Judgments."

86. Rohde et al., "Effects Can Precede Their Cause."

87. Imaizumi and Asai, "Dissociation of Agency."

88. Jörg et al., "How Responsiveness Affects Players' Perception."

89. Samaraweera et al., "Latency and Avatars in Virtual Environments."

90. Waltemate, "The Impact of Latency on Perceptual Judgments."

91. Marucci et al., "The Impact of Multisensory Integration."

92. 3D Systems, "Phantom Premium—Haptic Devices," https://www.3dsystems.com/haptics-devices/3d-systems-phantom-premium.

93. Patney et al., "Towards Foveated Rendering."

94. Fitts, "The Information Capacity of the Human Motor System."

95. Gallagher and Zahavi, "Primal Impression and Enactive Perception."

96. Warren and Verbrugge, "Auditory Perception of Breaking."

97. For a more detailed introduction to this field, see Neuhoff, "Ecological Psychoacoustics."

98. Gaver, "How Do We Hear in the World?"; and Gaver, "What in the World Do We Hear?"

99. Jensenius, *Sound Actions*.

100. Rath and Rocchesso, "Continuous Sonic Feedback."

101. Visell et al., "Continuous Auditory and Tactile Interaction Design."

102. Hallnäs and Redström, "From Use to Presence."

103. Hallnäs and Redström, "From Use to Presence."

104. Essl and O'Modhrain, "An Enactive Approach."

105. Visell et al., "Continuous Auditory and Tactile Interaction Design."

106. Williamson et al., "Shoogle."

107. Crease and Brewster, "Making Progress with Sounds."

108. Cabe and Pittenger, "Human Sensitivity to Acoustic Information."

109. Gallagher and Zahavi, *The Phenomenological Mind*.

110. Lee, "Guiding Movement by Coupling Taus."

111. Gallagher and Zahavi, *The Phenomenological Mind*; and Gallagher and Zahavi, "Primal Impression and Enactive Perception."

112. Gallagher and Zahavi, *The Phenomenological Mind*; and Gallagher and Zahavi, "Primal Impression and Enactive Perception."

113. Lee, "Guiding Movement by Coupling Taus."

114. Rodger et al., "Temporal Guidance of Musicians' Performance Movement," 2.

115. Lee et al., "Steering by Echolocation."

116. Lee et al., "Steering by Echolocation"; and Delafield-Butt et al., "A Perception–Action Strategy for Hummingbirds."

117. Lee et al., "Regulation of Gait in Long Jumping."

118. Kayed and Van der Meer, "A Longitudinal Study of Prospective Control."

119. Craig et al., "Guiding the Swing in Golf Putting"; and Schögler et al., "TauG-Guidance of Transients."

120. Iqbal and Bailey, "Leveraging Characteristics of Task Structure."

121. Watson et al., "Judging the 'Passability' of Dynamic Gaps."

122. Watson et al., "Judging the 'Passability' of Dynamic Gaps."

123. Dessing and Craig, "Bending It like Beckham."

124. Craig, "Understanding Perception and Action in Sport," 5.

125. Gallagher and Zahavi, *The Phenomenological Mind*; and Gallagher and Zahavi, "Primal Impression and Enactive Perception."

Chapter 5

1. Winston Churchill speech delivered to the House of Commons, October 28, 1943, cited in Blesser and Salter, *Spaces Speak, Are You Listening?*

2. Blesser and Salter, *Spaces Speak, Are You Listening?*

3. Blesser and Salter, *Spaces Speak, Are You Listening?*

4. Gibson, *The Senses Considered as Perceptual Systems*; and Gibson, *The Ecological Approach to Visual Perception.*

5. Ackerman, *A Natural History of the Senses*, 220.

6. Neidhardt et al., "Perceptual Matching of Room Acoustics."

7. Blesser and Salter, *Spaces Speak, Are You Listening?*

8. Campbell, *Past, Space, and Self*, 5.

9. Gallagher, *A Companion to Phenomenology and Existentialism.*

10. Gallagher, *A Companion to Phenomenology and Existentialism.*

11. Gallagher, *A Companion to Phenomenology and Existentialism.*

12. Merleau-Ponty, *Phenomenology of Perception*, 102.

13. Merleau-Ponty, *Phenomenology of Perception*, 91.

14. Gallagher, *A Companion to Phenomenology and Existentialism*, 350.

15. Gallagher, *A Companion to Phenomenology and Existentialism*, 354.

16. Campbell, *Past, Space, and Self.*

17. Campbell, *Past, Space, and Self.*

18. Klatzky, "Allocentric and Egocentric Spatial Representations."

19. Brain, "Visual Orientation."

20. Rizzolatti et al., "Afferent Properties of Periarcuate Neurons."

21. Serino, "Peripersonal Space (PPS) as a Multisensory Interface."

22. For a review of this research, see Serino, "Peripersonal Space (PPS) as a Multisensory Interface."

23. Iriki et al., "Coding of Modified Body Schema."

24. Iriki et al., "Self-Images in the Video Monitor."

25. Witt et al., "Tool Use Affects Perceived Distance."

26. Farnè et al., "Dynamic Size-Change of Peri-Hand Space."

27. Farnè et al., "Dynamic Size-Change of Peri-Hand Space."

28. Serino, "Peripersonal Space (PPS) as a Multisensory Interface."

29. Serino, "Peripersonal Space (PPS) as a Multisensory Interface."

30. Lin et al., "The Influence of Perceptual–Motor Variability." See also section 5.4.3 for further discussion of reaching in virtual environments.

31. Cutting and Vishton, "Perceiving Layout and Knowing Distances."

32. Mine et al., "Moving Objects in Space."

33. Cutting and Vishton, "Perceiving Layout and Knowing Distances."

34. Ängeslevä et al., "Body Mnemonics Portable Device Interaction Design Concept."

35. Berthoz, *The Brain's Sense of Movement*, 48.

36. Lewald and Karnath, "Vestibular Influence on Human Auditory Space Perception."

37. Turvey and Carello, "Obtaining Information by Dynamic (Effortful) Touching."

38. Runeson, "On the Possibility of 'Smart' Perceptual Mechanisms."

39. Turvey and Carello, "Obtaining Information by Dynamic (Effortful) Touching."

40. Runeson, "On the Possibility of 'Smart' Perceptual Mechanisms."

41. Zenner and Krüger, "Shifty."

42. Gibson, *The Senses Considered as Perceptual Systems*.

43. Fajen and Phillips, "Spatial Perception and Action."

44. Dessing and Craig, "Bending It like Beckham."

45. von Uexküll, *A Foray into the Worlds of Animals and Humans*.

46. For further discussion, see Turvey, *Lectures on Perception*.

47. Fajen and Phillips, "Spatial Perception and Action."

48. Warren and Whang, "Visual Guidance of Walking Through Apertures."

49. Warren, "Action-Scaled Information for the Visual Control of Locomotion."

50. Warren and Whang, "Visual Guidance of Walking Through Apertures."

51. Warren and Whang, "Visual Guidance of Walking Through Apertures."

52. Bhargava et al., "Revisiting Affordance Perception."

53. See Knapp and Loomis, "Limited Field of View of Head-Mounted Displays."

54. For this study, the virtual environment was rendered at ninety frames per second using the Unity 3D game engine and was displayed using an HTC Vive head-mounted display, with a 110-degree horizontal field of view.

55. Fajen, "Affordance Perception and the Visual Control of Locomotion."

56. Lee, "Visuo-Motor Coordination in Space–Time."

57. Fath and Fajen, "Static and Dynamic Visual Information."

58. Snyder and Cinelli, "Sensory Conflict Alters Visual Perception."

59. Watson et al., "Judging the 'Passability' of Dynamic Gaps." See also section 4.5.2.

60. Bootsma, "The (Current) Future Is Here!"

61. Warren and Whang, "Visual Guidance of Walking Through Apertures."

62. Fath and Fajen, "Static and Dynamic Visual Information."

63. Fajen and Matthis, "Direct Perception of Action-Scaled Affordance"; and Fajen and Matthis, "Visual and Non-Visual Contributions."

64. Fajen, "Affordance Perception and the Visual Control of Locomotion."

65. Fajen and Matthis, "Direct Perception of Action-Scaled Affordance."

66. Fajen, "Affordance Perception and the Visual Control of Locomotion," 89.

67. Fajen and Matthis, "Direct Perception of Action-Scaled Affordance."

68. Fajen and Matthis, "Visual and Non-Visual Contributions."

69. Fajen and Matthis, "Visual and Non-Visual Contributions."

70. Stefanucci et al., "Distances Appear Different on Hills."

71. Bhalla and Proffitt, "Visual–Motor Recalibration."

72. Fajen and Matthis, "Visual and Non-Visual Contributions."

73. Chardenon et al., "The Perceptual Control of Goal-Directed Locomotion."

74. Fajen and Phillips, "Spatial Perception and Action."

75. Cutting and Vishton, "Perceiving Layout and Knowing Distances."

76. Fink et al., "Catching Fly Balls in Virtual Reality."

77. Fink et al., "Catching Fly Balls in Virtual Reality."

78. Lee, "A Theory of Visual Control of Braking."

79. Bootsma, "The (Current) Future Is Here!"

80. Fajen and Phillips, "Spatial Perception and Action."

81. Choi and Mark, "Scaling Affordances."

82. Carello et al., "Visually Perceiving What Is Reachable."

83. Seifert et al., "The Perception of Nested Affordances."

84. Choi and Mark, "Scaling Affordances."

85. Wagman and Stoffregen, "It Doesn't Add Up."

86. Wagman and Morgan, "Nested Prospectivity in Perception."

87. Wagman and Morgan, "Nested Prospectivity in Perception."

88. Seifert et al., "The Perception of Nested Affordances."

89. Witt et al., "Tool Use Affects Perceived Distance."

90. Osiurak et al., "Tool Use and Perceived Distance."

91. Lin et al., "The Influence of Perceptual–Motor Variability."

92. Fink et al., "The Neural Basis of Vertical and Horizontal Line Bisection Judgments."

93. Ferroni et al., "The Remapping of Peripersonal Space."

94. Sengül et al., "Extending the Body to Virtual Tools."

95. Maravita et al., "Multisensory Integration and the Body Schema."

96. Dewez et al., "Towards 'Avatar-Friendly' 3D Manipulation Techniques."

97. Ebrahimi et al., "Effects of Anthropomorphic Fidelity."

98. Stroffregan and Pittenger, "Human Echolocation as a Basic Form of Perception and Action."

99. Carlile et al., "Spectral Information in Sound Localization."

100. Jourdain, *Diderot's Early Philosophical Works.*

101. Kolarik et al., "A Summary of Research Investigating Echolocation Abilities."

102. Bilsen, "Repetition Pitch."

103. Gordon and Rosenblum, "Perception of Sound-Obstructing Surfaces Using Body-Scaled Judgments."

104. Gibson, *The Ecological Approach to Visual Perception.*

105. Menounou et al., "Edge Wave on Axis Behind an Aperture or Disk."

106. Lee et al., "Common Principle of Guidance."

107. Stroffregan and Pittenger, "Human Echolocation as a Basic Form of Perception and Action."

108. Riehm et al., "Virtual Auditory Aperture Passability."

109. Riehm et al., "Virtual Auditory Aperture Passability."

110. Jacobs and Michaels, "Direct Learning."

111. Lee et al., "A Roadside Simulation of Road Crossing for Children."

112. Hassan, "Are Normally Sighted, Visually Impaired, and Blind Pedestrians Accurate and Reliable at Making Street Crossing Decisions?"

113. Emerson and Sauerburger, "Detecting Approaching Vehicles at Streets with No Traffic Control."

114. For an example of the literature on this subject, see Knoblauch et al., "Field Studies of Pedestrian Walking Speed."

115. Koutsoklenis and Papadopoulos, "Auditory Cues Used for Wayfinding."

116. Gordon and Rosenblum, "Effects of Intrastimulus Modality Change."

117. Vernat and Gordon, "Indirect Interception Actions."

118. Mieda and Kokubu, "Blind Footballers Direct Their Head."

119. The International Blind Sports Association is responsible for approving adaptive equipment used in all Paralympic sports, including blind football. See https://handilifesport.com/product/official-blind-football/.

120. Gibson, *The Senses Considered as Perceptual Systems.*

121. Haggard et al., "Embodiment and Disembodiment."

122. Stoffregan and Pittenger, "Human Echolocation as a Basic Form of Perception and Action."

123. Bernstein, *The Co-ordination and Regulation of Movements.*

124. Withagen and Michaels, "The Role of Feedback Information."

125. Chihak et al., "Synchronizing Self and Object Movement."

126. Wagman et al., "Attunement, Calibration, and Exploration."

127. Witt, "Action's Influence on Spatial Perception," 1.

128. Chow et al., "Nonlinear Pedagogy"; and Chow et al., "Coordination Changes in a Discrete Multi-Articular Action."

129. Kelly et al., "More than Just Perception–Action Recalibration."

130. Ponto et al., "Perceptual Calibration for Immersive Display Environments."

131. Leyrer et al., "The Influence of Eye Height and Avatars."

132. Huang et al., "Using Audio Reverberation."

133. Ebrahimi et al., "An Empirical Evaluation of Visuo-Haptic Feedback."

134. Altenhoff et al., "Effects of Calibration to Visual and Haptic Feedback."

135. Singapogu et al., "Perceptually Salient Haptic Rendering."

136. Slater, "Place Illusion and Plausibility."

137. Giudice, "Navigating Without Vision," 25.

138. Giudice, "Navigating Without Vision," 25.

139. Giudice, "Navigating Without Vision," 25.

140. Long and Giudice, "Establishing and Maintaining Orientation for Mobility."

141. Giudice, "Navigating Without Vision," 26.

142. Magnusson and Rassmus-Gröhn, "A Dynamic Haptic-Audio Traffic Environment."

143. Nam et al., "Haptic User Interfaces for the Visually Impaired."

144. Chai et al., "Hungry Cat"; and Kreimeier and Götzelmann, "Two Decades of Touchable and Walkable Virtual Reality."

145. Slater, "Place Illusion and Plausibility."

146. Siu et al., "Virtual Reality Without Vision."

147. Siu et al., "Virtual Reality Without Vision."

148. Montello, "Scale and Multiple Psychologies of Space."

149. Giudice, "Navigating Without Vision."

150. Giudice, "Navigating Without Vision."

151. Siu et al., "Virtual Reality Without Vision."

152. Gallagher and Zahavi, *The Phenomenological Mind.*

Chapter 6

1. Ishii and Ullmer, "Tangible Bits."

2. Collins Dictionary.com, "object," accessed December 20, 2024, https://www.collinsdictionary.com/us/dictionary/english/object.

3. Locke, "An Essay Concerning Human Understanding."

4. Locke, "An Essay Concerning Human Understanding."

5. Gallagher, *How the Body Shapes the Mind.*

6. Merleau-Ponty, *Phenomenology of Perception,* 223.

7. For a discussion of this, see Gallagher, *How the Body Shapes the Mind,* chap. 7.

8. Atkinson, *The Developing Visual Brain.*

9. Gallagher, *How the Body Shapes the Mind.*

10. See, e.g., Ferretti, "Two Visual Systems in Molyneux Subjects."

11. O'Regan and Noë, "A Sensorimotor Account of Vision and Visual Consciousness."

12. Gallagher, *How the Body Shapes the Mind.*

13. Gallagher, *How the Body Shapes the Mind,* 171.

14. Heidegger, *Sein und Zeit.*

15. Kaufer and Chemero, *Phenomenology.*

16. Verbeek, "Don Ihde."

17. Ihde, *Technology and the Lifeworld.*

18. Kirsh, "Embodied Cognition and the Magical Future of Interaction Design."

19. Culkin, "A Schoolman's Guide to Marshall McLuhan," 70.

20. Malafouris, *How Things Shape the Mind.*

21. Ihde and Malafouris, "*Homo faber* Revisited."

22. Verbeek, "Don Ihde."

23. Malafouris, "At the Potter's Wheel."

24. Aydin et al., "Technological Environmentality," 330.

25. Ihde, *Technology and the Lifeworld.*

26. Verbeek, "Don Ihde."

27. Gibson, *The Senses Considered as Perceptual Systems.*

28. Gibson, "Observations on Active Touch," 477.

29. Gibson, "Observations on Active Touch."

30. Lederman et al., "Cognitive Salience of Haptic Object Properties."

31. Lederman and Klatzky, "Hand Movements."

32. Massie and Salisbury, "The Phantom Haptic Interface."

33. O'Modhrain et al., "Designing Media for Visually-Impaired Users."

34. Lederman and Klatzky, "Hand Movements."

35. Noë, *Action in Perception*, 72–73.

36. Mossio and Taraborelli, "Action-Dependent Perceptual Invariants," 1331.

37. Gibson, *The Senses Considered as Perceptual Systems*, 3.

38. Solomon and Turvey, "Haptically Perceiving the Distances."

39. Alberti, *On Painting*.

40. Norman, "Affordance, Conventions, and Design."

41. For a comprehensive review of interpretations of theories of affordances within HCI, see Kaptelinin, *Affordances and Design*.

42. van Dijk, "Designing for Embodied Being-in-the-World,"12.

43. Ihde, *Technology and the Lifeworld*.

44. Kaptelinin and Nardi, "Affordances in HCI," 12.

45. Heidegger, *Sein und Zeit*.

46. Loomis, "Distal Attribution and Presence."

47. Dessing and Craig, "Bending It like Beckham."

48. van Beek et al., "Static Weight Perception."

49. Marshall et al., "Sensory Alignment in Immersive Entertainment."

50. Martin et al., "Multimodality in VR."

51. Lim et al., "A Systematic Review of Weight Perception."

52. Lederman and Klatzky, "Hand Movements."

53. van Beek et al., "Static Weight Perception."

54. Charpentier, "Analyse experimentale."

55. Harshfield and DeHardt, "Weight Judgment as a Function of Apparent Density of Objects."

56. Ellis and Lederman, "The Material–Weight Illusion Revisited."

57. Rinkenauer et al., "The Surface–Weight Illusion."

58. Lim et al., "A Systematic Review of Weight Perception."

59. van Beek et al., "Static Weight Perception."

60. Singapogu et al., "Perceptually Salient Haptic Rendering."

61. Hirao and Kawai, "Augmented Cross-Modality."

62. Lee et al., "Force Arrow 2."

63. Hirao and Kawai, "Augmented Cross-Modality."

64. Herbst and Stark, "Comparing Force Magnitudes by Means of Vibro-Tactile, Auditory, and Visual Feedback."

65. Singapogu et al., "Perceptually Salient Haptic Rendering."

66. Withagen and Michaels, "The Role of Feedback Information."

67. Mendes et al., "A Survey on 3D Virtual Object Manipulation."

68. For further information about Microsoft XBox, see https://www.xbox.com/en-US/accessories/controllers/xbox-wireless-controller.

69. Poupyrev et al., "The Go-Go Interaction Technique."

70. Araújo et al., "Mockup Builder."

71. See, e.g., Moehring and Froehlich, "Effective Manipulation of Virtual Objects."

72. Daniel Mendes et al., "The Benefits of DOF Separation."

73. Ishii and Ullmer, "Tangible Bits."

74. Uva et al., "Evaluating the Effectiveness of Spatial Augmented Reality."

75. Shavel, "MR Is Leaving AR in the Dust."

76. Baranyi et al., "Introducing the Concept of Internet of Digital Reality, Part I," 226.

77. Stewart et al., "Enaction and Engineering."

78. Gallagher, *How the Body Shapes the Mind*, chap. 7.

79. Held et al., "Revisiting the Molyneux Question."

80. Pham et al., "Scale Impacts Elicited Gestures for Manipulating Holograms."

81. Chalmers, "The Virtual and the Real," 327.

82. Pham et al., "Scale Impacts Elicited Gestures for Manipulating Holograms."

83. Slater, "Place Illusion and Plausibility," 3550.

84. Bailenson, *Experience on Demand*.

85. Baranyi et al., "Introducing the Concept of Internet of Digital Reality, Part I."

86. Deshpande and Kim, "The Effects of Augmented Reality."

87. Ängeslevä and Cooper, "Last Clock by Jussi Ängeslevä."

88. Rosenberg, "Virtual Fixtures."

89. Azuma et al., "Recent Advances in Augmented Reality."

90. Gibson, *The Ecological Approach to Visual Perception.*

91. Raja and Calvo, "Augmented Reality."

92. Slater, "Place Illusion and Plausibility."

93. Shavel, "MR is Leaving AR in the Dust."

94. Malafouris, "At the Potter's Wheel."

95. Piper et al., "Illuminating Clay."

96. Raja and Calvo, "Augmented Reality."

97. Adapted from Fishkin, "A Taxonomy for and Analysis of Tangible Interfaces."

98. Hornecker and Buur, "Getting a Grip on Tangible Interaction."

99. Jacob et al., "A Tangible Interface for Organizing Information."

100. See also Djajadiningrat et al., "Easy Doesn't Do It."

101. Hornecker and Buur, "Getting a Grip on Tangible Interaction."

102. Behrens et al., "Smart Citizen Sentiment Dashboard."

103. Hornecker and Buur, "Getting a Grip on Tangible Interaction."

104. Underkoffler and Ishii, "Urp."

105. Jordà et al., "The ReacTable."

106. Hornecker and Buur, "Getting a Grip on Tangible Interaction."

107. Ishii et al., "Radical Atoms."

108. Fishkin, "A Taxonomy for and Analysis of Tangible Interfaces."

109. Underkoffler and Ishii, "Urp."

110. Piper et al., "Illuminating Clay."

111. Holmquist et al., "Token-Based Access to Digital Information."

112. Kaptelinin and Nardi, *Acting with Technology.*

113. Dourish, *Where the Action Is.*

114. Hornecker and Dünser, "Of Pages and Paddles."

115. Hornecker, "Beyond Affordance."

116. Fitzmaurice et al., "Bricks."

117. Singapogu et al., "Perceptually Salient Haptic Rendering."

118. Nilsson et al., "Haptic Proxies for Virtual Reality."

119. Scopigno et al., "Digital Fabrication Techniques."

120. Granzow and Çamcı, "Recreating a Rare Instrument."

121. Nilsson et al., "Haptic Proxies for Virtual Reality."

122. Rasmussen et al., "Shape-Changing Interfaces."

123. Togler et al., "Living Interfaces."

124. Coelho et al., "Surflex."

125. Stanley and Okamura, "Controllable Surface Haptics." It is beyond the scope of this chapter to review the technical means for accomplishing such changes, but an excellent overview can be found in Alexander et al., "Grand Challenges in Shape-Changing Interface Research."

126. Rasmussen et al., "Shape-Changing Interfaces," 743.

127. Follmer et al., "inFORM."

128. Ishii et al., "Radical Atoms."

129. Follmer, "Dynamic Physical Affordances."

130. Essl and O'Modhrain, "An Enactive Approach."

131. Lederman and Klatzky, "Hand Movements."

132. O'Modhrain and Essl, "PebbleBox and CrumbleBag."

133. van Dijk, "Designing for Embodied Being-in-the-World."

134. Baranyi et al., "Introducing the Concept of Internet of Digital Reality, Part I."

135. van Dijk and Hummels, "Designing for Embodied Being-in-the-World."

136. van Dijk, "Designing for Embodied Being-in-the-World," 1.

137. van Dijk, "Designing for Embodied Being-in-the-World," 7.

138. van Dijk, "Designing for Embodied Being-in-the-World."

139. Merleau-Ponty, *Phenomenology of Perception.*

140. van Dijk, "Designing for Embodied Being-in-the-World," 12.

141. van Dijk, "Designing for Embodied Being-in-the-World."

142. van Dijk and Hummels, "Designing for Embodied Being-in-the-World."

143. van Dijk and Hummels, "Designing for Embodied Being-in-the-World," 52.

144. van Dijk and Hummels, "Designing for Embodied Being-in-the-World."

145. Dessing and Craig, "Bending It like Beckham."

Chapter 7

1. For more on the history of the Oculus Rift, see https://tech.facebook.com/reality-labs/2021/3/five-years-of-vr-an-oral-history-from-oculus-rift-to-quest-2/. For more on the history of the HTC Vive, see https://en.wikipedia.org/wiki/HTC_Vive.

2. Hummels et al., "Move to Get Moved."

3. Höök, *Designing with the Body*.

4. Loke and Robertson, "Moving and Making Strange."

5. Gallagher and Zahavi, *The Phenomenological Mind*; and Gallagher and Zahavi, "Primal Impression and Enactive Perception."

6. van Dijk and Hummels, "Designing for Embodied Being-in-the-World."

7. De Jaegher and Di Paolo, "Participatory Sense-Making," 488.

8. van Grunsven, "Disabled Body-Minds in Hostile Environments."

9. Criado Perez, *Invisible Women*.

10. Peck et al., "DiVRsify."

11. Munafo et al., "The Virtual Reality Head-Mounted Display Oculus Rift."

12. Spiel, "Body-Positive Computing."

13. Spiel, "The Bodies of TEI."

14. Adapted from Spiel, "The Bodies of TEI."

15. Höök, *Designing with the Body*; van Dijk and Hummels, "Designing for Embodied Being-in-the-World"; and van Dijk, "Designing for Embodied Being-in-the-World."

16. Kaufer and Chemero, *Phenomenology*.

17. Kolarik et al., "A Summary of Research Investigating Echolocation Abilities."

18. van Grunsven, "Disabled Body-Minds in Hostile Environments," 6.

19. van Grunsven, "Disabled Body-Minds in Hostile Environments," 6.

20. De Jaegher, "Embodiment and Sense-Making in Autism," 9.

21. Hamraie and Fritsch, "Crip Technoscience Manifesto."

22. ISO, "Guidelines for Standards Developers," 2.

23. Persson, "Universal Design, Inclusive Design."

24. Hamraie and Fritsch, "Crip Technoscience Manifesto."

25. Brown, "Changing America's Consciousness."

26. Charlton, *Nothing About Us Without Us*.

27. Spiel et al., "Nothing About Us Without Us."

28. Hamraie and Fritche, "Crip Technoscience Manifesto," 7.

29. Mankoff et al., "Disability Studies as a Source of Critical Inquiry."

30. "NSF Event Series to Spotlight Cross-Cutting Research."

31. van Grunsven, "Disabled Body-Minds in Hostile Environments."

32. For D4EB, see van Dijk and Hummels, "Designing for Embodied Being-in-the-World"; for soma design, see Höök, *Designing with the Body*.

33. Jensenius, *Sound Actions*.

34. Glinsky, *Theremin*.

35. Jensenius, *Sound Actions*.

36. Cooper, "Very Nervous System."

37. Waisvisz, "The Hands."

38. Essl and O'Modhrain, "An Enactive Approach."

39. Resnick, *Lifelong Kindergarten*.

40. Gillian et al., "A Machine Learning Toolbox."

41. Gillian et al., "A Machine Learning Toolbox."

42. Fiebrink and Cook, "The Wekinator." On AI, see Graf and Barthet, "Mixed Reality Musical Interface."

43. Jones, dir., *The Life of Brian*.

44. Fallman, "The New Good."

Bibliography

Ackerman, Diane. *A Natural History of the Senses*. Vintage, 1991.

Adgate, Brad. "Agencies Agree; 2021 Was a Record Year for Ad Spending, with More Growth Expected in 2022." *Forbes*, December 8, 2021. https://www.forbes.com/sites/bradadgate/2021/12/08/agencies-agree-2021-was-a-record-year-for-ad-spending-with-more-growth-expected-in-2022/.

Adolph, Karen E., Marion A. Eppler, and Eleanor J. Gibson. "Crawling Versus Walking Infants' Perception of Affordances for Locomotion over Sloping Surfaces." *Child Development* 64, no. 4 (1993): 1158–1174. https://doi.org/10.2307/1131332.

Alberti, Leon Battista. *On Painting*. 1435. Reprint, Penguin Books, 1991.

Alexander, Jason, Anne Roudaut, Jürgen Steimle, et al. "Grand Challenges in Shape-Changing Interface Research." In *Proceedings of the 2018 CHI Conference on Human Factors in Computing Systems*. ACM, 2018. https://doi.org/10.1145/3173574.3173873.

Almeida, Jorge, Bradford Z. Mahon, Ken Nakayama, and Alfonso Caramazza. "Unconscious Processing Dissociates Along Categorical Lines." *Proceedings of the National Academy of Sciences* 105, no. 39 (2008): 15214–15218. https://doi.org/10.1073/pnas.0805867105.

Altenhoff, Bliss M., Phillip E. Napieralski, Lindsay O. Long, et al. "Effects of Calibration to Visual and Haptic Feedback on Near-Field Depth Perception in an Immersive Virtual Environment." *Proceedings of the ACM Symposium on Applied Perception*. ACM, 2012. https://doi.org/10.1145/2338676.2338691.

Alvarez Igarzábal, Federico. *Time and Space in Video Games: A Cognitive-Formalist Approach*. Transcript Verlag, 2019. https://doi.org/10.1515/9783839447130.

Ängeslevä, Jussi, Ian Oakley, Stephen Hughes, and Sile O'Modhrain. "Body Mnemonics Portable Device Interaction Design Concept." *Proceedings of UIST*. ACM, 2003. https://uist.acm.org/archive/adjunct/2003/pdf/posters/p3-angesleva.pdf.

Ängeslevä, Jussi, and Ross Cooper. "Last Clock by Jussi Ängeslevä." Jussi Ängeslevä: Media Art and Design. Accessed January 5, 2025. http://angesleva.iki.fi/art/last/.

Araújo, Bruno R., Géry Casiez, Joaquim A. Jorge, and Martin Hachet. "Mockup Builder: 3D Modeling on and Above the Surface." *Computers and Graphics* 37, no. 3 (2013): 165–178. https://doi.org/10.1016/j.cag.2012.12.005.

Atkinson, Janette. *The Developing Visual Brain*. Oxford University Press, 2000.

Auvray, Malika, Sylvain Hanneton, Charles Lenay, and Kevin O'Regan. "There Is Something Out There: Distal Attribution in Sensory Substitution, Twenty Years Later." *Journal of Integrative Neuroscience* 4, no. 4 (2005): 505–521. https://doi.org/10.1142/s0219635205001002.

Aydin, Ciano, Margoth González Woge, and Peter-Paul Verbeek. "Technological Environmentality: Conceptualizing Technology as a Mediating Milieu." *Philosophy Technology* 32 (2019): 321–338. https://doi.org/10.1007/s13347-018-0309-3.

Azuma, Ronald, Yohan Baillot, Reinhold Behringer, Steven K. Feiner, Simon Julier, and Blair MacIntyre. "Recent Advances in Augmented Reality." *IEEE Computer Graphics and Applications* 21, no. 6 (2001): 34–47. https://doi.org/10.1109/38.963459.

Bach-y-Rita, Paul, Carter C. Collins, Frank A. Saunders, Benjamin White, and Lawrence Scadden. "Vision Substitution by Tactile Image Projection." *Nature* 221 (1969): 963–964. https://doi.org/10.1038/221963a0.

Baddeley, Alan. *Essentials of Human Memory*. Hove, 2013.

Baggs, Edward, and Anthony Chemero. "The Third Sense of Environment." In *Perception as Information Detection: Reflections on Gibson's Ecological Approach to Visual Perception*, edited by Jeffrey B. Wagman and Julia J. C. Blau. Routledge, 2019.

Bailenson, Jeremy. *Experience on Demand: What Virtual Reality Is, How It Works, and What It Can Do*. W. W. Norton & Company, 2016.

Baldassarre, Rocco. "Council Post: A Guide to Google Ads Smart Bidding." *Forbes*, January 7, 2022. https://www.forbes.com/sites/forbesagencycouncil/2022/01/07/a-guide-to-google-ads-smart-bidding/.

Baranyi, Péter, Ádám Csapó, Tamás Budai, and György Wersényi. "Introducing the Concept of Internet of Digital Reality, Part I." *Acta Polytechnica Hungarica* 18, no. 7 (2021): 225–240. https://doi.org/10.12700/APH.18.7.2021.7.12.

Behrens, Moritz, Nina Valkanova, Ava Fatah gen. Schieck, and Duncan P. Brumby. "Smart Citizen Sentiment Dashboard: A Case Study into Media Architectural Interfaces." In *Proceedings: 3rd ACM International Symposium on Pervasive Displays*. ACM, 2014. https://doi.org/10.1145/2611009.2611036.

Bell, Martha Ann, and Nathan A. Fox, "The Relations Between Frontal Brain Electrical Activity and Cognitive Development During Infancy." *Child Development* 63, no. 5 (1992): 1142–1163. https://doi.org/10.2307/1131523.

Benford, Steve, Holger Schnädelbach, Boriana Koleva, et al. "Expected, Sensed, and Desired: A Framework for Designing Sensing-Based Interaction." *ACM Transactions on Computer–Human Interaction (TOCHI)* 12, no. 1 (2005): 3–30. https://doi.org/10.1145/1057237.1057239.

Bennett, Mike, Kevin McCarthy, Sile O'Modhrain, and Barry Smyth. "SimpleFlow: Enhancing Gestural Interaction with Gesture Prediction, Abbreviation and Autocompletion." In *Human–Computer Interaction: INTERACT 2011. Lecture Notes in Computer Science*, Vol. 6946, edited by Pedro Campos, Nicholas Graham, Joaquim Jorge, Nuno Nunes, Philippe Palanque, and Marco Winckler. Springer, 2011. https://doi.org/10.1007/978-3-642-23774-4_47.

Bernstein, Nikolai A. *The Co-ordination and Regulation of Movements*. Pergamon Press, 1967.

Bernstein, Nikolai A. "A New Method of Mirror Cyclographie and Its Application Towards the Study of Labor Movements During Work on a Workbench." *Hygiene, Safety and Pathology of Labor* 5 (1930): 3–8.

Berthoz, Alain. *The Brain's Sense of Movement*. Harvard University Press, 2000.

Beus, Johannes. "Why (Almost) Everything You Knew About Google CTR Is No Longer Valid." Sistrix. July 14, 2020. https://www.sistrix.com/blog/why-almost-everything-you-knew-about-google-ctr-is-no-longer-valid/.

Bhalla, Mukul, and Dennis R. Proffitt. "Visual–Motor Recalibration in Geographical Slant Perception." *Journal of Experimental Psychology: Human Perception and Performance* 25, no. 4 (1999): 1076–1096. https://doi.org/10.1037//0096-1523.25.4.1076.

Bhargava, Ayush, Kathryn M. Lucaites, Leah S. Hartman, et al. "Revisiting Affordance Perception in Contemporary Virtual Reality." *Virtual Reality* 24, no. 4 (2020): 713–724. https://doi.org/10.1007/s10055-020-00432-y.

Bilsen, Frans A. "Repetition Pitch: Monaural Interaction of a Sound with the Repetition of the Same, but Phase Shifted, Sound." *Acta Acustica United with Acustica* 17, no. 5 (1966): 295–300.

Blesser, Barry, and Linda-Ruth Salter. *Spaces Speak, Are You Listening?* MIT Press, 2007.

Boehner, Kirsten, Rogério DePaula, Paul Dourish, and Phoebe Sengers. "How Emotion Is Made and Measured." *International Journal of Human–Computer Studies* 65, no. 4 (2007): 275–291. https://doi.org/10.1016/j.ijhcs.2006.11.016.

Bootsma, Reinoud J. "The (Current) Future Is Here!" *Perception* 38, no. 6 (2009): 851–858.

Bootsma, Reinoud J., Laure Fernandez, and Denis Mottet. "Behind Fitts' Law: Kinematic Patterns in Goal-Directed Movements." *International Journal of Human–Computer Studies* 61, no. 6 (2004): 811–821. https://doi.org/10.1016/j.ijhcs.2004.09.004.

Brain, W. Russell. "Visual Orientation with Special Reference to Lesions of the Right Cerebral Hemisphere." *Brain: A Journal of Neurology* 64 (1941): 244–272. https://doi.org/10.1093/brain/64.4.244.

Brebner, J. M. T., and A. T. Welford. "Introduction: An Historical Background Sketch." In *Reaction Times*, edited by A. T. Welford and J. M. T. Brebner. Academic Press, 1980.

Broughton, Mary C., and Catherine J. Stevens. "Analyzing Expressive Qualities in Movement and Stillness: Effort–Shape Analyses of Solo Marimbists' Bodily Expression." *Music Perception: An Interdisciplinary Journal* 29, no. 4 (2012): 339–357. https://doi.org/10.1525/mp.2012.29.4.339.

Brown, Steven E. "Changing America's Consciousness: A Brief History of the Independent Living Civil Rights Movement in the United States. In *The Routledge History of Disability*, edited by Roy Hanes, Ivan Brown, and Nancy E. Hansen. Routledge, 2017.

Brysbaert, Marc. "How Many Words Do We Read per Minute? A Review and Meta-Analysis of Reading Rate." *Journal of Memory and Language* 109 (2019): 104047. https://doi.org/10.1016/j.jml.2019.104047.

Cabe, Patrick A., and John B. Pittenger. "Human Sensitivity to Acoustic Information from Vessel Filling." *Journal of Experimental Psychology: Human Perception and Performance* 26, no. 1 (2000): 313–324. https://doi.org/10.1037//0096-1523.26.1.

Campbell. John. *Past, Space, and Self.* MIT Press, 1994.

Card, Stuart K., Thomas P. Moran, and Allen Newell. *The Psychology of Human–Computer Interaction*. Lawrence Erlbaum Associates, 1983.

Card, Stuart K., William K. English, and Betty J. Burr. "Evaluation of Mouse, Rate-Controlled Isometric Joystick, Step Keys, and Text Keys for Text Selection on a CRT." *Ergonomics* 31, no. 8 (1978): 601–613. https://doi.org/10.1080/00140137808931762.

Carello, Claudia, Alexis Grosofsky, Francene D. Reichel, H. Yosef Solomon, and Michael T. Turvey. "Visually Perceiving What Is Reachable." *Ecological Psychology* 1, no. 1 (1989): 27–54. https://doi.org/10.1207/s15326969eco0101_3.

Carello, Claudia, and Michael T. Turvey. "The Ecological Approach to Perception." In *The Encyclopedia of Cognitive Science*, edited by Lynn Nadel. Macmillan, 2002.

Carlile, Simon, Russell Martin, and Ken McAnally. "Spectral Information in Sound Localization." *International Review of Neurobiology* 70 (2005): 399–434. https://doi.org/10.1016/S0074-7742(05)70012-X.

Carôt, Alexander, and Christian Werner. "Fundamentals and Principles of Musical Telepresence." *Journal of Science and Technology of the Arts* 1, no. 1 (2009): 26–37. https://doi.org/10.7559/citarj.v1i1.6.

Chafe, Chris. "Tapping into the Internet as an Acoustical/Musical Medium." *Contemporary Music Review* 38, no. 4–5 (2009): 413–420. https://doi.org/10.1080/07494460903422362.

Chafe, Chris; Cáceres, Juan-Pablo; and Gurevich, Michael "Effect of Temporal Separation on Synchronization in Rhythmic Performance." *Perception* 39, no. 7 (2010): 982–992. https://doi.org/10.1068/p6465.

Chagnon, Marie-Julie, Louise Campbell, and Marcelo Wanderley. "On the Use of Laban-Bartenieff Techniques to Describe Ancillary Gestures of Clarinetists." Faculty of Music, McGill University. Accessed November 20, 2024. https://www-archive.idmil.org/_media/publications/2005/chagnon_2005_idmil.pdf.

Chai, Carmen, Bee Theng Lau, and Zheng Pan. "Hungry Cat—A Serious Game for Conveying Spatial Information to the Visually Impaired." *Multimodal Technologies and Interaction* 3, no. 1 (2019): 12. https://doi.org/10.3390/mti3010012.

Chalmers, David J. "The Virtual and the Real." *Disputatio* 9, no. 46 (2017): 309–352. https://doi.org/10.1515/disp-2017-0009.

Chardenon, A., Gilles Montagne, M. Laurent, and Reinoud J. Bootsma. "The Perceptual Control of Goal-Directed Locomotion: A Common Control Architecture for Interception and Navigation?" *Experimental Brain Research* 158 (2004): 100–108. https://doi.org/10.1007/s00221-004-1880-7.

Charpentier, A. "Analyse experimentale de quelques elements de la sensation de poids." *Archive de Physiologie Normale et Pathologiques* 3 (1891): 122–135.

Chemero, Anthony. "An Outline of a Theory of Affordances." *Ecological Psychology* 15, no. 2 (2003): 181–195. https://doi.org/10.1207/S15326969ECO1502_5.

Chemero, Anthony. *Radical Embodied Cognitive Science*. MIT Press, 2009.

Chen, Mingzhe, Walid Saad, and Changchuan Yin. "Virtual Reality over Wireless Networks: Quality-of-Service Model and Learning-Based Resource Management." *IEEE Transactions on Communications* 66, no. 11 (2018): 5621–5635. https://doi.org/10.1109/TCOMM.2018.2850303.

Chew, Elaine, Roger Zimmermann, Alexander A. Sawchuk, et al. "Musical Interaction at a Distance: Distributed Immersive Performance." Annual National Association of the Schools of Music. Accessed August 14, 2024. https://webspace.science.uu.nl/~fleis102/MusicNetwork-IMSC.pdf.

Chihak, Benjamin J., Jodie M. Plumert, Christine J. Ziemer, et al. "Synchronizing Self and Object Movement: How Child and Adult Cyclists Intercept Moving Gaps in a Virtual Environment." *Journal of Experimental Psychology: Human Perception and Performance* 36, no. 6 (2010): 1535–1552. https://doi.org/10.1037/a0020560.

Choi, Hyeg Joo, and Leonard S. Mark. "Scaling Affordances for Human Reach Actions." *Human Movement Science* 23, no. 6 (2004): 785–806. https://doi.org/10.1016/j.humov.2004.08.004.

Chow, Jia Yi, Keith Davids, Chris Button, and Michael Koh. "Coordination Changes in a Discrete Multi-Articular Action as a Function of Practice." *Acta Psychologica* 127, no. 1 (2008): 163–176. https://doi.org/10.1016/j.actpsy.2007.04.002.

Chow, Jia Yi, Keith Davids, Chris Button, Rick Shuttleworth, Ian Renshaw, and Duarte Araujo. 2006. "Nonlinear Pedagogy: A Constraints-Led Framework for Understanding Emergence of Game Play and Movement Skills." *Nonlinear Dynamics, Psychology, and Life Sciences* 10, no. 1 (2006): 71–103. https://europepmc.org/article/med/16393504.

Coelho, Marcelo, Hiroshi Ishii, and Pattie Maes. "Surflex: A Programmable Surface for the Design of Tangible Interfaces." In *CHI '08 Extended Abstracts on Human Factors in Computing Systems*. ACM, 2008. https://doi.org/10.1145/1358628.1358869.

Cole, Jonathan. *Pride and a Daily Marathon*. MIT Press, 1995.

Charlton, James. *Nothing About Us Without Us: Disability Oppression and Empowerment*. University of California Press, 1998. Online edition, California Scholarship Online, May 24, 2012. https://doi.org/10.1525/california/9780520207950.001.0001.

Cooper, Douglas. "Very Nervous System." *Wired*, March 1, 1995. https://www.wired.com/1995/03/rokeby/.

Craig, Cathy. "Understanding Perception and Action in Sport: How Can Virtual Reality Technology Help?" *Sports Technology* 6, no. 4 (2013): 161–169. https://doi.org/10.1080/19346182.2013.855224.

Craig, Cathy M., Cédric Goulon, Eric Berton, Guillaume Rao, Laure Fernandez, and Reinoud J. Bootsma. "Optic Variables Used to Judge Future Ball Arrival Position in Expert and Novice Soccer Players." *Attention, Perception, and Psychophysics* 71 (2009): 515–522. https://doi.org/10.3758/APP.71.3.515.

Craig, C. M., D. Delay, M. A. Grealy, and D. N. Lee. "Guiding the Swing in Golf Putting." *Nature* 405 (2000): 295–296. https://doi.org/10.1038/35012690.

Crease, Murray, and Stephen Brewster. "Making Progress with Sounds: The Design and Evaluation of an Audio Progress Bar." In *Proceedings of the 1998 International Conference on Auditory Display*. ACM, 1998. https://dl.acm.org/doi/10.5555/2227605.2227614.

Criado Perez, Caroline. *Invisible Women: Data Bias in a World Designed for Men*. Abrams, 2019.

Culkin, John M. "A Schoolman's Guide to Marshall McLuhan." *Saturday Review*, 1967.

Cummings, James J., and Jeremy N. Bailenson. "How Immersive Is Enough? A Meta-Analysis of the Effect of Immersive Technology on User Presence." *Media Psychology* 29, no. 2 (2015): 272–309. https://doi.org/10.1080/15213269.2015.1015740.

Cutting, James E. "Two Ecological Perspectives: Gibson vs. Shaw and Turvey." *American Journal of Psychology* 95, no. 2 (1982): 199–222. https://doi.org/10.2307/1422466.

Cutting, James E., and Peter M. Vishton. "Perceiving Layout and Knowing Distances: The Integration, Relative Potency, and Contextual Use of Different Information about Depth." In *Perception of Space and Motion*, edited by W. Epstein and S. J. Rogers. Academic Press, 1995.

Dawkins, Richard. "Why the Universe Seems so Strange." TED Talk, Oxford England, July 2005, 21 min., 42 sec. https://www.ted.com/talks/richard_dawkins_why_the_universe_seems_so_strange.

De Jaegher, Hanne. "Embodiment and Sense-Making in Autism." *Frontiers in Integrative Neuroscience* 7 (2013): 15. https://doi.org/10.3389/fnint.2013.00015.

De Jaegher, Hanne, and Ezequiel Di Paolo. "Participatory Sense-Making: An Enactive Approach to Social Cognition." *Phenomenology and the Cognitive Sciences* 6, no. 4 (2007): 485–507. https://doi.org/10.1007/s11097-007-9076-9.

Delafield-Butt, Jonathan, Andreas Galler, Benjaman Schögler, and David N. Lee. "A Perception–Action Strategy for Hummingbirds." *Perception* 39, no. 9 (2010): 1172–1174. https://doi.org/10.1068/p6708.

Descartes, René. *Meditations on First Philosophy: With Selections from the Objections and Replies*. Oxford University Press, 2008.

Deshpande, Abhiraj, and Inki Kim. "The Effects of Augmented Reality on Improving Spatial Problem Solving for Object Assembly." *Advanced Engineering Informatics* 38 (2018): 760–775. https://doi.org/10.1016/j.aei.2018.10.004.

Dessing, Joost C., and Cathy M. Craig. "Bending It like Beckham: How to Visually Fool the Goalkeeper." *PLoS One* 5, no. 10 (2010): e13161. https://doi.org/10.1371/journal.pone.0013161.

Dewez, Diane, Ludovic Hoyet, Anatole Lécuyer, and Ferran Argelaguet Sanz. "Towards 'Avatar-Friendly' 3D Manipulation Techniques: Bridging the Gap Between Sense of Embodiment and Interaction in Virtual Reality." In *Proceedings of the 2021 CHI Conference on Human Factors in Computing Systems*. ACM, 2021. https://doi.org/10.1145/3411764.3445379.

Dichgans, Johannes, and Thomas Brandt. "Visual–Vestibular Interaction: Effects on Self-Motion Perception and Postural Control." In *Perception: Handbook of Sensory Physiology*, Vol. 8, edited by Richard Held, Herschel W. Leibowitz, and Has-Lukas Teuber. Springer, 1978.

Dimitriou, Michael. "Enhanced Muscle Afferent Signals During Motor Learning in Humans." *Current Biology* 26, no. 8 (2016): 1062–1068. https://doi.org/10.1016/j.cub.2016.02.030.

Di Paolo, Ezequiel. "Autopoiesis, Adaptivity, Teleology, Agency." *Phenomenology and the Cognitive Sciences* 4, no. 4 (2005): 429–452. https://www.researchgate.net/publication/226935846_Autopoiesis_Adaptivity_Teleolog_Agency.

Dittrich, Winand H., Tom Troscianko, Stephen E. G. Lea, and D. O. Morgan. "Perception of Emotion from Dynamic Point-Light Displays Represented in Dance." *Perception* 25, no. 6 (1996): 727–738. https://doi.org/10.1068/p250727.

Djajadiningrat, Tom, Ben Matthews, and Marcelle Stienstra. "Easy Doesn't Do It: Skill and Expression in Tangible Aesthetics." *Personal and Ubiquitous Computing* 11, no. 8 (2007): 657–676. https://doi.org/10.1007/s00779-006-0137-9.

Dourish, Paul. "Epilogue: Where the Action Was, Wasn't, Should Have Been, and Might Yet Be." *ACM Transactions on Computer–Human Interaction (TOCHI)* 20, no. 1 (2013): 1–4. https://doi.org/10.1145/2442106.2442108.

Dourish, Paul. *Where the Action Is: Foundations of Embodied Interaction.* MIT Press, 2001. https://doi.org/10.7551/mitpress/7221.001.0001.

Dreyfus, Hubert L. *What Computers Still Can't Do: A Critique of Artificial Reason.* MIT Press, 1992.

Ebrahimi, Elham, Andrew Robb, Leah S. Hartman, Christopher C. Pagano, and Sabarish V. Babu. "Effects of Anthropomorphic Fidelity of Self-Avatars on Reach Boundary Estimation in Immersive Virtual Environments." In *Proceedings of the 15th ACM Symposium on Applied Perception.* ACM, 2018. https://doi.org/10.1145/3225153.3225170.

Ebrahimi, Elham, Sabarish V. Babu, Christopher C. Pagano, and Sophie Jörg. "An Empirical Evaluation of Visuo-Haptic Feedback on Physical Reaching Behaviors During 3D Interaction in Real and Immersive Virtual Environments." *ACM Transactions on Applied Perception (TAP)* 13, no. 4 (2016): 1–21. https://doi.org/10.1145/2947617.

Ellis, Robert R., and Susan J. Lederman. "The Material–Weight Illusion Revisited." *Perception and Psychophysics* 61, no. 8 (1999): 1564–1576. https://doi.org/10.3758/BF03213118.

Emerson, Robert Wall, and Dona Sauerburger. "Detecting Approaching Vehicles at Streets with No Traffic Control." *Journal of Visual Impairment and Blindness* 102, no. 12 (2008): 747–760. https://pmc.ncbi.nlm.nih.gov/articles/PMC2814321/.

Engbert, Kai, and Andreas Wohlschläger. "Intentions and Expectations in Temporal Binding." *Conscious Cognition* 16, no. 2 (2007): 255–264. https://doi.org/10.1016/j.concog.2006.09.010.

Essl, Georg, and Sile O'Modhrain. "An Enactive Approach to the Design of New Tangible Musical Instruments." *Organised Sound* 11, no. 3 (2006): 285–296. https://doi.org/10.1017/S135577180600152X.

Fagerberg, Petra, Anna Ståhl, and Kristina Höök. "Designing Gestures for Affective Input: An Analysis of Shape, Effort and Valence." In *Proceedings of the 2nd International Conference on Mobile and Ubiquitous Multimedia*. ACM, 2003. https://www.researchgate.net/publication/240838134_Designing_Gestures_for_Affective_Input_An_Analysis_of_Shape_Effort_and_Valence.

Fajen, Brett R. "Affordance Perception and the Visual Control of Locomotion." In *Human Walking in Virtual Environments: Perception, Technology, and Applications*, edited by Frank Steinicke, Yon Visell, Jennifer Campos, and Anatole Lécuyer. Springer, 2013. https://doi.org/10.1007/978-1-4419-8432-6_4.

Fajen, Brett, and Jonathan Matthis. "Direct Perception of Action-Scaled Affordances: The Shrinking Gap Problem." *Journal of Experimental Psychology: Human Perception and Performance* 37, no. 5 (2011): 1442–1457. https://doi.org/10.1037/a0023510.

Fajen, Brett, and Jonathan Matthis. "Visual and Non-Visual Contributions to the Perception of Object Motion During Self-Motion." *Journal of Vision* 11, no. 11 (2013): e55446. https://doi.org/10.1371/journal.pone.0055446.

Fajen, Brett R., and Flip Phillips. "Spatial Perception and Action." In *Handbook of Spatial Cognition*, edited by David Waller and Lynn Nadel. American Psychological Association, 2013. https://doi.org/10.1037/13936-004.

Fallman, Daniel. "In Romance with the Materials of Mobile Interaction: A Phenomenological Approach to the Design of Mobile Information Technology." Master's thesis, Umeå University, 2003.

Fallman, Daniel. "The New Good: Exploring the Potential of Philosophy of Technology to Contribute to Human–Computer Interaction." In *Proceedings of the SIGCHI Conference on Human Factors in Computing Systems*. ACM, 2011. https://doi.org/10.1145/1978942.1979099.

Fallman, Daniel, and John Waterworth. "Dealing with User Experience and Affective Evaluation in HCI Design: A Repertory Grid Approach." *International Conference on Human Factors in Computing Systems*. ACM, 2005. https://api.semanticscholar.org/CorpusID:13216344.

Farnè, Alessandro, Andrea Serino, and Elisabetta Làdavas. "Dynamic Size-Change of Peri-Hand Space Following Tool-Use: Determinants and Spatial Characteristics Revealed Through Cross-Modal Extinction." *Cortex* 43, no. 3 (2007): 436–443. https://doi.org/10.1016/S0010-9452(08)70468-4.

Farrer, Chlöé, G. Valentin, and Jean-Michel Hupé. "The Time Windows of the Sense of Agency." *Consciousness and Cognition* 22, no. 4 (2013): 1431–1441. https://doi.org/10.1016/j.concog.2013.09.010.

Fath, Aaron J., and Brett R. Fajen. "Static and Dynamic Visual Information About the Size and Passability of an Aperture." *Perception* 40, no. 8 (2011): 887–904. https://doi.org/10.1068/p6917.

Fdili Alaoui, Sarah, Thecla Schiphorst, Shannon Cuykendall, Kristin Carlson, Karen Studd, and Karen Bradley. "Strategies for Embodied Design: The Value and Challenges of Observing Movement." In *Proceedings of the 2015 ACM SIGCHI Conference on Creativity and Cognition*. ACM, 2015. https://doi.org/10.1145/2757226.2757238.

Ferretti, Gabriele. "Two Visual Systems in Molyneux Subjects." *Phenomenology and the Cognitive Sciences* 17, no. 4 (2018): 643–679. https://doi.org/10.1007/s11097-017-9533-z.

Ferroni, Francesca, Vittorio Gallese, Agata Marta Soccini, et al. "The Remapping of Peripersonal Space in a Real but Not in a Virtual Environment." *Brain Sciences* 12, no. 9 (2022). https://doi.org/10.3390/brainsci12091125.

Fiebrink, Rebecca, and Perry R. Cook. "The Wekinator: A System for Real-Time, Interactive Machine Learning in Music." In *Proceedings of The Eleventh International Society for Music Information Retrieval Conference*. IGITUR, 2010. https://citeseerx.ist.psu.edu/document?repid=rep1&type=pdf&doi=ced58510dbb49cc434d211092651d67aac030002.

Fink, Gereon R., John C. Marshall, Peter H. Weiss, and Karl Zilles. "The Neural Basis of Vertical and Horizontal Line Bisection Judgments: An FMRI Study of Normal Volunteers." *Neuroimage* 14, no. 1 (2001): S59–S67. https://doi.org/10.1006/nimg.2001.0819.

Fink, Philip W., Patrick S. Foo, and William H. Warren. 2009. "Catching Fly Balls in Virtual Reality: A Critical Test of the Outfielder Problem." *Journal of Vision* 9, no. 13 (2009): 14. https://doi.org/10.1167/9.13.14.

Fishkin, Kenneth P. "A Taxonomy for and Analysis of Tangible Interfaces." *Personal and Ubiquitous Computing* 8 (2004): 347–358. https://doi.org/10.1007/s00779-004-0297-4.

Fiske, Susan T. "Mind the Gap: In Praise of Informal Sources of Formal Theory." *Personality and Social Psychology Review* 8, no. 2 (2004): 132–137. https://doi.org/10.1207/s15327957pspr0802_6.

Fitts, Paul M. "The Information Capacity of the Human Motor System in Controlling the Amplitude of Movement." *Journal of Experimental Psychology* 47, no. 6 (1954): 381–391. https://doi.org/10.1037/h0055392.

Fitzmaurice, George W., Hiroshi Ishii, and William A. S. Buxton. "Bricks: Laying the Foundations for Graspable User Interfaces." In *Proceedings of the SIGCHI Conference on Human Factors in Computing Systems*. ACM, 1995. https://doi.org/10.1145/223904.223964.

Fleisig, Glenn S., Steve W. Barrentine, Nigel Zheng, Rafael F. Escamilla, and James R. Andrews. "Kinematic and Kinetic Comparison of Baseball Pitching Among Various Levels of Development." *Journal of Biomechanics* 32, no. 12 (1999): 1371–1413. https://doi.org/10.1016/S0021-9290(99)00127-X.

Fogtmann, Maiken Hillerup, Jonas Fritsch, and Karen Johanne Kortbek. "Kinesthetic Interaction: Revealing the Bodily Potential in Interaction Design." In *Proceedings of the 20th Australasian Conference on Computer–Human Interaction: Designing for Habitus and Habitat.* ACM, 2008. https://doi.org/10.1145/1517744.1517770.

Follmer, Sean. "Dynamic Physical Affordances for Shape-Changing and Deformable User Interfaces." PhD thesis, MIT, School of Architecture and Planning, Program in Media Arts and Sciences, 2015.

Follmer, Sean, Daniel Leithinger, Alex Olwal, Akimitsu Hogge, and Hiroshi Ishii. "inFORM: Dynamic Physical Affordances and Constraints Through Shape and Object Actuation." In *Proceedings of the 26th Annual ACM Symposium on User Interface Software and Technology.* ACM, 2013. https://doi.org/10.1145/2501988.2502032.

Gallagher, Shaun. *Action and Interaction.* Oxford University Press, 2020.

Gallagher, Shaun. *How the Body Shapes the Mind.* Oxford University Press, 2005. https://doi.org/10.1093/0199271941.001.0001.

Gallagher, Shaun. "The Intrinsic Spatial Frame of Reference." In *A Companion to Phenomenology and Existentialism,* edited by Hubert L. Dreyfus and Mark A. Wrathall. Wiley-Blackwell, 2006. https://doi.org/10.1002/9780470996508.ch24.

Gallagher, Shaun, and Dan Zahavi. *The Phenomenological Mind: An Introduction to Philosophy of Mind and Cognitive Science.* Routledge, 2008.

Gallagher, Shaun, and Dan Zahavi. "Primal Impression and Enactive Perception." In *Subjective Time: The Philosophy, Psychology, and Neuroscience of Temporality,* edited by Valtteri Arstila and Dan Lloyd. MIT Press, 2014. https://doi.org/10.7551/mitpress/8516.003.0010.

Gaver, William W. "How Do We Hear in the World? Explorations of Ecological Acoustics." *Ecological Psychology* 5, no. 1 (1993): 285–313. https://doi.org/10.1207/s15326969eco0504_2.

Gaver, William W. "Technology Affordances." In *Proceedings of the SIGCHI Conference on Human Factors in Computing Systems.* ACM, 1991. https://doi.org/10.1145/108844.108856.

Gaver, William W. "What in the World Do We Hear?: An Ecological Approach to Auditory Source Perception." *Ecological Psychology* 5, no. 1 (1993): 1–29. https://doi.org/10.1207/s15326969eco0501_1.

Gaver, William W., Jacob Beaver, and Steve Benford. "Ambiguity as a Resource for Design." In *Proceedings of the SIGCHI Conference on Human Factors in Computing Systems.* ACM, 2003. https://doi.org/10.1145/642611.64265.

Gibson, James. *The Ecological Approach to Visual Perception.* Psychology Press, 1979.

Gibson, James. "Observations on Active Touch." *Psychological Review* 69, no. 6 (1962): 477–491. https://doi.org/10.1037/h0046962.

Gibson, James. *Perception of the Visual World.* Houghton Mifflin, 1950.

Gibson, James. *The Senses Considered as Perceptual Systems.* Houghton Mifflin, 1966.

Gillespie, R. Brent, and Mark Cutkosky. 1996. "Stable User-Specific Haptic Rendering of the Virtual Wall." *ASME 1996 International Mechanical Engineering Congress and Exposition.* ASME, 1996. https://doi.org/10.1115/imece1996-0362.

Gillian, Nicholas. "Gesture Recognition for Musician Computer Interaction." PhD thesis, Queen's University Belfast, 2012.

Gillian, Nicholas, Benjamin Knapp, and Sile O'Modhrain. "A Machine Learning Toolbox for Musician Computer Interaction." In *Proceedings of the International Conference on New Interfaces for Musical Expression.* University of Oslo, 2011. https://nime.org/proc/gillian2011a/index.html.

Gillian, Nicholas, Benjamin Knapp, and Sile O'Modhrain. "Recognition of Multivariate Temporal Musical Gestures Using N-Dimensional Dynamic Time Warping." In *Proceedings of the International Conference on New Interfaces for Musical Expression.* University of Oslo, 2011. https://www.nime.org/proceedings/2011/nime2011_337.pdf.

Giudice, N. A. "Navigating Without Vision: Principles of Blind Spatial Cognition." In *Handbook of Behavioral and Cognitive Geography,* edited by D. R. Montello. Edward Elgar, 2018. https://doi.org/10.4337/9781784717544.00024.

Glinsky, Albert. *Theremin: Ether Music and Espionage.* University of Illinois Press, 2000.

Goel, Mayank, Jacob Wobbrock, and Shwetak Patel. "GripSense: Using Built-In Sensors to Detect Hand Posture and Pressure on Commodity Mobile Phones." In *Proceedings of the Twenty-fifth Annual ACM Symposium on User Interface Software and Technology.* ACM, 2012. https://doi.org/10.1145/2380116.2380184.

Goldstein, Bruce E. *Sensation and Perception.* 8th ed. Wadsworth, 2010.

Gordon, Michael S., and Lawrence D. Rosenblum. "Perception of Sound-Obstructing Surfaces Using Body-Scaled Judgments." *Ecological Psychology* 16, no. 2 (2004): 87–113. https://doi.org/10.1207/s15326969eco1602_1.

Gordon, Michael S., and Lawrence D. Rosenblum. "Effects of Intrastimulus Modality Change on Audiovisual Time-to-Arrival Judgments." *Perception and Psychophysics* 67, no. 4 (2005): 580–594. https://doi.org/10.3758/bf03193516.

Graf, Max, and Mathieu Barthet. "Mixed Reality Musical Interface: Exploring Ergonomics and Adaptive Hand Pose Recognition for Gestural Control." In *International Conference on New Interfaces for Musical Expression.* NIME, 2022. https://doi.org/10.21428/92fbeb44.56ba9b93.

Granzow, John, and Anil Çamcı. "Recreating a Rare Instrument Using VR and Fabrication: A Hyperreal Instrument Case Study." In *Forum Acusticum,* December 2020, Lyon, France, pp. 655–658. https://doi.org/10.48465/fa.2020.0847.

Haggard, Patrick, Tony Cheng, Brianna Beck, Francesca Fardo, and Georg Hajdu. "Embodiment and Disembodiment in Networked Music Performance." In *The Subject's Matter: Self-Consciousness and the Body*, edited by Frederique De Vignemont and Adrian J. T. Alsmith. MIT Press, 2017.

Hallnäs, Lars, and Johan Redström. "From Use to Presence: On the Expressions and Aesthetics of Everyday Computational Things." *ACM Transactions on Computer–Human Interaction* 9, no 2 (2002): 106–124. https://doi.org/10.1145/513665.513668.

Hamburg, Janet. "Coaching Athletes Using Laban Movement Analysis." *Journal of Physical Education, Recreation and Dance* 66, no. 2 (1995): 34–37. https://doi.org/10.1080/07303084.1995.10607040.

Hamraie, Aimi, and Kelly Fritsch. "Crip Technoscience Manifesto." *Catalyst* 5, no. 1 (2019). https://doi.org/10.28968/cftt.v5i1.29607.

Harrison, Steve, Deborah Tatar, and Phoebe Sengers. "The Three Paradigms of HCI." *AltCHI Session at the SIGCHI Conference on Human Factors in Computing Systems*. ACM, 2007. https://www.researchgate.net/publication/215835951_The_three_paradigms_of_HCI.

Harshfield, Stephen P., and Doris C. DeHardt. "Weight Judgment as a Function of Apparent Density of Objects." *Psychonomic Science* 20, no. 6 (1970): 365–366. https://doi.org/10.3758/BF03335692.

Hartmann, Björn, Maurizio Mancini, and Catherine Pelachaud. "Implementing Expressive Gesture Synthesis for Embodied Conversational Agents." In *Gesture in Human–Computer Interaction and Simulation*, edited by Sylvie Gibet, Nicolas Courty, and Jean-François Kamp. Springer, 2006. https://doi.org/10.1007/11678816_22.

Hassan, Shirin E. "Are Normally Sighted, Visually Impaired, and Blind Pedestrians Accurate and Reliable at Making Street Crossing Decisions?" *Investigative Ophthalmology and Visual Science* 53, no. 6 (2012): 2593–2600. https://doi.org/10.1167/iovs.11-9340.

Hausdorff, Jeffrey M. "Gait Dynamics in Parkinson's Disease: Common and Distinct Behavior Among Stride Length, Gait Variability, and Fractal-Like Scaling." *Chaos: An Interdisciplinary Journal of Nonlinear Science* 19, no. 2 (2009). https://doi.org/10.1063/1.3147408.

Hausdorff, Jeffrey M., Jenny Zitser, Anat Mirelman, and Nir Giladi. "Interaction Between Cognition and Gait in Patients with Parkinson's Disease." In *Dementia in Parkinson's Disease*, 2nd ed., edited by Murat Emre. Oxford University Press, 2009.

Heidegger, Martin. *Being and Time: A Translation of Sein und Zeit*. Translated by Joan Stambaugh. State University of New York Press, 1996.

Heidegger, Martin. *Sein und Zeit*. Max Niemeyer, 1957. Reprint, Max Niemeyer, 1993.

Held, Richard, Yuri Ostrovsky, Beatrice deGelder, and Pawan Sinha. "Revisiting the Molyneux Question." *Journal of Vision* 8, no. 6 (2008). https://doi.org/10.1167/8.6.523.

Hirao, Yutaro, and Takashi Kawai. "Augmented Cross-Modality: Translating the Physiological Responses, Knowledge and Impression to Audio-Visual Information in Virtual Reality." *Journal of Imaging Science and Technology* 62 (2019): jist0537. https://doi.org/10.2352/J.ImagingSci.Technol.2018.62.6.060402.

Holmquist, Lars Erik, Johan Redström, and Peter Ljungstrand. "Token-Based Access to Digital Information." In *Handheld and Ubiquitous Computing*, edited by Hans-W. Gellersen. Springer, 1999. https://doi.org/10.1007/3-540-48157-5_22.

Höök, Kristina. *Designing with the Body: Somaesthetic Interaction Design*. MIT Press, 2018.

Höök, Kristina, Sara Eriksson, Marie Louise Juul Søndergaard, et al. "Soma Design and Politics of the Body." In *Proceedings of the Halfway to the Future Symposium*. ACM, 2019. https://doi.org/10.1145/3363384.3363385.

Hornecker, Eva. "Beyond Affordance: Tangibles' Hybrid Nature." In *Proceedings of the Sixth International Conference on Tangible, Embedded and Embodied Interaction*. ACM, 2012. https://doi.org/10.1145/2148131.2148168.

Hornecker, Eva, and Andreas Dünser. "Of Pages and Paddles: Children's Expectations and Mistaken Interactions with Physical-Digital Tools." *Interacting with Computers* 21, no. 1–2 (2009): 95–107. https://doi.org/10.1016/j.intcom.2008.10.007.

Hornecker, Eva, and Jacob Buur. "Getting a Grip on Tangible Interaction: A Framework on Physical Space and Social Interaction." In *Proceedings of the SIGCHI Conference on Human Factors in Computing Systems*. ACM, 2006. https://doi.org/10.1145/1124772.1124838.

Huang, Yi-Hsuan, Roshan Venkatakrishnan, Rohith Venkatakrishnan, Sabarish V. Babu, and Wen-Chieh Lin. "Using Audio Reverberation to Compensate Distance Compression in Virtual Reality." *ACM Symposium on Applied Perception*. ACM, 2021. https://doi.org/10.1145/3474451.3476236.

Hudlicka, Eva. "Affective Game Engines: Motivation and Requirements." In *Proceedings of the 4th International Conference on Foundations of Digital Games*. ACM, 2009. https://doi.org/10.1145/1536513.1536565.

Hummels, Caroline, and Jelle van Dijk. "Seven Principles to Design for Embodied Sensemaking." In *Proceedings of the Ninth International Conference on Tangible, Embedded, and Embodied Interaction*. ACM, 2015. https://doi.org/10.1145/2677199.2680577.

Hummels Caroline, Kees C. J. Overbeeke, and Sietske Klooster. "Move to Get Moved: A Search for Methods, Tools and Knowledge to Design for Expressive and

Rich Movement-based Interaction." *Personal and Ubiquitous Computing* 11 (2007): 677–690. https://doi.org/10.1007/s00779-006-0135-y.

Hurley, Susan. "Perception And Action: Alternative Views." *Synthese* 129 (2001): 3–40. https://doi.org/10.1023/A:1012643006930.

Hurley, Susan, and Alva Noë. "Neural Plasticity and Consciousness." *Biology and Philosophy* 18, no. 1 (2003): 131–168. https://doi.org/10.1023/A:1023308401356.

Husserl, Edmund. *Phänomenologische Psychologie*. Vol. 9. Springer Science and Business Media, 1968.

Husserl, Edmund. *Phenomenological Psychology: Lectures, Summer Semester, 1925*. Springer, 1977. https://www.semanticscholar.org/paper/Phenomenological-Psychology%3A-Lectures%2C-Summer-1925-Husserl/0135227bc2811948a9a576436e2460ca5ea0af9c.

Husserl, Edmund. "The Phenomenology of Internal Time-Consciousness." Edited by Martin Heidegger. Indiana University Press, 1964.

Ihde, Don. *Technology and the Lifeworld: From Garden to Earth*. Indiana University Press, 1990.

Ihde, Don, and Lambros Malafouris. "*Homo faber* Revisited: Postphenomenology and Material Engagement Theory." *Philosophy and Technology* 32, no. 2 (2019): 195–214. https://doi.org/10.1007/s13347-018-0321-7.

Ihde, Don, Robert S. Cohen, and Marx W. Wartofsky. *Technics and Praxis*. Springer, 1979.

Imaizumi, Shu, and Tomohisa Asai. "Dissociation of Agency and Body Ownership Following Visuomotor Temporal Recalibration." *Frontiers in Integrative Neuroscience* 9, no. 35 (2015). https://doi.org/10.3389/fnint.2015.00035.

Iqbal, Shamsi T., and Brian P. Bailey. "Leveraging Characteristics of Task Structure to Predict the Cost of Interruption." In *Proceedings of the SIGCHI Conference on Human Factors in Computing Systems*. ACM, 2006. https://doi.org/10.1145/1124772.1124882.

Iriki, Atsushi, Michio Tanaka, Shigeru Obayashi, and Yoshiaki Iwamura. "Self-Images in the Video Monitor Coded by Monkey Intraparietal Neurons." *Neuroscience Research* 40, no. 2 (2001): 163–173. https://doi.org/10.1016/S0168-0102(01)00225-5.

Iriki, Atsushi, Michio Tanaka, and Yoshiaki Iwamura. "Coding of Modified Body Schema During Tool Use by Macaque Postcentral Neurones." *Neuroreport* 7, no. 14 (1996): 2325–2330. https://doi.org/10.1097/00001756-199610020-00010.

Isbister, Katherine, and Florian Mueller. "Guidelines for the Design of Movement-Based Games and Their Relevance to HCI 2015." *Human–Computer Interaction* 30, no. 3–4 (2015): 366–399. https://doi.org/10.1080/07370024.2014.996647.

Ishii, Hiroshi, Dávid Lakatos, Leonardo Bonanni, and Jean-Baptiste Labrune. "Radical Atoms: Beyond Tangible Bits, Toward Transformable Materials." *Interactions* 19, no. 1 (2012): 38–51. https://doi.org/10.1145/2065327.2065337.

Ishii, Hiroshi, and Brygg Ullmer. 1997. "Tangible Bits: Towards Seamless Interfaces Between People, Bits and Atoms." In *Proceedings of the ACM SIGCHI Conference on Human Factors in Computing Systems*. ACM, 1997. https://doi.org/10.1145/258549.258715.

ISO. "Guidelines for Standards Developers to Address the Needs of Older Persons and Persons with Disabilities." ISO/IEC Guide 71:2001(E). https://www.iso.org/files/live/sites/isoorg/files/archive/pdf/en/iso_iec_guide_71_2001.pdf.

Jacobs, David M., and Claire F. Michaels. "Direct Learning." *Ecological Psychology* 19, no. 4 (2007): 321–349. https://doi.org/10.1080/10407410701432337.

Jacobs, David M., Claire F. Michaels, and Sverker Runeson. "Learning to Perceive the Relative Mass of Colliding Balls: The Effects of Ratio Scaling and Feedback." *Perception and Psychophysics* 62, no. 7 (2000): 1332–1340. https://doi.org/10.3758/BF03212135.

Jacob, Robert J. K., Hiroshi Ishii, Gian Pangaro, and James Patten. "A Tangible Interface for Organizing Information Using a Grid." In *Proceedings of the SIGCHI Conference on Human Factors in Computing Systems*. ACM, 2002. https://doi.org/10.1145/503376.503437.

James, William. *The Principles of Psychology*. Vol. 1. Macmillan, 1890.

Jenison, Rick. "On Acoustic Information for Motion." *Ecological Psychology* 9, no. 2 (1997): 131–151. https://doi.org/10.1207/s15326969eco0902_2.

Jensen, Mads Vedel. "A Physical Approach to Tangible Interaction Design." In *Proceedings of the First International Conference on Tangible and Embedded Interaction*. ACM, 2007. https://doi.org/10.1145/1226969.1227018.

Jensenius, Alexander. *Sound Actions: Conceptualizing Musical Instruments*. MIT Press, 2022.

Jerald, Jason. *The VR Book: Human-Centered Design for Virtual Reality*. Morgan and Claypool, 2015.

Johansson, Gunnar. "Visual Perception of Biological Motion and a Model for Its Analysis." *Perception and Psychophysics* 14, no. 2 (1973): 201–211. https://doi.org/10.3758/BF03212378.

Jones, Terry, dir. *The Life of Brian*. HandMade Films, 1979.

Jönsson, Bodil. *Ten Thoughts About Time*. Robinson, 2003.

Jordà, Sergi, Günter Geiger, Marcos Alonso, and Martin Kaltenbrunner. "The ReacTable: Exploring the Synergy Between Live Music Performance and Tabletop Tangible

Interfaces." In *Proceedings of the 1st International Conference on Tangible and Embedded Interaction*. ACM, 2007. https://doi.org/10.1145/1226969.1226998.

Jörg, Sophie, Aline Normoyle, and Alla Safonova. "How Responsiveness Affects Players' Perception in Digital Games." In *Proceedings of the ACM Symposium on Applied Perception*. ACM, 2012. https://doi.org/10.1145/2338676.2338683.

Jourdain, Margaret. *Diderot's Early Philosophical Works*. Kessinger Publishing, 1916.

Ju, Wendy, and Leila Takayama. "Approachability: How People Interpret Automatic Door Movement as Gesture." *International Journal of Design* 3, no. 2 (2009): 77–86.

Kaetz, David. *Making Connections: Roots and Resonance in the Life and Teachings of Moshe Feldenkrais*. 2nd ed. River Centre, 2014.

Kant, Immanuel. *Critique of Pure Reason*. Translated and edited by Paul Guyer and Allen W. Wood. Cambridge University Press, 1998.

Kaptelinin, Victor. *Affordances and Design*. Interaction Design Foundation, 2014.

Kaptelinin, Victor, and Bonnie A. Nardi. *Acting with Technology: Activity Theory and Interaction Design*. MIT Press, 2009.

Kaptelinin, Victor, and Bonnie Nardi. "Affordances in HCI: Toward a Mediated Action Perspective." In *Proceedings of the SIGCHI Conference on Human Factors in Computing Systems*. ACM, 2012. https://doi.org/10.1145/2207676.2208541.

Kasahara, Shunichi, Shohei Nagai, and J. Rekimoto. "First Person Omnidirectional Video: System Design and Implications for Immersive Experience." In *Proceedings of the ACM International Conference on Interactive Experiences for TV and Online Video*. ACM, 2015. https://doi.org/10.1145/2745197.2745202.

Katsimbri, P. "The Biology of Normal Bone Remodelling." *European Journal of Cancer Care* 25, no. 5 (2017). https://doi.org/10.1111/ecc.12740.

Katta, Kathrin Gerling, Cynthia L. Bennett, Emeline Brulé, Rua M. Williams, Jennifer Rode, and Jennifer Mankoff. "Nothing About Us Without Us: Investigating the Role of Critical Disability." In *CHI EA '20: Extended Abstracts of the 2020 CHI Conference on Human Factors in Computing Systems*. ACM, 2020. https://doi.org/10.1145/3334480.3375150.

Kaufer, Stephan, and Anthony Chemero. *Phenomenology: An Introduction*. Polity, 2015.

Kay, Alan. *Doing with Images Makes Symbols: Communicating with Computers*. Video. Computer History Museum, 1987.

Kayed, Nanna Sønnichsen, and Audrey L. H. Van der Meer. "A Longitudinal Study of Prospective Control in Catching by Full-Term and Preterm Infants." *Experimental*

Brain Research 194, no. 2 (2009): 245–258. https://doi.org/10.1007/s00221-008-1692-2.

Kelly, Jonathan W., Lisa S. Donaldson, Lori A. Sjolund, and Jacob B. Freiberg. "More than Just Perception–Action Recalibration: Walking Through a Virtual Environment Causes Rescaling of Perceived Space." *Attention, Perception, and Psychophysics* 75, no. 7 (2013): 1473–1485. https://doi.org/10.3758/s13414-013-0503-4.

Kelso, J. A. Scott. "On the Self-Organizing Origins of Agency." *Trends in Cognitive Sciences* 20, no. 7 (2016): 490–499. https://doi.org/10.1016/j.tics.2016.04.004.

Kirsh, David. "Embodied Cognition and the Magical Future of Interaction Design." *ACM Transactions on Computer–Human Interaction* 20, no. 1 (2013): 1–30. https://doi.org/10.1145/2442106.2442109.

Kiverstein, Julian, and Michael Wheeler. *Heidegger and Cognitive Science.* Palgrave Macmillan, 2012.

Klatzky, Roberta L. "Allocentric and Egocentric Spatial Representations: Definitions, Distinctions, and Interconnections." In *Spatial Cognition: Lecture Notes in Computer Science*, Vol. 1404, edited by C. Freksa, C. Habel, and K. F. Wender. Springer, 1998. https://doi.org/10.1007/3-540-69342-4_1.

Klemmer, Scott R., Björn Hartmann, and Leila Takayama. "How Bodies Matter: Five Themes for Interaction Design." In *Proceedings of the Sixth Conference on Designing Interactive Systems*. ACM, 2006. https://doi.org/10.1145/1142405.1142429.

Klooster, Sietske, Richard Appleby, and Kees Overbeeke. "Design Education Moves: Movement as a Tool for Design Education, to Develop Sensitivity and Empathy." In *Second International Engineering and Product Design Education Conference*, September 2–3, 2004, Delft, The Netherlands, pp. 257–264. https://api.semanticscholar.org/CorpusID:110238985.

Knapp, Joshua M., and Jack M. Loomis. "Limited Field of View of Head-Mounted Displays Is Not the Cause of Distance Underestimation in Virtual Environments." *Presence: Teleoperators and Virtual Environments* 13, no. 5 (2004): 572–577. https://doi.org/10.1162/1054746042545238.

Knoblauch, Richard L., Martin T. Pietrucha, and Marsha Nitzburg. "Field Studies of Pedestrian Walking Speed and Start-Up Time." *Transportation Research Record* 1538, no. 1 (1996): 27–38. https://doi.org/10.1177/0361198196153800104.

Knowles, Tom. "I'm So Sorry, Says Inventor of Endless Online Scrolling." *The Times*, April 27, 2019. https://www.thetimes.co.uk/article/i-m-so-sorry-says-inventor-of-endless-online-scrolling-9lrv59mdk.

Knudson, Duane V., and Craig S. Morrison. *Qualitative Analysis of Human Movement.* Human Kinetics, 2002.

Kolarik, Andrew J., Silvia Cirstea, Shahina Pardhan, and Brian C. J. Moore. "A Summary of Research Investigating Echolocation Abilities of Blind and Sighted Humans." *Hearing Research* 310 (2014): 60–68. https://doi.org/10.1016/j.heares.2014.01.010.

Koutsoklenis, Athanasios, and Konstantinos Papadopoulos. "Auditory Cues Used for Wayfinding in Urban Environments by Individuals with Visual Impairments." *Journal of Visual Impairment and Blindness* 105, no. 10 (2011): 703–714.

Kreimeier, Julian, and Timo Götzelmann. "Two Decades of Touchable and Walkable Virtual Reality for Blind and Visually Impaired People: A High-Level Taxonomy." *Multimodal Technologies and Interaction* 4, no. 4 (2020). https://doi.org/10.3390/mti4040079.

Kudrle, Sara, Michel Proulx, Pascal Carrieres, and Marco Lopez. "Fingerprinting for Solving A/V Synchronization Issues within Broadcast Environments." *SMPTE Motion Imaging Journal* 120, no 5 (2010): 36–46. https://doi.org/10.5594/j18059.

Kugler, Peter N., J. A. Scott Kelso, and M. T. Turvey. "On the Concept of Coordinative Structures as Dissipative Structures: Theoretical Lines of Convergence." *Advances in Psychology* 1 (1980): 3–47. https://doi.org/10.1016/S0166-4115(08)61936-6.

Kuo, Arthur D. "The Six Determinants of Gait and the Inverted Pendulum Analogy: A Dynamic Walking Perspective." *Human Movement Science* 26, no. 4 (2007): 617–656. https://doi.org/10.1016/j.humov.2007.04.003.

Laban, Rudolf, and F. C. Lawrence. *In Effort: Economy of Human Movement*. Macdonald and Evans, 1974.

Lakoff, George, and Mark Johnson. *Metaphors We Live By*. University of Chicago Press, 2003.

Larssen, Astrid Twenebowa, Lian Loke, T. J. Robertson, and Jenny Edwards. "Understanding Movement as Input for Interaction: A Study of Two EyeToy™ Games." *Australian Computer Human Interaction Conference* (2004).

Lashley, K. S. "Integrative Functions of the Cerebral Cortex." *Physiological Reviews* 13, no. 1 (1933): 1–42.

Latash, Mark L. "The Bliss (Not the Problem) of Motor Abundance (Not Redundancy)." *Experimental Brain Research* 217, no. 1 (2012): 1–5. https://doi.org/10.1007/s00221-012-3000-4.

Latash, Mark L. *Synergy*. Oxford University Press, 2008.

Lederman, Susan J., Craig Summers, and Roberta L. Klatzky. "Cognitive Salience of Haptic Object Properties: Role of Modality-Encoding Bias." *Perception* 25, no. 8 (1996): 983–998. https://doi.org/10.1068/p250983.

Lederman, Susan, and Roberta Klatzky. "Hand Movements: A Window into Haptic Object Recognition." *Cognitive Psychology* 19, no. 3 (1987): 342–368. https://doi.org/10.1016/0010-0285(87)90008-9.

Lee, David N. "Visuo-Motor Coordination in Space–Time." In *Advances in Psychology*, Vol. 1, edited by George E. Stelmach, and Jean Requin. Elsevier, 1980. https://doi.org/10.1016/S0166-4115(08)61951-2.

Lee, David N. "Guiding Movement by Coupling Taus." *Ecological Psychology* 10, no. 3–4 (1998): 221–250. https://doi.org/10.1207/s15326969eco103&4_4.

Lee, David N. "How Movement Is Guided." University of Edinburgh. Accessed April 4, 2022. https://www.pmarc.ed.ac.uk/ideas/pdf/HowMovtGuided100311.pdf.

Lee, David N. "A Theory of Visual Control of Braking Based on Information About Time-to-Collision." *Perception* 5, no. 4 (1976): 437–459. https://doi.org/10.1068/p050437.

Lee, David N., Apostolos P. Georgopoulos, Martyn J. Clark, Cathy M. Craig, and Nicholas Port. "Guiding Contact by Coupling the Taus of Gaps." *Experimental Brain Research* 139 (2001): 151–159. https://doi.org/10.1007/s002210100725.

Lee, David N., J. Roly Lishman, and James A. Thomson. "Regulation of Gait in Long Jumping." *Journal of Experimental Psychology: Human Perception and Performance* 8, no. 3 (1982): 448–459. https://doi.org/10.1037/0096-1523.8.3.448.

Lee, David N., and Reddish, Paul E. "Plummeting Gannets: A Paradigm of Ecological Optics." *Nature* 293 (1981): 293–294. https://doi.org/10.1038/293293a0.

Lee, D. N., J. A. Simmons, P. A. Saillant, and F. Bouffard. "Steering by Echolocation: A Paradigm of Ecological Acoustics." *Journal of Comparative Physiology A* 176, no. 3 (1995): 347–354. https://doi.org/10.1007/BF00219060.

Lee David N., Frederikus Ruud van der Weel, Tris Hitchcock, Eddie Matejowsky, and Jack D. Pettigrew. "Common Principle of Guidance by Echolocation and Vision." *Journal of Comparative Physiology A* 171, no. 5 (1992): 563–571. https://doi.org/10.1007/BF00194105.

Lee, David N., David S. Young, and Carmel M. McLaughlin. "A Roadside Simulation of Road Crossing for Children." *Ergonomics* 27, no. 12 (1984): 1271–1281. https://doi.org/10.1080/00140138408963608.

Lee, Jun, Jee-In Kim, and HyungSeok Kim. "Force Arrow 2: A Novel Pseudo-Haptic Interface for Weight Perception in Lifting Virtual Objects." In *2019 IEEE International Conference on Big Data and Smart Computing (BigComp)*. IEEE, 2019. https://doi.org/10.1109/BIGCOMP.2019.8679400.

Lewald, Jörg, and Hans-Otto Karnath. "Vestibular Influence on Human Auditory Space Perception." *Journal of Neurophysiology* 84, no. 2 (2000): 1107–1111. https://doi.org/10.1152/jn.2000.84.2.1107.

Leyrer, Markus, Sally A. Linkenauger, Heinrich H. Bülthoff, Uwe Kloos, and Betty Mohler. "The Influence of Eye Height and Avatars on Egocentric Distance Estimates in Immersive Virtual Environments." In *Proceedings of the ACM SIGGRAPH*

Symposium on Applied Perception in Graphics and Visualization. ACM, 2011. https://doi.org/10.1145/2077451.2077464.

Lim, Woan Ning, Kian Meng Yap, Yunli Lee, Chyanna Wee, and Ching Chiuan Yen. "A Systematic Review of Weight Perception in Virtual Reality: Techniques, Challenges, and Road Ahead." *IEEE Access* 9 (2021): 163253–163283. https://doi.org/10.1109/ACCESS.2021.3131525.

Lin, Lisa P. Y., Neil M. McLatchie, and Sally A. Linkenauger. "The Influence of Perceptual–Motor Variability on the Perception of Action Boundaries for Reaching." *Journal of Experimental Psychology: Human Perception and Performance* 46, no. 5 (2020): 474–488. https://doi.org/10.1037/xhp0000724.

Locke, John. "An Essay Concerning Human Understanding." In *The Works of John Locke in Nine Volumes*, 12th ed., Vol. 1. Rivington, 1824. https://oll.libertyfund.org/title/locke-the-works-vol-1-an-essay-concerning-human-understanding-part-1.

Loke, Lian, Astrid T. Larssen, Toni Robertson, and Jenny Edwards. "Understanding Movement for Interaction Design: Frameworks and Approaches." *Personal and Ubiquitous Computing* 11, no. 8 (2007): 691–701. https://doi.org/10.1007/s00779-006-0132-1.

Loke, Lian, and Toni Robertson. "Moving and Making Strange: An Embodied Approach to Movement-Based Interaction Design." *ACM Transactions on Computer–Human Interaction (TOCHI)* 20, no. 1 (2013): 1–25. https://doi.org/10.1145/2442106.2442113.

Long, R. G., and N. A. Giudice. "Establishing and Maintaining Orientation for Mobility." In *Foundations of Orientation and Mobility*, 3rd ed., Vol. 1, History and Theory, edited by William R. Wiener, Richard W. Welsh, and Bruce B. Blasch. American Foundation for the Blind, 2010.

Loomis, Jack M. "Distal Attribution and Presence." *Presence: Teleoperators and Virtual Environments* 1, no. 1 (1992): 113–119. https://doi.org/10.1162/pres.1992.1.1.113.

Magnusson, Charlotte, and Kirsten Rassmus-Gröhn. "A Dynamic Haptic-Audio Traffic Environment." In *Proceedings of EuroHaptics 2004*. Interaction Design Foundation, 2004. https://portal.research.lu.se/en/publications/a-dynamic-haptic-audio-traffic-environment.

Malafouris, Lambros. "At the Potter's Wheel: An Argument for Material Agency." In *Material Agency: Towards a Non-Anthropocentric Approach*, edited by Carl Knappett and Lambros Malafouris. Springer, 2008. https://doi.org/10.1007/978-0-387-74711-8_2.

Malafouris, Lambros. *How Things Shape the Mind: A Theory of Material Engagement*. MIT Press, 2013. https://doi.org/10.7551/mitpress/9476.001.0001.

Mankoff, Jennifer, Gillian R. Hayes, and Devva Kasnitz. "Disability Studies as a Source of Critical Inquiry for the Field of Assistive Technology." In *Proceedings of the 12th International ACM SIGACCESS Conference on Computers and Accessibility*. ACM, 2010. https://doi.org/10.1145/1878803.1878807.

Maravita, Angelo, Charles Spence, and Jon Driver. "Multisensory Integration and the Body Schema: Close to Hand and Within Reach." *Current Biology* 13, no. 13 (2003): R531–539. https://doi.org/10.1016/S0960-9822(03)00449-4.

Marshall, Joe, Steve Benford, Richard Byrne, and Paul Tennent. "Sensory Alignment in Immersive Entertainment." In *Proceedings of the 2019 CHI Conference on Human Factors in Computing Systems*. ACM, 2019. https://doi.org/10.1145/3290605.3300930.

Martin, Daniel, Sandra Malpica, Diego Gutierrez, Belen Masia, and Ana Serrano. "Multimodality in VR: A Survey." *ACM Computing Surveys* 54, no. 10 (2022): 1–36. https://doi.org/10.1145/3508361.

Marucci, Matteo, Gianluca Di Flumeri, Gianluca Borghini, et al. "The Impact of Multisensory Integration and Perceptual Load in Virtual Reality Settings on Performance, Workload and Presence." *Scientific Reports* 11, no. 1 (2021): 1–15. https://doi.org/10.1038/s41598-021-84196-8.

Massie, T. H., and J. K. Salisbury. "The Phantom Haptic Interface: A Device for Probing Virtual Objects." In *Proceedings of the ASME Winter Annual Meeting, Symposium on Haptic Interfaces for Virtual Environment and Teleoperator Systems*. ASME, 1994.

McBeath, Michael K., Dennis M. Shaffer, and Mary K. Kaiser. "How Baseball Outfielders Determine Where to Run to Catch Fly Balls." *Science* 268, no. 5210 (1995): 569–573. https://doi.org/10.1126/science.772510.

McGann, Marek. "'Enactive' Cognition: A Cognition Briefing." euCognition: The European Network for the Advancement of Artificial Cognitive Systems. Accessed March 17, 2016. http://www.vernon.eu/euCognition/cognition_briefing_enactive_cognition.htm.

McGann, Marek. "Perceptual Modalities: Modes of Presentation or Modes of Interaction?" *Journal of Consciousness Studies* 17, no. 1–2 (2020): 72–94. https://www.researchgate.net/publication/233510440_Perceptual_ModalitiesModes_of_Presentation_or_Modes_of_Interaction.

Menary, Richard. "Intentionality, Cognitive Integration and the Continuity Thesis." *Topoi* 28, no. 31–43 (2009). https://doi.org/10.1007/s11245-008-9044-1.

Mendes, D., F. M. Caputo, A. Giachetti, A. Ferreira, and J. Jorge. "A Survey on 3D Virtual Object Manipulation: From the Desktop to Immersive Virtual Environments." *Computer Graphics Forum* 38, no. 1 (2019): 21–45. https://doi.org/10.1111/cgf.13390.

Mendes, Daniel, Filipe Relvas, Alfredo Ferreira, and Joaquim Jorge. "The Benefits of DOF Separation in Mid-Air 3D Object Manipulation." In *Proceedings of the 22nd ACM Conference on Virtual Reality Software and Technology*. ACM, 2016. https://dl.acm.org/doi/10.1145/2993369.2993396.

Menounou, Penelope, Michael R. Bailey, and David T. Blackstock. 2000. "Edge Wave on Axis Behind an Aperture or Disk Having a Ragged Edge." *The Journal of the Acoustical Society of America* 107, no. 1 (2000): 103–111. https://doi.org/10.1121/1.428343.

Merleau-Ponty, Maurice. *Phenomenology of Perception*. Routledge and Kegan Paul, 1962.

Michaels, Claire F., and Raoul R. D. Oudejans. "The Optics and Actions of Catching Fly Balls: Zeroing out Optical Acceleration." *Ecological Psychology* 4, no. 4 (1992): 199–222. https://doi.org/10.1207/s15326969eco0404_1.

Mieda, Takumi, and Masahiro Kokubu. "Blind Footballers Direct Their Head Towards an Approaching Ball During Ball Trapping." *Scientific Reports* 10, no. 1 (2020): 1–12.

Miller, Robert B. "Response Time in Man–Computer Conversational Transactions." In *Proceedings of the December 9–11, 1968, Fall Joint Computer Conference, Part 1*. ACM, 1968. https://doi.org/10.1145/1476589.1476628.

Mine, Mark R., Frederick P. Brooks Jr., and Carlo H. Sequin. 1997. "Moving Objects in Space: Exploiting Proprioception in Virtual-Environment Interaction." In *Proceedings of the 24th Annual Conference on Computer Graphics and Interactive Techniques*. ACM, 1997. https://doi.org/10.1145/258734.258747.

Moehring, M., and B. Froehlich. "Effective Manipulation of Virtual Objects Within Arm's Reach." In *Proceedings of the 2011 IEEE Virtual Reality Conference*. IEEE, 2011. https://doi.org/10.1109/VR.2011.5759451.

Montello, Daniel R. "Scale and Multiple Psychologies of Space." In *Spatial Information Theory: A Theoretical Basis for GIS*, edited by Andrew U. Frank and Irene Campari. Springer, 1993. https://doi.org/10.1007/3-540-57207-4_21.

Mossio, Matteo, and Dario Taraborelli. "Action-Dependent Perceptual Invariants: From Ecological to Sensorimotor Approaches." *Consciousness and Cognition* 17, no. 4 (2008): 1324–1340. https://doi.org/10.1016/j.concog.2007.12.003.

Mottet, Denis, and Reinoud J. Bootsma. "The Dynamics of Goal-Directed Rhythmical Aiming." *Biological Cybernetics* 80, no. 4 (1999): 235–245. https://doi.org/10.1007/s004220050521.

Mottet, Denis, Yves Guiard, Thierry Ferrand, and Reinoud J. Bootsma. "Two-Handed Performance of a Rhythmical Fitts Task by Individuals and Dyads." *Journal of Experimental Psychology: Human Perception and Performance* 27, no. 6 (2001): 1275–1286. https://doi.org/10.1037/0096-1523.27.6.1275.

Munafo, Justin, Meg Diedrick, and Thomas A. Stoffregen. "The Virtual Reality Head-Mounted Display Oculus Rift Induces Motion Sickness and Is Sexist in Its Effects." *Experimental Brain Research* 235, no. 3 (2017): 889–901. https://doi.org/10.1007/s00221-016-4846-7.

Nam, Chang S., Yueqing Li, Takehiko Yamaguchi, and Tonya L. Smith-Jackson. "Haptic User Interfaces for the Visually Impaired: Implications for Haptically Enhanced Science Learning Systems." *International Journal of Human–Computer Interaction* 28, no. 12 (2012): 784–798. https://doi.org/10.1080/10447318.2012.661357.

Neidhardt, Annika, Christian Schneiderwind, and Florian Klein. "Perceptual Matching of Room Acoustics for Auditory Augmented Reality in Small Rooms: Literature Review and Theoretical Framework." *Trends in Hearing* 26 (2022). https://doi.org 10.1177/23312165221092919.

Neuhoff, John G. "Ecological Psychoacoustics: Introduction and History." In *Ecological Psychoacoustics*, edited by John G. Neuhoff. Brill, 2004. https://doi.org/10.1163/9780080477442_004.

Nilsson, Niels Christian, Andre Zenner, Adalberto L. Simeone, Donald Degraen, and Florian Daiber. "Haptic Proxies for Virtual Reality: Success Criteria and Taxonomy." In *CHI 2021 Workshop on Everyday Proxy Objects for Virtual Reality*, May 8–13, 2021, Yokohama, Japan. http://epo4vr.dfki.de/.

Nöe, Alva. *Action in Perception*. MIT Press, 2006.

Norman, Donald A. 1999. "Affordance, Conventions, and Design." *Interactions* 6, no. 3 (1999): 38–43. https://doi.org/10.1145/301153.301168.

Norman, Donald, and Stephen Draper. *User Centered System Design: New Perspectives on Human–Computer Interaction*. Lawrence Erlbaum Associates, 1986.

"NSF Event Series to Spotlight Cross-Cutting Research by Persons with Disabilities." US National Science Foundation. July 9, 2024. https://new.nsf.gov/sbe/updates/event-series-spotlight-cross-cutting-research.

Nusseck, Manfred, and Marcelo M. Wanderley. "Music and Motion: How Music-Related Ancillary Body Movements Contribute to the Experience of Music." *Music Perception: An Interdisciplinary Journal* 26, no. 4 (2009): 335–353. https://doi.org/10.1525/mp.2009.26.4.335.

Ojeda, Lauro, and Johann Borenstein. "Non-GPS Navigation for Security Personnel and First Responders." *Journal of Navigation* 60, no. 3 (2007): 391–407. https://doi.org/10.1017/S0373463307004286.

O'Modhrain, Sile, and Georg Essl. "PebbleBox and CrumbleBag: Tactile Interfaces for Granular Synthesis." In *Proceedings of the 2004 Conference on New Interfaces for Musical Expression*. National University of Singapore, 2004.

O'Modhrain, Sile, Nicholas A. Giudice, John A. Gardner, and Gordon E. Legge. "Designing Media for Visually-Impaired Users of Refreshable Touch Displays: Possibilities and Pitfalls." *IEEE Transactions on Haptics* 8, no. 3 (2015): 248–257. https://doi.org/10.1109/TOH.2015.2466231.

O'Modhrain, Sile, Stefania Serafin, Chris Chafe, and Julius O. Smith. "Qualitative and Quantitative Assessment of a Virtual Bowed String Instrument." *International Computer Music Association* 2000 (2000): 66–69. http://hdl.handle.net/2027/spo.bbp2372.2000.199.

O'Regan, J. Kevin. *Why Red Doesn't Sound Like a Bell: Understanding the Feel of Consciousness*. Oxford University Press, 2011.

O'Regan, J. Kevin, and Alva Noë. "A Sensorimotor Account of Vision and Visual Consciousness." *Behavioral and Brain Sciences* 24, no. 5 (2001): 939–973. https://doi.org/10.1017/S0140525X01000115.

Osiurak, François, Nicolas Morgado, and Richard Palluel-Germain. "Tool Use and Perceived Distance: When Unreachable Becomes Spontaneously Reachable." *Experimental Brain Research* 218 (2012): 331–339. https://doi.org/10.1007/s00221-012-3036-5.

Oulasvirta, Antti, Sakari Tamminen, Virpi Roto, and Jaana Kuorelahti. "Interaction in 4-Second Bursts: The Fragmented Nature of Attentional Resources in Mobile HCI." In *Proceedings of the SIGCHI Conference on Human Factors in Computing Systems*. ACM, 2005. https://doi.org/10.1145/1054972.1055101.

Patney, Anjul, Marco Salvi, Joohwan Kim, et al. "Towards Foveated Rendering for Gaze-Tracked Virtual Reality." *ACM Transactions on Graphics* 35, no. 6 (2016): 1–12. https://doi.org/10.1145/2980179.2980246.

Peck, Tabitha C., Kyla A. McMullen, and John Quarles. "Break the Cycle and Develop VR for Everyone." *IEEE Computer Graphics and Applications* 41, no. 6 (2021): 133–142. https://doi.org/ 10.1109/MCG.2021.3113455.

Penny, Simon. *Making Sense*. MIT Press, 2018.

Persson, Hans, Henrik Åhman, Alexander Arvei Yngling, and Jan Gulliksen. "Universal Design, Inclusive Design, Accessible Design, Design for All: Different Concepts—One Goal? On The Concept of Accessibility—Historical, Methodological and Philosophical Aspects." *Universal Access in the Information Society* 14 (2015): 505–526. https://doi.org/10.1007/s10209-014-0358-z.

Pham, Tran, Jo Vermeulen, Anthony Tang, and Lindsay MacDonald Vermeulen. "Scale Impacts Elicited Gestures for Manipulating Holograms: Implications for AR Gesture Design." In *Proceedings of the 2018 Designing Interactive Systems Conference*. ACM, 2018. https://doi.org/10.1145/3196709.3196719.

Piaget, Jean. *The Construction of Reality in the Child*. Basic Books, 1952.

Picard, Rosalind W. *Affective Computing*. MIT Press, 2000.

Piper, Ben, Carlo Ratti, and Hiroshi Ishii. "Illuminating Clay: A 3-D Tangible Interface for Landscape Analysis." In *Proceedings of the SIGCHI Conference on Human Factors in Computing Systems*. ACM, 2002. https://doi.org/10.1145/503376.503439.

Plater, Alan. *Only a Matter of Time*. Radio Play. January 21, 2008. https://www.bbc.co.uk/programmes/b007qrqk.

Pockett, Susan. "How Long Is 'Now'? Phenomenology and the Specious Present." *Phenomenology and the Cognitive Sciences* 2, no. 1 (2003): 55–68. https://doi.org/10.1023/A:1022960122740.

Ponto, Kevin, Michael Gleicher, Robert G. Radwin, and Hyun Joon Shin. "Perceptual Calibration for Immersive Display Environments." *IEEE Transactions on Visualization and Computer Graphics* 19, no. 4 (2013): 691–700.

Pöppel, Ernst. *Mindworks: Time and Conscious Experience*. Springer, 1988.

Poupyrev, Ivan, Mark Billinghurst, Suzanne Weghorst, and Tadao Ichikawa. "The Go-Go Interaction Technique: Non-Linear Mapping for Direct Manipulation in VR." In *Proceedings of the 9th Annual ACM Symposium on User Interface Software and Technology*. ACM, 2006. https://doi.org/10.1145/237091.237102.

Prilutsky, Boris I., and Vladimir M. Zatsiorsky. "Tendon Action of Two-Joint Muscles: Transfer of Mechanical Energy Between Joints During Jumping, Landing, and Running." *Journal of Biomechanics* 27, no. 1 (1994): 25–34. https://doi.org/10.1016/0021-9290(94)90029-9.

Raja, Vicente, and Paco Calvo. "Augmented Reality: An Ecological Blend." *Cognitive Systems Research* 42 (2017): 58–72. https://doi.org/10.1016/j.cogsys.2016.11.009.

Rallis, Ioannis, Nikolaos Bakalos, Nikolaos Doulamis, Anastasios Doulamis, and Athanasios Voulodimos. "Bidirectional Long Short-Term Memory Networks and Sparse Hierarchical Modeling for Scalable Educational Learning of Dance Choreographies." *The Visual Computer* 37 (2021): 47–62. https://doi.org/10.1007/s00371-019-01741-3.

Rasmussen, Majken K., Esben W. Pedersen, Marianne G. Petersen, and Kasper Hornbæk. "Shape-Changing Interfaces: A Review of the Design Space and Open Research Questions." In *Proceedings of the SIGCHI Conference on Human Factors in Computing Systems*. ACM, 2012. https://doi.org/10.1145/2207676.2207781.

Rath, Matthias, and Davide Rocchesso. "Continuous Sonic Feedback from a Rolling Ball." *IEEE MultiMedia* 12, no. 2 (2005): 60–69. https://doi.org/10.1109/MMUL.2005.24.

Resnick, Mitchel. *Lifelong Kindergarten: Cultivating Creativity Through Projects, Passion, Peers, and Play*. MIT Press, 2017.

Riehm, Christopher, Anthony Chemero, Paula L. Silva, and Kevin Shockley. "Virtual Auditory Aperture Passability." *Experimental Brain Research* 237, no. 1 (2019): 191–200. https://doi.org/10.1007/s00221-018-5407-z.

Rinkenauer, Gerhard, Stefan Mattes, and Rolf Ulrich. "The Surface–Weight Illusion: On the Contribution of Grip Force to Perceived Heaviness." *Perception and Psychophysics* 61, no. 1 (1999): 23–30. https://doi.org/10.3758/BF03211946.

Rizzolatti, Giacomo, Cristiana Scandolara, Massimo Matelli, and Maurizio Gentilucci. "Afferent Properties of Periarcuate Neurons in Macaque Monkeys. I. Somatosensory Responses." *Behavioural Brain Research* 2, no. 2 (1981): 125–146.

Robertson, Toni, and Lian Loke. "Designing Situations." In *Proceedings of the 21st Annual Conference of the Australian Computer–Human Interaction Special Interest Group*. ACM, 2009. https://doi.org/10.1145/1738826.1738828.

Rocchesso, Davide, Pietro Polotti, and Stefano Delle Monache. "Designing Continuous Sonic Interaction." *International Journal of Design* 3, no. 3 (2009). https://www.ijdesign.org/index.php/IJDesign/article/view/620/271.

Rodger, Matthew, Sile O'Modhrain, and Cathy M. Craig. "Temporal Guidance of Musicians' Performance Movement Is an Acquired Skill." *Experimental Brain Research* 226, no. 2 (2013): 221–230. https://doi.org/10.1007/s00221-013-3427-2.

Rohde, Marieke, Meike Scheller, and Marc O. Ernst. "Effects Can Precede Their Cause in the Sense of Agency." *Neuropsychologia* 65 (2014): 191–196. https://doi.org/10.1016/j.neuropsychologia.2014.10.011.

Rosenberg, L. B. "Virtual Fixtures: Perceptual Tools for Telerobotic Manipulation." In *Proceedings of the 1993 IEEE Virtual Reality Annual International Symposium*. IEEE, 1993. https://doi.org/10.1109/VRAIS.1993.380795.

Rosenblum, Lawrence D., Claudia Carello, and Richard E. Pastore. "Relative Effectiveness of Three Stimulus Variables for Locating a Moving Sound Source." *Perception* 16, no. 2 (1987): 175–186. https://doi.org/10.1068/p160175.

Rosenblum, Lawrence D., James W. Dias, and Josh Dorsi. "The Supramodal Brain: Implications for Auditory Perception." *Journal of Cognitive Psychology* 29, no. 1 (2017): 65–87. https://doi.org/10.1080/20445911.2016.1181691.

Rowan, Daniel, Timos Papadopoulos, David Edwards, et al. "Identification of the Lateral Position of a Virtual Object Based on Echoes by Humans." *Hearing Research* 300 (2013): 56–65. https://doi.org/10.1016/j.heares.2013.03.005.

Ruggieri, Vezio, and Alexander Katsnelson. "An Analysis of a Performance by the Violinist D. Oistrakh: The Hypothetical Role of Postural Tonic-Static and Entourage Movements." *Perceptual and Motor Skills* 82, no. 1 (1996): 291–300.

Runeson, Sverker. "On the Possibility of 'Smart' Perceptual Mechanisms." *Scandinavian Journal of Psychology* 18, no. 1 (1977): 172–179. https://doi.org/10.1111/j.1467-9450.1977.tb00274.x.

Runeson, Sverker, and Gunilla Frykholm. "Visual Perception of Lifted Weight." *Journal of Experimental Psychology: Human Perception and Performance* 7, no. 4 (1981): 733–740. https://doi.org/10.1037/0096-1523.7.4.733.

Samaraweera, Gayani, Rongkai Guo, and John Quarles. "Latency and Avatars in Virtual Environments and the Effects on Gait for Persons with Mobility

Impairments." In *2013 IEEE Symposium on 3D User Interfaces*. IEEE, 2013. https://doi.org/10.1109/3DUI.2013.6550192.

Schiphorst, Thecla. "A Case Study of Merce Cunningham's Use of the Lifeforms Computer Choreographic System in the Making of Trackers." Master's thesis, Simon Fraser University, 1986.

Schögler, Benjaman, Gert-Jan Pepping, and David N. Lee. "TauG-Guidance of Transients in Expressive Musical Performance." *Experimental Brain Research* 189, no. 3 (2008): 361–372. https://doi.org/10.1007/s00221-008-1431-8.

Scopigno, R., P. Cignoni, N. Pietroni, M. Callieri, and M. Dellepiane. "Digital Fabrication Techniques for Cultural Heritage: A Survey." *Computer Graphics Forum* 36, no. 1 (2017): 6–21. https://doi.org/10.1111/cgf.12781.

Seifert, Ludovic, Matt Dicks, Frieder Wittmann, and Peter Wolf. "The Perception of Nested Affordances: An Examination of Expert Climbers." *Psychology of Sport and Exercise* 52 (2021): 101843. https://doi.org/10.1016/j.psychsport.2020.101843.

Sellen, Abigail J., and Richard H. R. Harper. *The Myth of the Paperless Office*. MIT Press, 2001.

Sengül, Ali, Michiel Van Elk, Giulio Rognini, Jane Elizabeth Aspell, Hannes Bleuler, and Olaf Blanke. "Extending the Body to Virtual Tools Using a Robotic Surgical Interface: Evidence from the Crossmodal Congruency Task." *PLoS One* 7, no. 12 (2012): e49473. https://doi.org/10.1371/journal.pone.0049473.

Seow, Steven C. *Designing and Engineering Time: The Psychology of Time Perception in Software*. Addison-Wesley, 2008.

Seow, Steven C. "Information Theoretic Models of HCI: A Comparison of the Hick–Hyman Law and Fitts' Law." *Human–Computer Interaction* 20, no. 3 (2005): 315–352. https://doi.org/10.1207/s15327051hci2003_3.

Serino, Andrea. "Peripersonal Space (PPS) as a Multisensory Interface Between the Individual and the Environment, Defining the Space of the Self." *Neuroscience and Biobehavioral Reviews* 99 (2019): 138–159. https://doi.org/10.1016/j.neubiorev.2019.01.016.

Serlin, Ilene A. "Maxine Sheetstone: The Primacy of Movement—Expanded Second Edition (Advances in Consciousness Research)." *American Journal of Dance Therapy* 35 (2013): 201–206. https://doi.org/10.1007/s10465-013-9164-z.

Shaffer, Dennis M., Scott M. Krauchunas, Marianna Eddy, and Michael K. McBeath. "How Dogs Navigate to Catch Frisbees." *Psychological Science* 15, no. 7 (2004): 437–441. https://doi.org/10.1111/j.0956-7976.2004.00698.x.

Shannon, Claude, and Warren Weaver. *The Mathematical Theory of Communication*. University of Illinois Press, 1949.

Shapiro, Lawrence. *Embodied Cognition*. Routledge, 2011.

Shavel, Tatyana. "MR Is Leaving AR in the Dust: So Say These Mixed Reality Examples." Iflexion, March 4, 2020. https://www.iflexion.com/blog/mixed-reality-examples.

Sheets-Johnstone, Maxine. "Maxine Sheets-Johnstone: Dance, Philosophy and Movement." Interview by Serge Prengel. Relational Implicit. Active Pause, May 2010. https://activepause.com/zug/transcripts/Sheets-Johnstone-2010-05.pdf.

Sheets-Johnstone, Maxine. *The Primacy of Movement*. John Benjamins, 2011.

Shin, Kang G., and Parameswaran Ramanathan. "Real-Time Computing: A New Discipline of Computer Science and Engineering." *Proceedings of IEEE, Special Issue on Real-Time Systems* 82, no. 1 (January 1994): 6–24. https://doi.org/10.1109/5.259423.

Shneiderman, Ben. "Response Time and Display Rate in Human Performance with Computers." *ACM Computing Surveys* 16, no. 3 (1984): 265–285. https://doi.org/10.1145/2514.2517.

Simons, Daniel J., and Christopher F. Chabris. "Gorillas in Our Midst: Sustained Inattentional Blindness for Dynamic Events." *Perception* 28 (1999): 1059–1074.

Singapogu, Ravikiran B., Christopher C. Pagano, Timothy C. Burg, Paul G. Dorn, Ron Zacharia, and DongBin Lee. "Perceptually Salient Haptic Rendering for Enhancing Kinesthetic Perception in Virtual Environments." *Journal on Multimodal User Interfaces* 8, no. 3 (2014): 319–331. https://doi.org/10.1007/s12193-014-0164-1.

Siu, Alexa F., Mike Sinclair, Robert Kovacs, Eyal Ofek, Christian Holz, and Edward Cutrell. "Virtual Reality Without Vision: A Haptic and Auditory White Cane to Navigate Complex Virtual Worlds." In *Proceedings of the 2020 CHI Conference on Human Factors in Computing Systems*. ACM, 2020. https://doi.org/10.1145/3313831.3376353.

Slater, Mel. "Place Illusion and Plausibility Can Lead to Realistic Behaviour in Immersive Virtual Environments." *Philosophical Transactions of the Royal Society, B: Biological Sciences* 364, no. 1535 (2009): 3549–3557. https://doi.org/10.1098/rstb.2009.0138.

Smith, Linda B., and Esther Thelen, eds. *A Dynamic Systems Approach to Development*. MIT Press, 1993.

Smith, Linda B., Esther Thelen, Robert Titzer, and Dewey McLin. "Knowing in the Context of Acting: The Task Dynamics of the A-not-B Error." *Psychological Review* 106, no. 2 (1999): 235–260. https://doi.org/10.1037/0033-295X.106.2.235.

Snyder, Natalie, and Michael Cinelli. 2020. "Sensory Conflict Alters Visual Perception of Action Capabilities During Crossing of a Closing Gap in Virtual Reality." *Quarterly Journal of Experimental Psychology* 73 (12): 2309–2316. https://doi.org/10.1177/1747021820942212.

Solomon, H. Y., and M. T. Turvey. "Haptically Perceiving the Distances Reachable with Hand-Held Objects." *Journal of Experimental Psychology: Human Perception and Performance* 14, no. 3 (1988): 404–427. https://doi.org/10.1037/0096-1523.14.3.404.

Soukoreff, R. William, and I. Scott MacKenzie. "Towards a Standard for Pointing Device Evaluation, Perspectives on 27 Years of Fitts' Law Research in HCI 2004." *International Journal of Human–Computer Studies* 61, no. 6 (2004): 751–789. https://doi.org/10.1016/j.ijhcs.2004.09.001.

Souza, Vinicius, Anderson Maciel, Luciana Nedel, and Regis Kopper. "Measuring Presence in Virtual Environments: A Survey." *ACM Computing Surveys (CSUR)* 54, no. 8 (2022). https://doi.org/10.1145/3466817.

Spiel, Katta. "The Bodies of TEI: Investigating Norms and Assumptions in the Design of Embodied Interaction." In *Proceedings of the Fifteenth International Conference on Tangible, Embedded, and Embodied Interaction*. ACM, 2021. https://doi.org/10.1145/3430524.3440651.

Spiel, Katta. "Body-Positive Computing as a Means to Counteract Normative Biases in Fitness Trackers." *XRDS: Crossroads, The ACM Magazine for Students* 25, no. 4 (2019): 34–37. https://dl.acm.org/doi/10.1145/3331065.

Spiel, Katta, Kathrin Gerling, Cynthia L. Bennett, et al. "Nothing About Us Without Us: Investigating the Role of Critical Disability Studies in HCI." In *Extended Abstracts of the 2020 CHI Conference on Human Factors in Computing Systems*. ACM, 2020. https://doi.org/10.1145/3334480.3375150.

Stanley, Andrew A., and Allison M. Okamura. "Controllable Surface Haptics via Particle Jamming and Pneumatics." *IEEE Transactions on Haptics* 8, no. 1 (2015): 20–30. https://doi.org/10.1109/TOH.2015.2391093.

Stauffert, Jan-Philipp, Florian Niebling, and Marc Erich Latoschik. "Latency and Cybersickness: Impact, Causes, and Measures. A Review." *Frontiers in Virtual Reality* 1 (2020). https://doi.org/10.3389/frvir.2020.582204.

Stefanucci, Jeanine K., Dennis R. Proffitt, Tom Banton, and William Epstein. "Distances Appear Different on Hills." *Perception and Psychophysics* 67, no. 6 (2005): 1052–1060.

Stetson, Chess, Matthew P. Fiesta, and David M. Eagleman. "Does Time Really Slow Down During a Frightening Event?" *PLoS One* 2, no. 12 (2007): e1295. https://doi.org/10.1371/journal.pone.0001295.

Stewart, John, Armen Khatchatourov, and Charles Lenay. "Enaction and Engineering." Communication Presented at the *Interdisciplines Virtual Workshop Enactive Network of Excellence* 20 (2004). https://www.academia.edu/1817582/Enaction_and_engineering.

Stroffregen, Thomas A., and John B. Pittenger. "Human Echolocation as a Basic Form of Perception and Action." *Ecological Psychology* 7, no. 3 (1995): 181–216. https://doi.org/10.1207/s15326969eco0703_2.

Suchman, Lucy. *Human–Machine Reconfigurations: Plans and Situated Actions.* Cambridge University Press, 2007.

Suchman, Lucy. *Plans and Situated Actions: The Problem of Human–Machine Communication.* Cambridge University Press, 1987.

Sum, Cella M., Rahaf Alharbi, Franchesca Spektor, et al. "Dreaming Disability Justice in HCI." In *CHI Conference on Human Factors in Computing Systems Extended Abstracts.* ACM, 2022. https://doi.org/10.1145/3491101.3503731.

Svanæs, Dag. "Interaction Design for and with the Lived Body: Some Implications of Merleau-Ponty's Phenomenology." *ACM Transactions on Computer–Human Interaction* 20, no. 1 (2013): 1–30. https://doi.org/10.1145/2442106.2442114.

Teixeira, Fabricio, and Caio Braga. "The State of UX in 2024." UX Trends. Accessed September 15, 2024. https://trends.uxdesign.cc/2024.

Thompson, Evan, and Mog Stapleton. "Making Sense of Sense-Making: Reflections on Enactive and Extended Mind Theories." *Topoi* 28, no. 1 (2009): 23–30. https://doi.org/10.1007/s11245-008-9043-2.

Thornson, Carol A., Brian F. Goldiez, and Huy Le. "Predicting Presence: Constructing the Tendency Toward Presence Inventory." *International Journal of Human–Computer Studies* 67, no. 1 (2009): 62–78. https://doi.org/10.1016/j.ijhcs.2008.08.006.

Togler, Jonas, Fabian Hemmert, and Reto Wettach. "Living Interfaces: The Thrifty Faucet." In *Proceedings of the 3rd International Conference on Tangible and Embedded Interaction.* ACM, 2009. https://doi.org/10.1145/1517664.1517680.

Turvey, Michael T. "Affordances and Prospective Control: An Outline of the Ontology." *Ecological Psychology* 4, no. 3 (1992): 173–187. https://doi.org/10.1207/s15326969eco0403_3.

Turvey, Michael T. *Lectures on Perception: An Ecological Perspective.* Routledge, 2018.

Turvey, Michael T., and Claudia Carello. "Obtaining Information by Dynamic (Effortful) Touching." *Philosophical Transactions of the Royal Society B: Biological Sciences* 366, no. 1581 (2011): 3123–3132. https://doi.org/10.1098/rstb.2011.0159.

Turvey, M. T., Hollis Fitch, and Betty Tuller. "The Bernstein Perspective: I. The Problems of Degrees of Freedom and Context-Conditioned Variability." In *Human Motor Behavior: An Introduction*, edited by J. A. Scott Kelso. Psychology Press, 1982.

Uchida, Thomas K., and Scott L. Delp. *Biomechanics of Movement: The Science of Sports, Robotics, and Rehabilitation.* MIT Press, 2021.

Underkoffler, John, and Hiroshi Ishii. "Illuminating Light: An Optical Design Tool with a Luminous-Tangible Interface." *Proceedings of the SIGCHI Conference on Human Factors in Computing Systems*. ACM, 1998. https://doi.org/10.1145/274644.274717.

Underkoffler, John, and Hiroshi Ishii. "Urp: A Luminous-Tangible Workbench for Urban Planning and Design." In *Proceedings of the SIGCHI Conference on Human Factors in Computing Systems*. ACM, 1999. https://doi.org/10.1145/302979.303114.

Uva, Antonio E., Michele Gattullo, Vito M. Manghisi, Daniele Spagnulo, Giuseppe L. Cascella, and Michele Fiorentino. "Evaluating the Effectiveness of Spatial Augmented Reality in Smart Manufacturing: A Solution for Manual Working Stations." *International Journal of Advanced Manufacturing Technology* 94 (2018): 509–521. https://doi.org/10.1007/s00170-017-0846-4.

van Beek, Femke E., Raymond J. King, Casey Brown, Massimiliano Di Luca, and Sean Keller. "Static Weight Perception Through Skin Stretch and Kinesthetic Information: Detection Thresholds, JNDs, and PSEs." *IEEE Transactions on Haptics* 14, no. 1 (2021): 20–31. https://doi.org/10.1109/TOH.2020.3009599.

Van Damme, Sam, Maria Torres Vega, and Filipe De Turck. "Human-centric Quality Management of Immersive Multimedia Applications." In *Sixth IEEE Conference on Network Softwarization (NetSoft)*. ACM, 2020. https://doi.org/10.1109/NetSoft48620.2020.9165335.

van de Langenberg, Rolf, Idsart Kingma, and Peter J. Beek. "The Perception of Limb Orientation Depends on the Center of Mass." *Journal of Experimental Psychology: Human Perception and Performance* 34, no. 3 (2008): 624–639. https://doi.org/10.1037/0096-1523.34.3.624 34.

van Dijk, Jelle. "Cognition Is Not What It Used to Be: Reconsidering Usability from an Embodied Embedded Cognition Perspective." *Human Technology: An Interdisciplinary Journal on Humans in ICT Environments* 5, no. 1 (2009): 29–46. http://www.humantechnology.jyu.fi.

van Dijk, Jelle. "Designing for Embodied Being-in-the-World: A Critical Analysis of the Concept of Embodiment in the Design of Hybrids." *Multimodal Technologies and Interaction* 2, no. 1 (2018). https://doi.org/10.3390/mti2010007.

van Dijk, Jelle, and Caroline Hummels. "Designing for Embodied Being-in-the-World: Two Cases, Seven Principles and One Framework." In *Proceedings of the Eleventh International Conference on Tangible, Embedded, and Embodied Interaction*. ACM, 2017. https://doi.org/10.1145/3024969.3025007.

van Grunsven, Janna. "Disabled Body-Minds in Hostile Environments: Disrupting an Ableist Cartesian Sociotechnical Imagination with Enactive Embodied Cognition and Critical Disability Studies." *Topoi* 44 (2024): 505–515. https://doi.org/10.1007/s11245-024-10080-5.

Varela, Francisco J. "The Specious Present: A Neurophenomenology of Time Consciousness." In *Naturalizing Phenomenology: Issues in Contemporary Phenomenology and Cognitive Science*, edited by Jean Petitot, Francisco J. Varela, Bernard Pachoud, and Jean-Michel Roy. Stanford University Press, 1999.

Varela, Francisco J., Evan Thompson, and Eleanor Rosch. *The Embodied Mind: Cognitive Science and Human Experience*, rev. ed. MIT Press, 1991.

Verbeek, Peter P. C. C. "Don Ihde: The Technological Lifeworld." *American Philosophy of Technology: The Empirical Turn*, edited by H. J. Achterhuis. Indiana University Press, 2001. https://research.utwente.nl/en/publications/don-ihde-the-technological-lifeworld.

Vernat, Jean-Phillipe, and Michael S. Gordon. 2011. "Indirect Interception Actions by Blind and Sighted Perceivers: The Role of Modality and Tau." *Scandinavian Journal of Psychology* 52, no. 1 (2011): 83–92.

Vincent, James. *Beyond Measure: The Hidden History of Measurement from Cubits to Quantum Constants*. W. W. Norton & Company, 2022.

Visell, Yon, Roderick Murray-Smith, Stephen Brewster, and John Williamson. "Continuous Auditory and Tactile Interaction Design." In *Sonic Interaction Design*, edited by Karmen Franinović and Stefania Serafin. MIT Press, 2013. https://doi.org/10.7551/mitpress/8555.003.0005.

Viviani, Paolo, and Natalie Stucchi. "Biological Movements Look Uniform: Evidence of Motor-Perceptual Interactions." *Journal of Experimental Psychology: Human Perception and Performance* 18, no. 3 (1992): 603–623. https://doi.org/10.1037//0096-1523.18.3.603.

Viviani, Paolo, and Tamar Flash. "Minimum-Jerk, Two-Thirds Power Law, and Isochrony: Converging Approaches to Movement Planning." *Experimental Psychology: Human Perception and Performance* 21, no. 1 (1995): 32–53. https://doi.org/10.1037/0096-1523.21.1.32.

von Helmholtz, Hermann. *Handbuch der physiologischen Optik: mit 213 in den Text eingedruckten Holzschnitten und 11 Tafeln*. L. Voss, 1867.

von Uexküll, Jakob. *A Foray into the Worlds of Animals and Humans with a Theory of Meaning*. 1934. Reprint, University of Minnesota Press, 2010.

Wagman, Jeffrey B., Kevin Shockley, Michael A. Riley, and M. T. Turvey. "Attunement, Calibration, and Exploration in Fast Haptic Perceptual Learning." *Journal of Motor Behavior* 33, no. 4 (2001): 323–327. https://doi.org/10.1080/00222890109601917.

Wagman, Jeffrey B., and Lydia L. Morgan. "Nested Prospectivity in Perception: Perceived Maximum Reaching Height Reflects Anticipated Changes in Reaching Ability." *Psychonomic Bulletin and Review* 17, no. 6 (2010): 905–909. https://doi.org/10.3758/PBR.17.6.905.

Wagman, Jeffrey B., and Thomas A. Stoffregen. "It Doesn't Add Up: Nested Affordances for Reaching Are Perceived as a Complex Particular." *Attention, Perception, and Psychophysics* 82 (2020): 3832–3841. https://doi.org/10.3758/s13414-020-02108-w.

Waisvisz, Michel. "The Hands: A Set of Remote MIDI Controllers." In *Proceedings of the International Computer Music Conference*. International Computer Music Association, 1985. http://hdl.handle.net/2027/spo.bbp2372.1985.049.

Walker, Jamie. "Curse of a Sunburnt Land." *The Australian*, February 9, 2009. https://web.archive.org/web/20090210180628/http://www.theaustralian.news.com.au/story/0,25197,25025610-28737,00.html.

Waltemate, Thomas, Irene Senna, Felix Hülsmann, et al. "The Impact of Latency on Perceptual Judgments and Motor Performance in Closed-Loop Interaction in Virtual Reality." In *Proceedings of the 22nd ACM Conference on Virtual Reality Software and Technology*. ACM, 2016. https://doi.org/10.1145/2993369.2993381.

Wanderley, Marcelo M., Bradley W. Vines, Neil Middleton, Cory McKay, and Wesley Hatch. "The Musical Significance of Clarinetists' Ancillary Gestures: An Exploration of the Field." *Journal of New Music Research* 34, no. 1 (2005): 97–113. https://doi.org/10.1080/09298210500124208.

Ward, Nicholas. *Effortful Interaction: A New Paradigm for the Design of Digital Musical Instruments*. Queen's University Belfast, 2013.

Warren, William H. "Action-Scaled Information for the Visual Control of Locomotion." In *Closing the Gap: The Scientific Writings of David N. Lee*, edited by Gert-Jan Pepping and Madeleine A. Grealy. Psychology Press, 2007.

Warren, William, and Robert Verbrugge. "Auditory Perception of Breaking and Bouncing Events: A Case Study in Ecological Acoustics." *Journal of Experimental Psychology: Human Perception and Performance* 10, no. 5 (1984): 704–712. https://doi.org/10.1037/0096-1523.10.5.704.

Warren Jr., William H., and Suzanne Whang. "Visual Guidance of Walking Through Apertures: Body-Scaled Information for Affordances." *Journal of Experimental Psychology: Human Perception and Performance* 13, no. 3 (1987): 371–383. https://doi.org/10.1037/0096-1523.13.3.371.

Watson, Gareth, Sebastien Brault, Richard Kulpa, Benoit Bideau, Joe Butterfield, and Cathy Craig. "Judging the 'Passability' of Dynamic Gaps in a Virtual Rugby Environment." *Human Movement Science* 30, no. 5 (2011): 942–956. https://doi.org/10.1016/j.humov.2010.08.004.

Weber, Andreas, and Francisco J. Varela. "Life after Kant: Natural Purposes and the Autopoietic Foundations of Biological Individuality." *Phenomenology and the Cognitive Sciences* 1, no. 2 (2002): 97–125. https://doi.org/10.1023/A:1020368120174.

Wechsung, Ina, and Katrien De Moor. "Quality of Experience Versus User Experience." In *Quality of Experience: Advanced Concepts, Applications and Methods*, edited by Sebastian Möller and Alexander Raake. Springer, 2014. https://doi.org/10.1007/978-3-319-02681-7_3.

Wessel, David, and Matthew Wright. "Problems and Prospects for Intimate Musical Control of Computers." *Computer Music Journal* 26, no. 3 (2002): 11–22. https://www.jstor.org/stable/3681975.

West, Mick. "Programming Responsiveness." *Game Developer* (blog). July 9, 2008. https://www.gamedeveloper.com/programming/programming-responsiveness.

Wheeler, Michael. "The Revolution Will Not Be Optimised: Radical Enactivism, Extended Functionalism and the Extensive Mind." *Topoi* 36, no. 3 (2017): 457–472. https://doi.org/10.1007/s11245-015-9356-x.

Williamson, John, Roderick Murray-Smith, and Stephen Hughes. "Shoogle: Multimodal Excitatory Interaction on Mobile Devices." In *Proceedings of the SIGCHI Conference on Human Factors in Computing Systems*. ACM, 2007. https://doi.org/10.1145/1240624.1240642.

Wilson, Robert A., and Lucia Foglia. "Embodied Cognition." In *The Stanford Encyclopedia of Philosophy*, spring 2017 edition, edited by Edward N. Zalta. Accessed July 25, 2011. https://plato.stanford.edu/archives/spr2017/entries/embodied-cognition/.

Winograd, Terry, and Fernando Flores. *Understanding Computers and Cognition: A New Foundation for Design*. Ablex, 1986.

Withagen, Rob, and Claire F. Michaels. "The Role of Feedback Information for Calibration and Attunement in Perceiving Length by Dynamic Touch." *Journal of Experimental Psychology: Human Perception and Performance* 31, no. 6 (2005): 1379–1390. https://doi.org/10.1037/0096-1523.31.6.1379.

Withagen, Rob, Harjo J. de Poel, Duarte Araújo, and Gert-Jan Pepping. "Affordances Can Invite Behavior: Reconsidering the Relationship Between Affordances and Agency." *New Ideas in Psychology* 30, no. 2 (2012): 250–258. https://doi.org/10.1016/j.newideapsych.2011.12.003.

Witt, Jessica K. "Action's Influence on Spatial Perception: Resolution and a Mystery." *Current Opinion in Psychology* 32 (2020): 153–157.

Witt, Jessica K., Dennis R. Proffitt, and William Epstein. "Tool Use Affects Perceived Distance, but Only When You Intend to Use It." *Journal of Experimental Psychology: Human Perception and Performance* 31, no 5 (2005): 880–888. https://doi.org/10.1037/0096-1523.31.5.

Wittmann, Mark. "Embodied Time: The Experience of Time, the Body, and the Self." In *Subjective Time: The Philosophy, Psychology, and Neuroscience of Temporality*, edited by V. Arstila and D. Lloyd. Boston Review, 2014.

Yang, Jing-Lan, Shiauyee Chen, Mei-Hwa Jan, Yeong-Fwu Lin, and Jiu-jenq Lin. "Proprioception Assessment in Subjects with Idiopathic Loss of Shoulder Range of Motion: Joint Position Sense and a Novel Proprioceptive Feedback Index." *Journal of Orthopaedic Research* 26, no. 9 (2008): 1218–1224. https://doi.org/10.1002/jor.20627.

Young, William R., Matthew W. M. Rodger, and Cathy M. Craig. "Auditory Observation of Stepping Actions Can Cue Both Spatial and Temporal Components of Gait in Parkinson's Disease Patients." *Neuropsychologia* 57 (2014): 140–153. https://doi.org/10.1016/j.neuropsychologia.2014.03.009.

Zadra, Jonathan R., and Dennis R. Proffitt. "Optic Flow Is Calibrated to Walking Effort." *Psychonomic Bulletin and Review* 23, no. 5 (2016): 1531–5320. https://doi.org/10.3758/s13423-016-1017-6.

Zahavi, Dan. "Inner (Time-) Consciousness." In *On Time: New Contributions to the Husserlian Phenomenology of Time*, edited by Dieter Lohmar and Ichiro Yamaguchi. Springer, 2010. https://doi.org/10.1007/978-90-481-8766-9_16.

Zahrai, Kseniia, Ekant Veer, Paul William Ballantine, Huibert Peter de Vries, and Girish Prayag. "Either You Control Social Media or Social Media Controls You: Understanding the Impact of Self-Control on Excessive Social Media Use from the Dual-System Perspective." *Journal of Consumer Affairs* 56, no. 2 (2022): 806–848. https://doi.org/10.1111/joca.12449.

Zenner, Andre, and Antonio Krüger. "Shifty: A Weight-Shifting Dynamic Passive Haptic Proxy to Enhance Object Perception in Virtual Reality." *IEEE Transactions on Visualization and Computer Graphics* 23, no. 4 (2017): 1285–1294. https://doi.org/10.1109/TVCG.2017.2656978.

Zhu, Qin, and Geoffrey P. Bingham. "Learning to Perceive the Affordance for Long-Distance Throwing: Smart Mechanism or Function Learning?" *Journal of Experimental Psychology: Human Perception and Performance* 36, no. 4 (2010): 862–875. https://doi.org/10.1037/a0018738.

Index

Page numbers followed by an *f* denote a figure, a *t* denotes a table.

Publisher contact:
The MIT Press
Massachusetts Institute of Technology
77 Massachusetts Avenue, Cambridge, MA 02139
mitpress.mit.edu

EU Authorised Representative:
Easy Access System Europe, Mustamäe tee 50,
10621 Tallinn, Estonia
gpsr.requests@easproject.com

Printed by Integrated Books International,
United States of America